AF780635

# Precise Point Positioning with GPS, GLONASS, BeiDou, and Galileo

# Precise Point Positioning with GPS, GLONASS, BeiDou, and Galileo

Editors

**Shengfeng Gu**
**Xiaopeng Gong**
**Yidong Lou**
**Chuang Shi**

Basel • Beijing • Wuhan • Barcelona • Belgrade • Novi Sad • Cluj • Manchester

*Editors*
Shengfeng Gu
Wuhan University
Wuhan
China

Xiaopeng Gong
Wuhan University
Wuhan
China

Yidong Lou
Wuhan University
Wuhan
China

Chuang Shi
Beihang University
Beijing
China

*Editorial Office*
MDPI
St. Alban-Anlage 66
4052 Basel, Switzerland

This is a reprint of articles from the Special Issue published online in the open access journal *Remote Sensing* (ISSN 2072-4292) (available at: https://www.mdpi.com/journal/remotesensing/special_issues/precise_point_positioning_with_gps_glonass_beidou_galileo).

For citation purposes, cite each article independently as indicated on the article page online and as indicated below:

Lastname, A.A.; Lastname, B.B. Article Title. *Journal Name* **Year**, *Volume Number*, Page Range.

**ISBN 978-3-0365-8986-2 (Hbk)**
**ISBN 978-3-0365-8987-9 (PDF)**
**doi.org/10.3390/books978-3-0365-8987-9**

© 2023 by the authors. Articles in this book are Open Access and distributed under the Creative Commons Attribution (CC BY) license. The book as a whole is distributed by MDPI under the terms and conditions of the Creative Commons Attribution-NonCommercial-NoDerivs (CC BY-NC-ND) license.

# Contents

*remote sensing*

MDPI

*Article*

# Performance Assessment of BDS-3 PPP-B2b/INS Loosely Coupled Integration

**Xiaofei Xu [1], Zhixi Nie [1,*], Zhenjie Wang [1], Boyang Wang [1] and Qinghuai Du [2]**

[1] College of Oceanography and Space Informatics, China University of Petroleum, Qingdao 266580, China; b19010058@s.upc.edu.cn (X.X.); sdwzj@upc.edu.cn (Z.W.); b19010077@s.upc.edu.cn (B.W.)
[2] Shengli Branch, Sinopec Geophysical Corporation, Dongying 257100, China; dqh062@163.com
* Correspondence: niezhixi@upc.edu.cn

**Abstract:** The BeiDou global navigation satellite system (BDS-3) has been officially providing a real-time precise point positioning (PPP) augmentation service, known as the PPP-B2b service, since 2020. Decimeter-level positioning accuracy is expected to be achieved based on the PPP-B2b service. It shows great potential for global navigation satellite system (GNSS) real-time applications, including, for example, vehicle positioning on land. However, the application of the PPP-B2b service is still full of challenges in the urban environment because of GNSS signal blockage. The inertial navigation system (INS) is a popular technology which can provide continuous positions under GNSS challenging scenarios. In this study, we constructed a BDS-3 PPP-B2b/INS loosely coupled integration system for vehicle positioning and evaluated its performance through two automotive experiments. In the first experiment, four periods of 30 s GNSS outages were simulated to evaluate the performance of PPP-B2b/INS loosely coupled integration during GNSS outages. During the simulated GNSS outages, PPP-B2b positioning did not work. Nevertheless, PPP-B2b/INS loosely coupled integration provided continues solution through INS mechanization. The averaged positioning errors at the last epoch of outages were 300.6/498.0/41.0 cm for PPP-B2b/MEMS-IMU and 18.6/21.8/6.1 cm for PPP-B2b/Tactical-IMU loosely coupled integration, in the east, north and up directions, respectively. In the second experiment, we drove the land vehicle in a complex urban environment for 15 min. During this period, two GNSS signal interruptions occurred due to the occlusion of bridges, lasting 15 s and 5 s, respectively. The results show that the improvement of positioning accuracy in the east, north, and up components were 64.1%, 77.8%, and 73.8% respectively for PPP-B2b/MEMS-IMU loosely coupled integration, and 63.9%, 79.5%, and 74.4% respectively for PPP-B2b/Tactical-IMU loosely coupled integration, as compared to the positioning accuracy of PPP-B2b only.

**Keywords:** GNSS; INS; PPP-B2b service; loosely coupled integration

**Citation:** Xu, X.; Nie, Z.; Wang, Z.; Wang, B.; Du, Q. Performance Assessment of BDS-3 PPP-B2b/INS Loosely Coupled Integration. *Remote Sens.* **2022**, *14*, 2957. https://doi.org/10.3390/rs14132957

Academic Editors: Chuang Shi, Shengfeng Gu, Yidong Lou and Xiaopeng Gong

Received: 27 May 2022
Accepted: 19 June 2022
Published: 21 June 2022

**Publisher's Note:** MDPI stays neutral with regard to jurisdictional claims in published maps and institutional affiliations.

**Copyright:** © 2022 by the authors. Licensee MDPI, Basel, Switzerland. This article is an open access article distributed under the terms and conditions of the Creative Commons Attribution (CC BY) license (https://creativecommons.org/licenses/by/4.0/).

## 1. Introduction

Real-time kinematic (RTK) and precise point positioning (PPP) are usually conducted to satisfy the high-precision positioning requirements of autonomous cars and unmanned aerial vehicles. RTK technology can archive centimeter-level positioning accuracy with short initialization time [1,2], but it needs regional corrections from a reference station or reference network. The concept of PPP technology was proposed in 1997 [3,4], and it could obtain decimeter- even to centimeter-level positioning accuracy using standalone global navigation satellite system (GNSS) equipment [5]. Over the decades, PPP has become a widely used high-precision GNSS positioning technology. Precise GNSS satellite products are indispensable for the implementation of PPP. Motivated by the requirements of GNSS real-time applications, the international GNSS service (IGS) has been providing real-time service (RTS) through Internet communication since 2013 [6,7]. The RTS corrections, including precise orbit correction, clock offset correction, and code biases, are sent to users based on the Networked Transport of RTCM via Internet Protocol (NTRIP) [8]. By applying these corrections, the users can compute precise satellite orbits and clocks, and then carry out real-time PPP. However, RTS may be interrupted when the Internet is unavailable [9].

The commissioning of the BeiDou global navigation satellite system (BDS-3) was announced on 31 July 2020 [10,11]. Several featured services are provided by BDS-3, including short message communication, international search and rescue, and the PPP-B2b service, in addition to global positioning, navigation and time (PNT) services [12]. The BDS-3 PPP-B2b service can support real-time PPP based on PPP-B2b corrections broadcast by 3 BDS-3 geostationary earth orbit (GEO) satellites [13,14]. The PPP-B2b corrections include satellite orbit correction, clock correction, and code bias correction. At present, the BDS-3 PPP-B2b service only provides corrections for BDS-3/GPS satellites. The BDS-3 PPP-B2b service can be obtained free of charge. In addition, unlike IGS-RTS, Internet communication is not required since the BDS-3 PPP-B2b service is a satellite-based service. Therefore, the PPP-B2b service has great potential for real-time GNSS applications. Recently, publications have presented the precision of PPP-B2b corrections and positioning performance based on the PPP-B2b service [15–19]. These results indicate that the signal-in-space accuracy of precise ephemeris calculated with PPP-B2b corrections is at the decimeter- even to centimeter-level. In term of positioning performance, centimeter-level accuracy in the static mode and decimeter-level in the kinematic mode could be achieved. However, positioning performance based on the BDS-3 PPP-B2b service would be significantly degraded in severe environments, due to GNSS signal blockages. To overcome this weakness, GNSS should be integrated with other sensors [20]. The inertial navigation system (INS) is an autonomous and spontaneous navigation system, which has potential to overcome degradation of GNSS positioning performance when GNSS signal outage occurs. Many valuable studies in PPP/INS integration have been published in the past decades, including loosely coupled (LC) model and tightly coupled (TC) integration [21–25]. In these publications, the precise ephemeris derived from IGS final precise products or from the IGS RTS service is adopted for PPP/INS integration processing. However, there is a considerable delay obtaining IGS final precise products. As for the IGS RTS service, the PPP/INS integration is restricted when there is a lack of Internet communication. It is worth mentioned that the above disadvantages can be overcome when the BDS-3 PPP-B2b service has implemented PPP/INS integration.

To our knowledge, there has until now been no published research on BDS-3 PPP-B2b/INS integration. In this paper, we focus on performance assessment of BDS-3 PPP-B2b/INS loosely coupled integration. Two vehicle kinematic experiments were carried out to assess the performance of PPP-B2b/INS loosely coupled integration. The rest of the paper is organized as follows. In Section 2, the methodology is introduced, including PPP positioning based on PPP-B2b corrections, the INS model, and their loosely coupled integration. In Section 3, the experimental set-ups and data processing strategies are presented in detail. Section 4 presents the performance of PPP-B2b/INS loosely coupled integration in urban environments. Finally, conclusions and perspectives are illustrated in Section 5.

## 2. Methodology

BDS-3 PPP-B2b/INS loosely coupled integration can be divided into three parts, including PPP positioning based on the PPP-B2b service, the INS model, and the integration of BDS-3 PPP-B2b and INS. In this section, these three parts are described in detail.

### 2.1. PPP Positioning Based on PPP-B2b Service

As is known, precise satellite orbit and clock products play a key role in PPP processing. For PPP-B2b service, the satellite orbit corrections and clock offset corrections are broadcast by BDS-3 GEO satellites. Up to now, the PPP-B2b service has provided corrections only for BDS-3/GPS satellites. According to the PPP-B2b Interface Control Document (ICD) from China Satellite Navigation Office (CSNO) [26], the PPP-B2b orbit corrections are given in the radial $(\delta O_{radial})$, along-track $\left(\delta O_{along}\right)$, and cross-track $(\delta O_{cross})$ directions.

Hence, the orbit correction vector $\delta \boldsymbol{O}_{B2b} = \begin{bmatrix}\delta O_{radial} & \delta O_{along} & \delta O_{cross}\end{bmatrix}^T$ should first be transformed to the Earth-Center Earth-Fixed (ECEF) frame. This is because the satellite positions derived from CNAV1/LNAV broadcast ephemeris are based on the ECEF frame. This transformation can be described as follows:

$$\delta \boldsymbol{X}^{sat} = \begin{bmatrix}\boldsymbol{e}_{radial} & \boldsymbol{e}_{along} & \boldsymbol{e}_{cross}\end{bmatrix} \cdot \delta \boldsymbol{O}_{B2b} \tag{1}$$

with

$$\begin{cases} \boldsymbol{e}_{radial} = \frac{\boldsymbol{r}}{|\boldsymbol{r}|} \\ \boldsymbol{e}_{cross} = \frac{\boldsymbol{r} \times \dot{\boldsymbol{r}}}{|\boldsymbol{r} \times \dot{\boldsymbol{r}}|} \\ \boldsymbol{e}_{along} = \boldsymbol{e}_{cross} \times \boldsymbol{e}_{radial} \end{cases} \tag{2}$$

where $\delta \boldsymbol{X}^{sat} = \begin{bmatrix}\delta O_x & \delta O_y & \delta O_z\end{bmatrix}^T$ represents the PPP-B2b orbit correction vector in the ECEF frame. $\boldsymbol{r}$ and $\dot{\boldsymbol{r}}$ are the satellite position and velocity vectors derived from broadcast ephemeris. Then, the precise satellite position vector $\boldsymbol{X}_{prec}$ can be computed by

$$\boldsymbol{X}_{prec} = \boldsymbol{X}_{brdc} - \delta \boldsymbol{X}^{sat} \tag{3}$$

where $\boldsymbol{X}_{brdc}$ is the satellite position vector derived from broadcast ephemeris. As described in ICD document, the PPP-B2b clock correction parameter is defined as offset to the broadcast ephemeris clock in meters. The precise satellite clock offset is given by

$$dt^{sat}_{prec} = dt^{sat}_{brdc} - \frac{C_0}{c} \tag{4}$$

where $C_0$ represents the PPP-B2b clock correction parameter; $dt^{sat}_{brdc}$ is the satellite clock offset derived from broadcast ephemeris; $dt^{sat}_{prec}$ denotes the precise PPP-B2b clock offset; $c$ is the velocity of light in a vacuum.

The GNSS raw code and carrier-phase measurements can read as [27,28]

$$\begin{cases} P_i = \rho + c \cdot (dt_r - dt^s) + T + \frac{f_1^2}{f_i^2} \cdot I_1 + B_{r,i} - B_i^s + \varepsilon_{P_i} \\ L_i = \rho + c \cdot (dt_r - dt^s) + T - \frac{f_1^2}{f_i^2} \cdot I_1 + \lambda_i (N_i + b_{r,i} - b_i^s) + \varepsilon_{L_i} \end{cases} \tag{5}$$

where $i$ represents the frequency number; $P_i$ and $L_i$ are raw code and phase measurements; $\rho$ denotes the geometric distance from satellite to receiver; $dt_r$ and $dt^s$ are the clock offsets at the receiver and satellite, respectively; $f_i$ is the frequency value, $T$ is the tropospheric delay, and $I_1$ denotes the ionospheric delay for $L_1$; $B_{r,i}$ and $B_i^s$ represent the code bias at the receiver/satellite end; $b_{r,i}$ and $b_i^s$ denote the phase bias at the receiver/satellite end; $\varepsilon_{P_i}$ and $\varepsilon_{L_i}$ are the unmodelled errors of code/phase measurements. The relativistic effect, Sagnac effect, Shapiro time delay [29], site displacements [30], and phase windup [31] should be corrected according to the corresponding models.

The ionosphere-free code and phase combinations are usually adopted by PPP to eliminate the first-order ionospheric delay. The ionospheric-free code and phase combinations read as [32]:

$$\begin{cases} P_{IF} = \alpha \cdot P_i + (1-\alpha) \cdot P_j = \rho + c \cdot d\bar{t}_r - c \cdot dt^s - B^s_{IF} + T + \varepsilon_{P_{IF}} \\ L_{IF} = \alpha \cdot L_i + (1-\alpha) \cdot L_j = \rho + c \cdot d\bar{t}_r - c \cdot dt^s + T + \lambda_{IF} \cdot \overline{N}_{IF} + \varepsilon_{L_{IF}} \end{cases} \tag{6}$$

where $\alpha = f_i^2 / \left(f_i^2 - f_j^2\right)$ and $\lambda_{IF} \cdot \overline{N}_{IF} = \lambda_{IF} N_{IF} - b^s_{IF} + b_{r,IF} - B_{r,IF}$, $b^s_{IF}$ and $b_{r,IF}$ are phase bias of ionospheric-free phase combination at the satellite/receiver end, $B^s_{IF}$ and $B_{r,IF}$ represent code bias of ionospheric-free code combination at the satellite/receiver end, $\lambda_{IF}$ and $N_{IF}$ are the ionospheric-free wavelength and the ionospheric-free ambiguity, respectively; $d\bar{t}_r$ is the recombined receiver clock offset, which absorbs the receiver code bias of ionospheric-free code combination; $\varepsilon_{P_{IF}}$ and $\varepsilon_{L_{IF}}$ denote the unmodelled errors of ionospheric-free code and phase combinations. It is noted that the ionospheric-free code bias at the satellite end should be corrected by applying PPP-B2b differential code bias corrections [15].

When the recovered PPP-B2b precise satellite orbits and clock offsets have been applied, the satellite orbit and clock errors are considered eliminated. The tropospheric delay can be divided into the dry and wet parts. The Saastamoinen model is usually used to correct the dry part, and the wet part must be estimated as unknown. Then the ionospheric-free code and phase combinations can be linearized as

$$\begin{cases} p_{IF} = -\boldsymbol{e}\cdot\boldsymbol{x} + c\cdot d\overline{t}_r + M_W\cdot zwd + \varepsilon_{P_{IF}} \\ l_{IF} = -\boldsymbol{e}\cdot\boldsymbol{x} + c\cdot d\overline{t}_r + M_W\cdot zwd + \lambda_{IF}\cdot\overline{N}_{IF} + \varepsilon_{L_{IF}} \end{cases} \tag{7}$$

where $\boldsymbol{e}$, $\boldsymbol{x}$ represent the unit vector from receiver to satellite, and the vector of position increment, respectively; $zwd$ denotes the zenith wet delay and $M_W$ is the corresponding mapping function. In this equation, the remaining unknown parameters include only the receiver position increment vector $\boldsymbol{x}$, the receiver clock offset $d\overline{t}_r$, the zenith wet delay $zwd$ and the ionosphere-free ambiguity $\lambda_{IF}\overline{N}_{IF}$. By adopting a Kalman filter, the unknown parameters can be exactly estimated. It should be noted that in this paper the ionospheric-free Doppler combination is used to derive the velocity [23].

### 2.2. INS Model

The mechanization of INS in the navigation frame ($n$-frame) can be expressed as an integral process for the following equation [33,34]:

$$\begin{bmatrix} \dot{\boldsymbol{p}}^n_{INS} \\ \dot{\boldsymbol{v}}^n_{INS} \\ \dot{\boldsymbol{C}}^n_b \end{bmatrix} = \begin{bmatrix} \boldsymbol{D}^{-1}\boldsymbol{v}^n_{INS} \\ \boldsymbol{C}^n_b\boldsymbol{f}^b - (2\boldsymbol{\omega}^n_{ie} + \boldsymbol{\omega}^n_{en}) \times \boldsymbol{v}^n_{INS} + \boldsymbol{g}^n \\ \boldsymbol{C}^n_b\left(\boldsymbol{\omega}^b_{ib}\times\right) - (\boldsymbol{\omega}^n_{ie} + \boldsymbol{\omega}^n_{en}) \times \boldsymbol{C}^n_b \end{bmatrix} \tag{8}$$

where $\boldsymbol{D}^{-1}$ is the diagonal matrix to transform the rectangular coordinates into geodetic coordinates; $\boldsymbol{p}^n_{INS}$ and $\boldsymbol{v}^n_{INS}$ denote the INS position vector and INS velocity vector in $n$-frame; $\boldsymbol{C}^n_b$ represents the transformation matrix from body-frame ($b$-frame) to $n$-frame; $\boldsymbol{f}^b$ is the inertial measurement unit (IMU) specific force measurement projected in $b$-frame; $\boldsymbol{\omega}^b_{ib}$ denotes the IMU angular rate measurement expressed in $b$-frame; $\boldsymbol{\omega}^n_{ie}$ is the vector of earth rotation rate expressed in $n$-frame; $\boldsymbol{g}^n$ is the gravity vector presented in $n$-frame; $\boldsymbol{\omega}^n_{en}$ represents the rotation rate of ECEF frame relative to that of $n$-frame expressed in $n$-frame; The symbol "$\times$" represents the cross-product operator. Obviously, the INS update of position, velocity, and attitude can be implement based on Equation (8).

The perturbation of position, velocity and attitude must be considered in the INS data processing. In this paper, the error model of attitude, velocity, and position is defined as the psi-angle error model, which reads as

$$\begin{bmatrix} \delta\dot{\boldsymbol{p}}^n_{INS} \\ \delta\dot{\boldsymbol{v}}^n_{INS} \\ \dot{\boldsymbol{\phi}} \end{bmatrix} = \begin{bmatrix} -\boldsymbol{\omega}^n_{en} \times \delta\boldsymbol{p}^n_{INS} + \delta\boldsymbol{v}^n_{INS} \\ -(2\boldsymbol{\omega}^n_{ie} + \boldsymbol{\omega}^n_{en}) \times \delta\boldsymbol{v}^n_{INS} + \boldsymbol{C}^n_b\boldsymbol{f}^b \times \boldsymbol{\phi} + \delta\boldsymbol{g}^n + \boldsymbol{C}^n_b\delta\boldsymbol{f}^b \\ -(\boldsymbol{\omega}^n_{ie} + \boldsymbol{\omega}^n_{en}) \times \boldsymbol{\phi} - \boldsymbol{C}^n_b\delta\boldsymbol{\omega}^b_{ib} \end{bmatrix} \tag{9}$$

where $\delta\boldsymbol{p}^n_{INS}$ and $\delta\boldsymbol{v}^n_{INS}$ are the error vector of INS position and velocity, respectively; $\boldsymbol{\phi}$ indicates the vector of misalignment angles; $\delta\boldsymbol{g}^n$ is the gravity error; $\delta\boldsymbol{f}^b$ is error vector of specific force measurement; $\delta\boldsymbol{\omega}^b_{ib}$ denotes error vector of the IMU angular rate. Generally, the error vectors of specific force and angular rate measurement include scale factor errors, bias errors, and white noise [35]. Here, only bias errors are considered and the bias errors of IMU sensors are described as the first Gauss-Markov procedure [34]:

$$\begin{bmatrix} \delta\dot{\boldsymbol{B}}_a \\ \delta\dot{\boldsymbol{B}}_g \end{bmatrix} = -\frac{1}{\tau}\begin{bmatrix} \delta\boldsymbol{B}_a \\ \delta\boldsymbol{B}_g \end{bmatrix} + \begin{bmatrix} \boldsymbol{w}_a \\ \boldsymbol{w}_g \end{bmatrix} \tag{10}$$

where $\delta B_g$ and $\delta B_a$ represent the error vector of gyro biases and accelerometer biases, respectively; $\tau$ is the correction time; and $w_a$, $w_g$ represents the driving noise for accelerometer and gyro, respectively.

### 2.3. PPP-B2b/INS Loosely Coupled Integration Model

A typical state vector of 15 states is used in PPP-B2b/INS loosely coupled integration, which can be expressed as

$$X_k = \begin{bmatrix} \delta p^n_{INS} & \delta v^n_{INS} & \phi & \delta B_a & \delta B_g \end{bmatrix}^T \tag{11}$$

The system model of PPP-B2b/INS loosely coupled integration in the discrete form can be simplified as

$$X_k = (I + F \cdot \Delta t) X_{k-1} + \Gamma_{k-1} w_{k-1}, w_{k-1} \sim N(0, Q_{k-1}) \tag{12}$$

where $X_{k-1}$ and $X_k$ are the vector of state parameter at epoch $k-1$ and $k$, respectively; $I$ is the identify matrix; $\Delta t$ denotes the time interval between two adjacent epochs; $w_{k-1}$ represents the vector of the process noise at epoch $k-1$, and $\Gamma_{k-1}$ denotes the matrix of the noise distribution; $Q_{k-1}$ represents the covariance matrix of process noise; $F$ is the dynamic matrix, which can be derived from the INS error model as follows:

$$F = \begin{bmatrix} F_{rr} & F_{rv} & 0 & 0 & 0 \\ F_{vr} & F_{vv} & \left(f^b \times\right) & C^n_b & 0 \\ F_{\phi r} & F_{\phi v} & -(\omega^n_{ie} + \omega^n_{en} \times) & 0 & -C^n_b \\ 0 & 0 & 0 & -\frac{1}{\tau} I & 0 \\ 0 & 0 & 0 & 0 & -\frac{1}{\tau} I \end{bmatrix} \tag{13}$$

where $\mathbf{0}$ denotes the zero matrix; $F_{rr}$, $F_{rv}$, $F_{vr}$, $F_{vv}$, $F_{\phi r}$, and $F_{\phi v}$ are sub-matrices of dynamic matrix, which can be derived from Equation (9). Then the Kalman time update can be expressed as [36]

$$\begin{cases} X^-_k = \Phi_{k,k-1} X^+_{k-1} \\ P^-_k = \Phi_{k,k-1} P^+_{k-1} \Phi^T_{k,k-1} + Q_{k-1} \end{cases} \tag{14}$$

where the superscript "+" and "−" denote the estimated and predicted information; $\Phi_{k,k-1} = (I + F \cdot \Delta t)$ represents the state transition matrix; $P_k$ represents the covariance matrix of the states.

When the PPP-B2b positioning solution and INS predicted solution are available at the same epoch, the Kalman filter measurement update can be operated. However, the positions and velocities from INS are referred to the IMU center, while those from PPP-B2b positioning are typically based on the GNSS antenna phase center. Therefore, the corresponding lever-arm offsets must be compensated. Here, the accurate lever-arm offsets are measured and directly applied to the position and velocity from INS. Then, the observation model for PPP-B2b/INS loosely coupled integration can be expressed as

$$Z_k = H_k X_k + \begin{bmatrix} \varepsilon_p \\ \varepsilon_v \end{bmatrix}, \begin{bmatrix} \varepsilon_p \\ \varepsilon_v \end{bmatrix} \sim N(0, R_k) \tag{15}$$

with

$$Z_k = \begin{bmatrix} p^n_{GNSS} \\ v^n_{GNSS} \end{bmatrix} - \begin{bmatrix} p^n_{INS} \\ v^n_{INS} \end{bmatrix} \tag{16}$$

$$H_k = \begin{bmatrix} I_{3\times3} & 0_{3\times3} & \left(C^n_b l^b \times\right) & 0_{3\times3} & 0_{3\times3} \\ 0_{3\times3} & I_{3\times3} & H_\phi & 0_{3\times3} & -C^n_b\left(l^b \times\right) \end{bmatrix} \tag{17}$$

where $H_\phi = \left[ (\omega_{ie}^n + \omega_{en}^n) \times C_b^n \left( l^b \times \right) + C_b^n \left( l^b \times \omega_{ib}^b \right) \times \right]$; $p_{GNSS}^n$, $v_{GNSS}^n$ denote the vector of position and velocity obtained from PPP-B2b positioning; $\begin{bmatrix} \varepsilon_p & \varepsilon_v \end{bmatrix}^T$ denotes the vector of measurement noise of position and velocity and $R_k$ represents its related covariance matrix; $l^b$ represents the vector of lever-arm offsets; $H_k$ is the design matrix; $Z_k$ is defined as the innovation vector, which consists of the difference of positions and velocities between PPP-B2b positioning solution and INS predicted solution. Finally, the Kalman measurement update can be given as

$$\begin{cases} K_k = P_k^- H_k^T \left( H_k P_k^- H_k^T + R_k \right)^{-1} \\ X_k^+ = X_k^- + K_k \left( Z_k - H_k X_k^- \right) \\ P_k^+ = (I - K_k H_k) P_k^- \end{cases} \tag{18}$$

where $K_k$ is the Kalman gain.

## 3. Description of Experiments

In order to assess the performance of BDS-3 PPP-B2b/INS loosely coupled integration, two land vehicle experiments in open-sky scenery and urban canyon scenery (hereinafter referred to as Experiment A and Experiment B, respectively) were carried out in Qingdao, China. In these two experiments, a tactical-grade IMU (ISA100C) [37] and a MEMS-grade IMU (ADIS-16505) [38] were equipped to evaluate the impact of IMU grade on BDS-3 PPP-B2b/INS loosely coupled integration. Two GNSS antennas were mounted on the roof of the land vehicle, and connected to a NovAtel PwrPak7 receiver [39] and a FRII-PLUS PPP-B2b receiver (http://www.femtomes.com, accessed on 20 October 2021). The FRII-PLUS receiver was only used to collected PPP-B2b messages, while all GNSS observations used in the evaluation were collected by the PwrPak7 receiver. The sampling rate of GNSS observations was 1 Hz. Figure 1 shows the installation of experimental equipment. Relevant specific information for the IMU sensors is listed in Table 1.

**Figure 1.** Experimental equipment mounted on the vehicle.

**Table 1.** Parameters of the IMU sensors used in the experiments.

| IMU Sensor | Grade | Sampling Rate | Bias | | Random Walk | |
|---|---|---|---|---|---|---|
| | | | Gyro. (°/h) | Acc. (mGal) | Angular (°/$\sqrt{h}$) | Velocity (m/s/$\sqrt{h}$) |
| IAS100C | Tactical | 200 Hz | 0.75 | 1000 | 0.03 | 0.1 |
| ADIS-16505 | MEMS | 100 Hz | 10 | 1500 | 0.34 | 0.18 |

During these two experiments, a GNSS base station was established on the roof of Engineering-C building in the China University of Petroleum (East China). It should be noted that the distance between the base and rover stations was less than 5 km. Using GNSS observations collected by PwrPak7, and IMU data from ISA100C, together with GNSS measurements at the base station, the smoothed RTK/INS tightly coupled solution was obtained by commercial Inertial Explorer software (IE 8.90) from the NovAtel company, and this solution was used as reference.

In PPP-B2b positioning, the B1C and B2a signals of BDS-3, L1 and L2 signals of GPS were selected to form the ionospheric-free combination, and the cut-off elevation angle was set to 7 degrees. The remaining processing strategies are shown in detail in Table 2.

**Table 2.** Strategies of PPP-B2b positioning.

| Item | Model/Strategy |
|---|---|
| Satellite orbit and clock | Applying the PPP-B2b precise products |
| Weight method | Elevation-dependent weight [40] |
| Phase windup | Corrected [31] |
| Sagnac effect | Corrected [29] |
| Special relativistic effect | Corrected [29] |
| Shapiro time delay | Corrected [29] |
| Tidal effects | Corrected according to IERS Conventions 2010 [30] |
| Troposphere | Saastamoinen model [41] is adopted to correct hydrostatic delay, the zenith wet delay is estimated for each epoch as a random walk noise, and Global Mapping Function [42] is used. |
| Phase ambiguity | Estimated as a constant for each ambiguity arc |
| Estimator | Kalman filter |

## 4. Result and Discussion

In this section, the performance of BDS-3 PPP-B2b/INS loosely coupled integration was investigated in both open-sky and complex urban environments.

### 4.1. Experiment A

Experiment A was conducted from 16:30:40 to 16:59:59 on 3 December 2021 in GPS time. The trajectory of the land vehicle is shown in Figure 2.

During Experiment A, the velocities of the land vehicle were within ±20 m/s in the east and north directions and within ±1 m/s in the vertical direction. The number of visible BDS-3/GPS satellites and the position dilution of precision (PDOP) values are shown in Figure 3. The number of visible satellites ranged from 9 to 15 with an average value of 12.5, and the PDOP varied from 1.28 to 2.70 with an average value of 1.59.

Figure 4 shows the positioning errors of PPP-B2b only, PPP-B2b/MEMS-IMU loosely coupled integration, and PPP-B2b/tactical-IMU loosely coupled integration, respectively. As with traditional PPP processing, a convergence period is indispensable in PPP-B2b positioning and PPP-B2b/INS loosely coupled integration. In this study, the convergence time was defined for horizontal/vertical positioning accuracy better than 30 cm/60 cm, with such a positioning accuracy for at least 60 continuous epochs. The statistics of the positioning results are summarized in Table 3. It should be noted that positioning results before full convergence were excluded when calculating positioning accuracy. The biases in the east, north and up directions were 6.7/23.7/26.5 cm for PPP-B2b only, and the corresponding values for PPP-B2b/tactical-IMU and PPP-B2b/MEMS-IMU loosely coupled integration schemes were 6.6/23.3/26.0 cm and 6.2/23.5/25.8 cm, respectively. In terms of the root mean square (RMS) value of positioning errors, the improvement of PPP-B2b/INS loosely coupled integration was not significant compared to PPP-B2b only. This is because Experiment A was carried out within an open-sky environment and thus PPP-B2b on its own was already able to obtain high-precision positioning accuracy.

**Figure 2.** The trajectory of land vehicle in Experiment A.

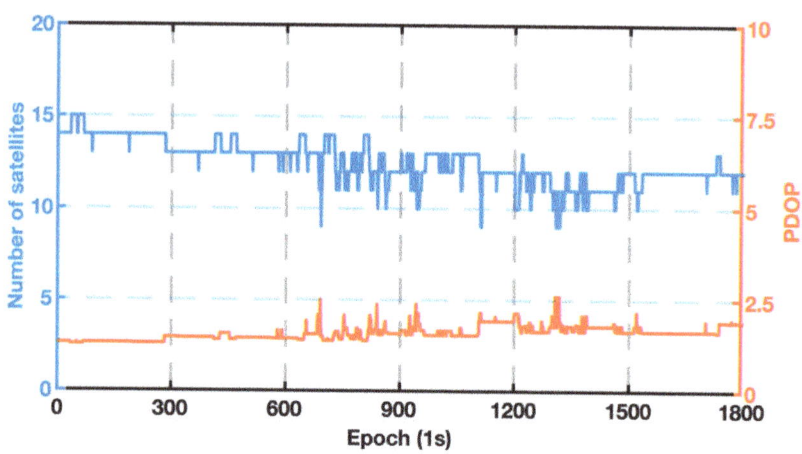


**Figure 3.** Number of visible BDS-3/GPS satellites and PDOPs in Experiment A.

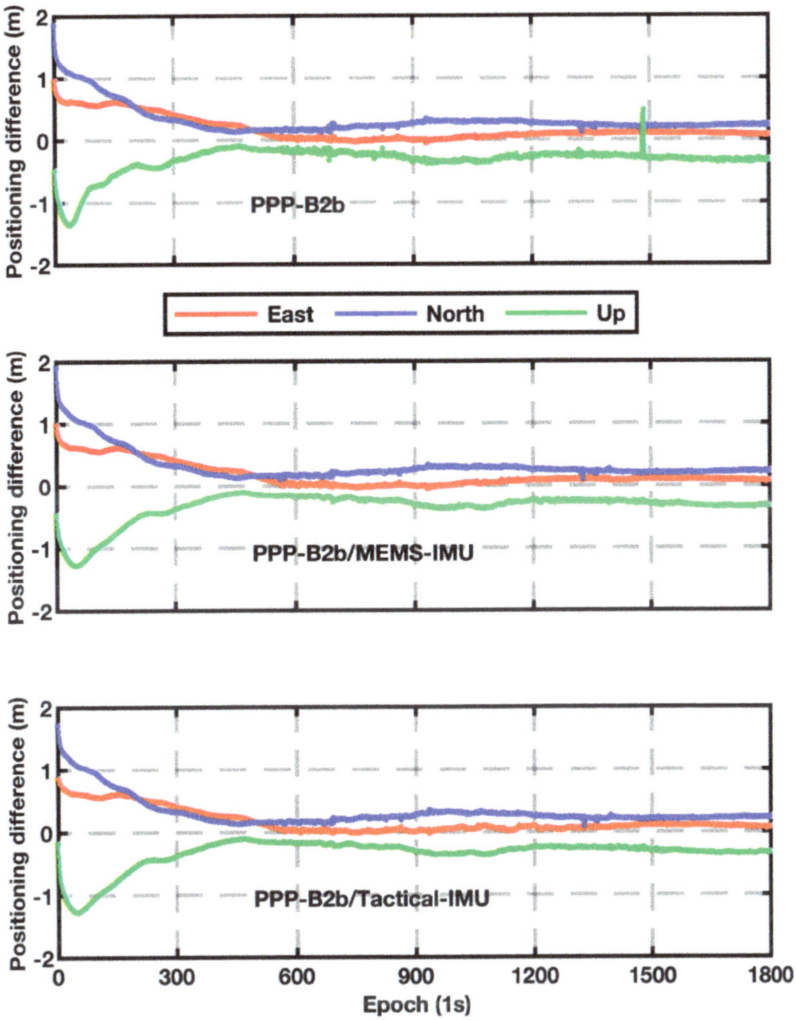


**Figure 4.** Positioning errors of PPP-B2b only (**top** panel), PPP-B2b/MEMS-IMU loosely coupled integration (**middle** panel), and PPP-B2b/tactical−IMU loosely coupled integration (**bottom** panel), for Experiment A.

**Table 3.** Statistics of the positioning errors (unit: cm).

| Schemes | Bias | | | RMS | | |
|---|---|---|---|---|---|---|
| | East | North | Up | East | North | Up |
| PPP-B2b | 6.7 | 23.7 | 26.5 | 8.1 | 24.2 | 27.5 |
| PPP-B2b/MEMS-IMU | 6.6 | 23.3 | 26.0 | 7.9 | 24.0 | 27.0 |
| PPP-B2b/tactical-IMU | 6.2 | 23.5 | 25.8 | 7.1 | 23.8 | 26.5 |

It is recognized that PPP-B2b only is unable to obtain a high-precision solution during GNSS signal outages, while PPP-B2b/INS loosely coupled integration continues to work by INS mechanization. Four periods of GNSS signal outage were simulated to present the superiority of PPP-B2b/INS loosely coupled integration. The periods of GNSS outage were simulated from the 600th/900th/1200th/1500th epoch, and each period lasted 30 s. The corresponding positioning errors of PPP-B2b/MEMS-IMU and PPP-B2b/tactical-IMU loosely coupled integration are plotted in Figures 5 and 6, respectively. Clearly, the positioning accuracy become progressively worse with increased time of GNSS outages in both PPP-B2b/MEMS-IMU and PPP-B2b/tactical-IMU loosely coupled integration schemes. When the GNSS signal outage time increased to 30 s, the positioning errors in the east, north, and up directions decreased to 300.0 cm, 498.0 cm, and 41.0 cm, respectively, for PPP-B2b/MEMS-IMU loosely coupled integration. The results for the PPP-B2b/tactical-IMU loosely coupled integration scheme dropped to 18.6 cm, 21.8 cm, and 6.1 cm in the east, north, and up components, respectively. Table 4 lists the statistics results, showing that the positioning performance of PPP-B2b/tactical-IMU loosely coupled integration is significantly better than that of PPP-B2b/MEMS-IMU loosely coupled integration.

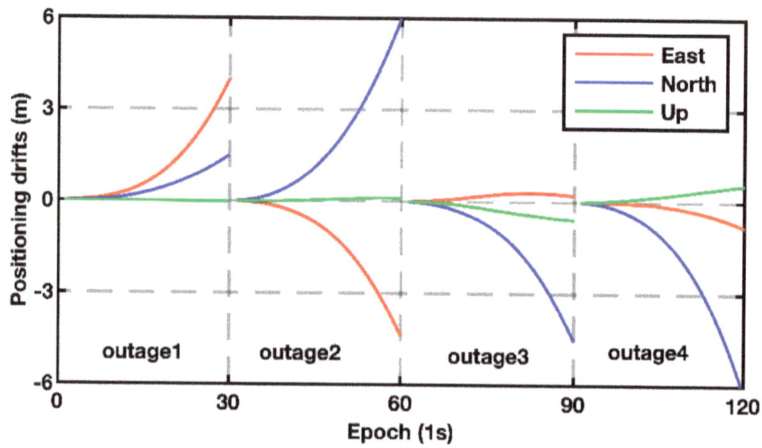


**Figure 5.** Positioning errors of the PPP-B2b/MEMS−IMU loosely coupled integration during the simulated GNSS signal outages.

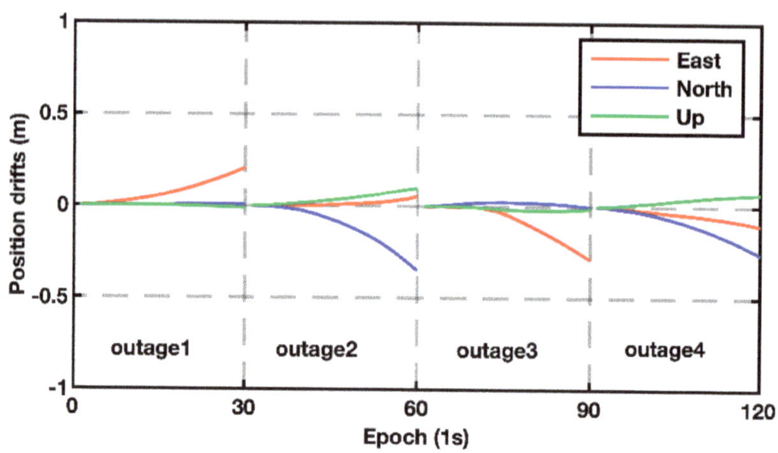


**Figure 6.** Positioning errors of the PPP-B2b/tactical−IMU loosely coupled integration during the simulated GNSS signal outages.

**Table 4.** Statistics of positioning errors for PPP-B2b/INS loosely coupled integration during GNSS outages (unit: cm).

| Schemes | Direction | 5 s | 10 s | 15 s | 20 s | 25 s | 30 s |
|---|---|---|---|---|---|---|---|
| PPP-B2b/tactical-IMU | East | 0.5 | 1.8 | 3.9 | 7.7 | 12.7 | 18.6 |
| | North | 0.8 | 2.5 | 5.3 | 9.3 | 14.6 | 21.8 |
| | Up | 0.4 | 1.3 | 2.5 | 3.7 | 5.0 | 6.1 |
| PPP-B2b/MEMS-IMU | East | 3.8 | 17.4 | 45.7 | 97.4 | 180.5 | 300.6 |
| | North | 8.0 | 35.2 | 87.3 | 174.6 | 308.9 | 498.0 |
| | Up | 1.8 | 7.2 | 15.7 | 24.7 | 33.4 | 41.0 |

### 4.2. *Experiment B*

To fully demonstrate the performance of BDS-3 PPP-B2b/INS loosely coupled integration in an urban scenario, we conducted Experiment B in a complex urban environment. Experiment B was carried out from 09:00:00 to 09:49:59 on 5 December 2021 in GPS time. The trajectory of the land vehicle is shown in Figure 7.

**Figure 7.** The trajectory of the land vehicle in Experiment B.

Figure 8 presents the number of visible BDS-3/GPS satellites and the values of PDOP. The number of available BDS-3/GPS satellites dropped frequently, and at the same time the PDOP value significantly increased as several typical city features such as tall buildings, trees, and over-bridges appeared along the vehicle route.

During the first 35 min, there were enough satellites for the implementation of PPP-B2b positioning. The positioning errors of PPP-B2b only, PPP-B2b/MEMS-IMU loosely coupled integration and PPP-B2b/tactical-IMU loosely coupled integration are plotted in Figure 9. It can be found that all three schemes provided similar positioning accuracy in this 35-min period. The RMS values of positioning errors were 20.7 cm, 17.3 cm, and 47.6 cm in the east, north, and up directions for PPP-B2b only. For PPP-B2b/INS loosely coupled integration, the corresponding RMS values were 20.8 cm, 17.4 cm, and 47.5 cm for the PPP-B2b/MEMS-IMU scheme, and 20.3 cm, 17.4 cm, and 47.5 cm for the PPP-B2b/tactical-IMU scheme. It should be noted that a convergence period of 15 min was removed when we computed the positioning accuracy. Clearly, the equipped IMU sensors could not improve the performance of the PPP-B2b service when there was no GNSS signal blockage.

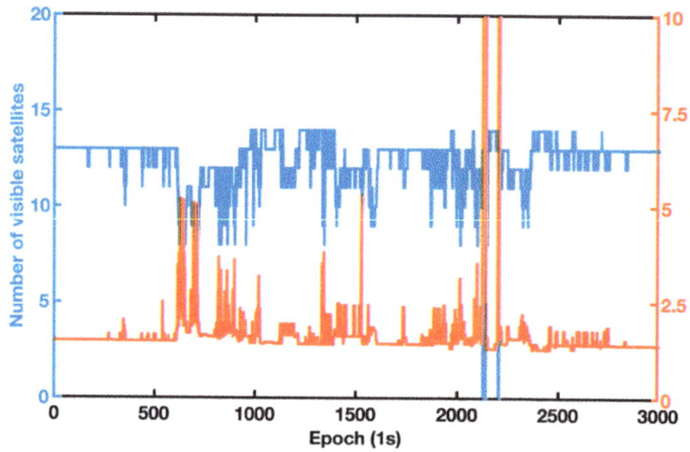


**Figure 8.** Number of visible BDS-3/GPS satellites and PDOPs in Experiment B.

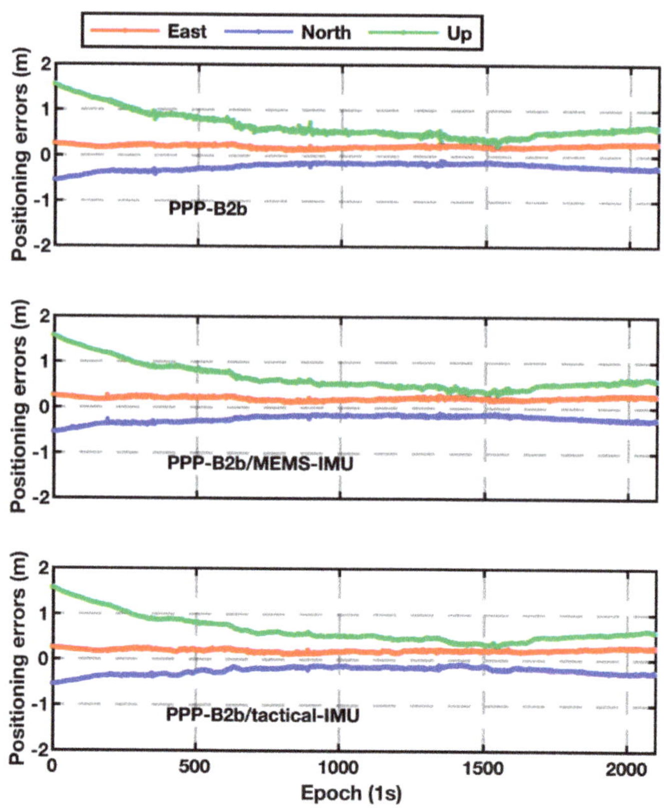


**Figure 9.** Positioning errors of PPP-B2b only (**top** panel), PPP-B2b/MEMS-IMU loosely coupled integration (**middle** panel), and PPP-B2b/tactical−IMU loosely coupled integration (**bottom** panel) during the period from 09:00:00 to 09:34:59.

From 09:35:00 to the end of Experiment B, the land vehicle crossed two over-bridges, as shown in Figure 10. Two GNSS signal outages appeared when the land vehicle crossed the bridges, lasting 15 s and 5 s, respectively. During these two GNSS signal outages, PPP-B2b positioning could be implemented, because the number of visible satellites was less than four. Therefore, the PPP-B2b positioning was re-converged after crossing the bridges. Figure 11 shows the positioning errors of PPP-B2b only, PPP-B2b/MEMS-IMU loosely coupled integration, and PPP-B2b/tactical-IMU loosely coupled integration. It can be seen that the positioning accuracy of PPP-B2b/INS loosely coupled integration exhibited better performance than PPP-B2b positioning during the re-convergence phase. The overall positioning accuracies for the three schemes were calculated from 09:35:00 to 09:49:59 and are presented in Table 5. The RMS values of positioning errors for the east, north, and up components were 114.2 cm, 200.0 cm, and 244.3 cm for PPP-B2b only. These figures improved to 41.0 cm, 44.5 cm, and 64.0 cm for PPP-B2b/MEMS-IMU loosely coupled integration, and to 41.2 cm, 41.0 cm, and 62.5 cm for PPP-B2b/tactical-IMU loosely coupled integration. Obviously, the performance of PPP-B2b/INS loosely coupled integration was significantly improved compared to PPP-B2b only in the last 15-min period of Experiment B. The positioning accuracy of PPP-B2b/tactical-IMU loosely coupled integration was slightly better than that of PPP-B2b/MEMS-IMU loosely coupled integration.

**Figure 10.** Two bridge scenarios during the period from 09:35:00 to 09:49:59.

**Table 5.** The positioning accuracy during the period from 09:35:00 to 09:49:59 (unit: cm).

| Schemes | Bias | | | RMS | | |
|---|---|---|---|---|---|---|
| | **East** | **North** | **Up** | **East** | **North** | **Up** |
| PPP-B2b | 14.9 | 48.1 | 12.9 | 114.2 | 200.0 | 244.3 |
| PPP-B2b/MEMS-IMU | 31.7 | 25.1 | 25.2 | 41.0 | 44.5 | 64.0 |
| PPP-B2b/tactical-IMU | 32.1 | 24.2 | 25.1 | 41.2 | 41.0 | 62.5 |

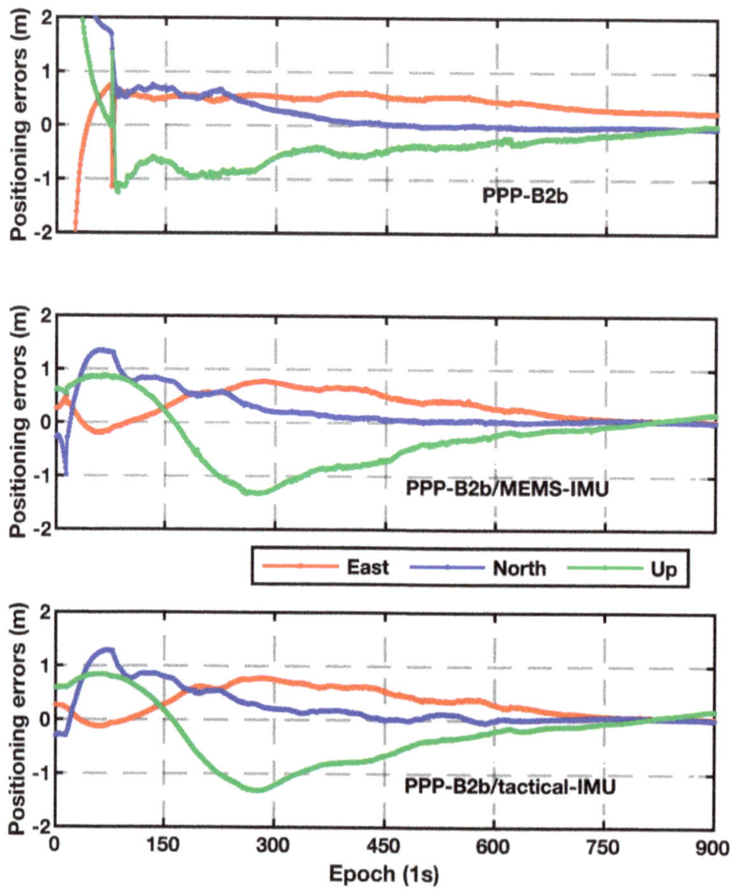


**Figure 11.** Positioning errors of PPP-B2b only (**top** panel), PPP-B2b/MEMS-IMU loosely coupled integration (**middle** panel), and PPP-B2b/tactical−IMU loosely coupled integration (**bottom** panel) during the period from 09:35:00 to 09:49:59.

## 5. Conclusions

Since July 2020, PPP-B2b has served as a featured service of BDS-3 to support real-time PPP. Compared to the IGS RTS service, PPP-B2b is a satellite-based service, which is not limited by Internet communication. Its high-precision positioning performance shows great potential in real-time GNSS applications. However, the BDS-3 PPP-B2b service continues to encounter challenges in urban environments due to GNSS signal outages. In order to overcome the drawbacks of the BDS-3 PPP-B2b service in urban environments, we set up a BDS-3 PPP-B2b/INS loosely coupled integration system and evaluated its performance in different urban scenarios.

The experimental results indicate that PPP-B2b/INS loosely coupled integration cannot improve the positioning performance in open-sky environments where there are enough GPS/BDS-3 satellites for PPP-B2b positioning. However, PPP-B2b/INS loosely coupled integration can show its superiority when GNSS signal outage occurs. During GNSS signal outages, the INS mechanization can provide continuous positioning. Therefore, the performance of PPP-B2b positioning can be improved by adopting PPP-B2b/INS loosely coupled integration in GNSS blockage environments. In the first experiment, we simulated four 30-s periods of GNSS signal outages. At the last epoch of the simulated outages, the averaged positioning errors in the east, north, and up directions were 300.0 cm, 498.0 cm, and 41.0 cm, respectively, for PPP-B2b/MEMS-IMU loosely coupled integration, and 18.6 cm, 21.8 cm, and 6.1 cm, respectively, for PPP-B2b/tactical-IMU loosely coupled integration. In addition, we also evaluated the performance of PPP-B2b/INS loosely coupled integration in a real complex urban environment. When the land vehicle crossed bridges, two GNSS signal interruptions appeared that lasted 15 s and 5 s, respectively. Compared to PPP-B2b only, the respective improvement of positioning accuracy in the east, north, and up components was 64.1%, 77.8%, and 73.8% for PPP-B2b/MEMS-IMU loosely coupled integration, and 63.9%, 79.5%, and 74.4% for PPP-B2b/Tactical-IMU loosely coupled integration.

It can be concluded that when there are sufficient valid satellites for PPP-B2b positioning, the positioning accuracy of PPP-B2b/INS loosely coupled integration depends mainly on the accuracy of PPP-B2b. In urban environments, positioning performance can be significantly improved compared to PPP–B2b alone, by adopting PPP-B2b/INS loosely coupled integration especially in GNSS signal blockage environments.

**Author Contributions:** X.X.; Conceptualization, methodology, software, visualization, and writing—original draft preparation. Z.N.; Conceptualization, software, writing—review and editing, and funding acquisition. Z.W.; writing—review and editing, supervision, project administration, and funding acquisition. B.W.; validation and formal analysis. Q.D.; validation and formal analysis. All authors have read and agreed to the published version of the manuscript.

**Funding:** This study was supported by the National Natural Science Foundation of China (No. 42104011), the Shandong Provincial Natural Science Foundation (No. ZR2021QD069), the National Key Research and Development Program of China (No. 2019YFC1509205), and Sinopec Science and Technology Entry Program (JP210004).

**Data Availability Statement:** The datasets used in this study are managed by the College of Oceanography and Space Informatics, China University of Petroleum, Qingdao, China and can be available on request from the corresponding author.

**Conflicts of Interest:** The authors declare no conflict of interest.

## References


1. Bock, Y.; Nikolaidis, R.M.; de Jonge, P.J.; Bevis, M. Instantaneous geodetic positioning at medium distances with the Global Positioning System. *J. Geophys. Res. Solid Earth* **2020**, *105*, 28223–28253. [CrossRef]
2. Rizos, C. Network RTK research and implementation-a geodetic perspective. *J. Glob. Posit. Syst.* **2009**, *1*, 144–151. [CrossRef]
3. Kouba, J.; Héroux, P. Precise point positioning using IGS orbit and clock products. *GPS Solut.* **2001**, *5*, 12–28. [CrossRef]
4. Zumberge, J.; Heflin, M.; Jefferson, D.; Watkins, M.; Webb, F. Precise point positioning for the efficient and robust analysis of GPS data from large networks. *J. Geophys. Res. Solid Earth* **1997**, *102*, 5005–5017. [CrossRef]
5. Teunissen, P.; Montenbruck, O. *Springer Handbook of Global Navigation Satellite Systems*; Springer International Publishing: New York, NY, USA, 2017.
6. Hadas, T.; Bosy, J. IGS RTS precise orbits and clocks verification and quality degradation over time. *GPS Solut.* **2015**, *19*, 93–105. [CrossRef]
7. Nie, Z.; Gao, Y.; Wang, Z.; Ji, S.; Yang, H. An approach to GPS clock prediction for real-time PPP during outages of RTS stream. *GPS Solut.* **2018**, *22*, 14. [CrossRef]
8. Elsobeiey, M.; Al-Harbi, S. Performance of real-time Precise Point Positioning using IGS real-time service. *GPS Solut.* **2016**, *20*, 565–571. [CrossRef]
9. Nie, Z.; Wang, B.; Wang, Z.; He, K. An offshore real-time precise point positioning technique based on a single set of BeiDou short-message communication devices. *J. Geod.* **2020**, *94*, 1–11. [CrossRef]
10. Cai, H.; Meng, Y.; Geng, C.; Gao, W.; Zhang, T.; Li, G.; Shao, B.; Xin, J.; Lu, H.; Mao, Y. BDS-3 performance assessment: PNT, SBAS, PPP, SMC and SAR. *Acta Geod. Cartogr. Sinica.* **2021**, *50*, 427–435.
11. Yang, Y.; Liu, L.; Li, J.; Yang, Y.; Zhang, T.; Mao, Y.; Sun, B.; Ren, X. Featured services and performance of BDS-3. *Sci. Bull.* **2021**, *66*, 2135–2143. [CrossRef]
12. China Satellite Navigation Office (CSNO). BeiDou Navigation Satellite System Open Service Performance Standard (Version 2.0). Available online: http://www.beidou.gov.cn/xt/gfxz/201812/P020181227529449178798.pdf (accessed on 30 September 2020).
13. Liu, C.; Gao, W.; Liu, T.; Wang, D.; Yao, Z.; Gao, Y.; Nie, X.; Wang, W.; Li, D.; Zhang, W.; et al. Design and implementation of a BDS precise point positioning service. *Navig. J. Instit. Navig.* **2020**, *67*, 875–891. [CrossRef]
14. Yang, Y.; Gao, W.; Guo, S.; Mao, Y.; Yang, Y. Introduction to BeiDou-3 navigation satellite system. *Navigation* **2019**, *66*, 7–18. [CrossRef]

15. Nie, Z.; Xu, X.; Wang, Z.; Du, J. Initial Assessment of BDS PPP-B2b Service: Precision of Orbit and Clock Corrections, and PPP Performance. *Remote Sens.* **2021**, *13*, 2050. [CrossRef]
16. Zhang, W.; Lou, Y.; Song, W.; Sun, W.; Zou, X.; Gong, X. Initial assessment of BDS-3 precise point positioning service on GEO B2b signal. *Adv. Space Res.* **2021**, *69*, 690–700. [CrossRef]
17. Ren, Z.; Gong, H.; Peng, J.; Tang, C.; Huang, X.; Sun, G. Performance assessment of real-time precise point positioning using BDS PPP-B2b service signal. *Adv. Space Res.* **2021**, *68*, 3242–3254. [CrossRef]
18. Tao, J.; Liu, J.; Hu, Z.; Zhao, Q.; Chen, G.; Ju, B. Initial Assessment of the BDS-3 PPP-B2b RTS compared with the CNES RTS. *GPS Solut.* **2021**, *25*, 1–16. [CrossRef]
19. Xu, Y.; Yang, Y.; Li, J. Performance evaluation of BDS-3 PPP-B2b precise point positioning service. *GPS Solut.* **2021**, *25*, 1–14. [CrossRef]
20. Cox, D.B. Integration of GPS with Inertial Navigation Systems. *Navig. J. Inst. Navig.* **1978**, *25*, 236–245. [CrossRef]
21. Abd Rabbou, M.; El-Rabbany, A. Tightly coupled integration of GPS precise point positioning and MEMS-based inertial systems. *GPS Solut.* **2015**, *19*, 601–609. [CrossRef]
22. Li, X.; Li, X.; Huang, J.; Shen, Z.; Wang, B.; Yuan, Y.; Zhang, K. Improving PPP-RTK in urban environment by tightly coupled integration of GNSS and INS. *J. Geod.* **2021**, *95*, 1–18. [CrossRef]
23. Liu, S.; Sun, F.; Zhang, L.; Li, W.; Zhu, X. Tight integration of ambiguity-fixed PPP and INS: Model description and initial results. *GPS Solut.* **2016**, *20*, 39–49. [CrossRef]
24. Roesler, G.; Martell, H. Tightly coupled processing of precise point positioning (PPP) and INS data. In Proceedings of the 22nd International Technical Meeting of the Satellite Division of The Institute of Navigation (ION GNSS 2009), Savannah, GA, USA, 22–25 September 2009; pp. 1898–1905.
25. Zhang, X.; Zhu, F.; Zhang, Y.; Mohamed, F.; Zhou, W. The improvement in integer ambiguity resolution with INS aiding for kinematic precise point positioning. *J. Geod.* **2019**, *93*, 993–1010. [CrossRef]
26. China Satellite Navigation Office (CSNO). BeiDou Navigation Satellite System Signal in Space Interface Control Document Precise Point Positioning Service Signal PPP-B2b (Version 1.0). Available online: http://www.beidou.gov.cn/xt/gfxz/202008/P020200803362062482940.pdf (accessed on 30 September 2020).
27. Leick, A.; Rapoport, L.; Tatarnikov, D. *GPS Satellite Surveying*, 4th ed.; John Wiley & Sons: Hoboken, NJ, USA, 2015.
28. Zhang, B.; Hou, P.; Liu, T. PPP-RTK functional models formulated with undifferenced and un combined GNSS observations. *Satell. Navig.* **2022**, *3*, 1–15. [CrossRef]
29. Ashby, N. Relativity in the Global Positioning System Imprint/Terms of Use. *Living Rev. Relativ.* **2003**, *6*, 1–42. [CrossRef]
30. Petit, G.; Luzum, B. *IERS Conventions (2010)*; Bureau International Des Poids et Mesures Sevres: Frankfurt, Germany, 2010.
31. Wu, J.-T.; Wu, S.C.; Hajj, G.A.; Bertiger, W.I.; Lichten, S.M. Effects of antenna orientation on GPS carrier phase. In Proceedings of the AAS/AIAA Astrodynamics Conference, Durango, CO, USA, 19–22 August 1991; pp. 1647–1660.
32. Geng, J.; Meng, X.; Dodson, A.; Teferle, F. Integer ambiguity resolution in precise point positioning: Method comparison. *J. Geod.* **2010**, *84*, 569–581. [CrossRef]
33. Niu, X.; Zhang, Q.; Gong, L.; Liu, C.; Zhang, H.; Shi, C.; Wang, J.; Coleman, M. Development and evaluation of GNSS/INS data processing software for position and orientation systems. *Surv. Rev.* **2015**, *47*, 87–98. [CrossRef]
34. Shin, E.H. Estimation Techniques for Low-Cost Inertial Navigation. Ph.D. Thesis, University of Calgary, Calgary, AB, Canada, 2005.
35. Groves, P.D. *Principles of GNSS, Inertial, and Multisensor Integrated Navigation Systems*, 2nd ed.; Artech House: Norwood, MA, USA, 2013.
36. Yang, Y. *Adaptive Navigation and Kinematic Positioning*; Surveying and Mapping Press: Beijing, China, 2006.
37. NovAtel Inc. IMU-ISA-100C Product Sheet. Available online: https://hexagondownloads.blob.core.windows.net/public/Novatel/assets/Documents/Papers/IMUISA100CD19539v2/IMU-ISA-100CProductSheet.pdf (accessed on 15 October 2021).
38. Analog Devices Inc. ADIS16505 Data Sheet. Available online: https://www.analog.com/media/en/technical-documentation/data-sheets/adis16505.pdf (accessed on 20 October 2021).
39. NovAtel Inc. PwrPak7 Product Sheet. Available online: https://hexagondownloads.blob.core.windows.net/public/Novatel/assets/Documents/Papers/PwrPak7-Product-Sheet/PwrPak7-Product-Sheet.pdf (accessed on 15 October 2021).
40. Ge, M.; Gendt, G.; Rothacher, M.; Shi, C.; Liu, J. Resolution of GPS carrier-phase ambiguities in precise point positioning (PPP) with daily observations. *J. Geod.* **2008**, *82*, 389–399. [CrossRef]
41. Saastamoinen, J. Atmospheric correction for the troposphere and stratosphere in radio ranging satellites. *J. Geophys. Res. Atmosph.* **1972**, *15*, 247–251. [CrossRef]
42. Böhm, J.; Niell, A.; Tregoning, P.; Schuh, H. Global Mapping Function (GMF): A new empirical mapping function based on numerical weather model data. *Geophys. Res. Lett.* **2006**, *33*, 1–4. [CrossRef]

remote sensing

MDPI

*Article*

# Performance of Multi-GNSS in the Asia-Pacific Region: Signal Quality, Broadcast Ephemeris and Precise Point Positioning (PPP)

**Mengyuan Li [1], Guanwen Huang [1,2,*], Le Wang [1], Wei Xie [1] and Fan Yue [1]**

1 College of Geology Engineering and Geomatics, Chang'an University, Xi'an 710054, China; limy@chd.edu.cn (M.L.); wangle18@chd.edu.cn (L.W.); chdxiewei@chd.edu.cn (W.X.); south_wind@chd.edu.cn (F.Y.)
2 Key Laboratory of Ecological Geology and Disaster Prevention, Ministry of Natural Resources, Xi'an 710054, China
* Correspondence: guanwen@chd.edu.cn; Tel.: +86-13636801167

**Abstract:** Since BeiDou Navigation Satellite System (BDS) and Japan's Quasi-Zenith Satellite System (QZSS) have more visible satellites in the Asia-Pacific region, and navigation satellites of Global Positioning System (GPS), Galileo satellite navigation system (Galileo), and GLONASS satellite navigation system (GLONASS) are uniformly distributed globally, the service level of multi-mode Global Navigation Satellite System (GNSS) in the Asia-Pacific region should represent the best service capability. Based on the observation data of 10 Multi-GNSS Experiment (MGEX) stations, broadcast ephemeris and precision ephemeris from 13 to 19 October 2021, this paper comprehensively evaluated the service capability of multi-GNSS in the Asia-Pacific region from three aspects of observation data quality, broadcast ephemeris performance, and precision positioning level. The results show that: (1) the carrier-to-noise-density ratio (C/N0) quality of the GPS and Galileo is the best, followed by BDS and GLONASS, and QZSS is the worst. GPS, BDS-2, GLONASS, and QZSS pseudorange multipath values range from 0 to 0.6 m, while Galileo system and BDS-3 pseudorange multipath values range from 0 to 0.8 m. (2) In terms of broadcast ephemeris accuracy, BDS-3 broadcast ephemeris has the best orbit, and the three-dimensional (3D) Root Mean Square (RMS) is 0.21 m; BDS-2 was the worst, with a 3D RMS of 1.99 m. The broadcast ephemeris orbits of GPS, Galileo, QZSS, and GLONASS have 3D RMS of 0.60 m, 0.62 m, 0.83 m, and 1.27 m, respectively. For broadcast ephemeris clock offset: Galileo has the best performance, 0.61 ns, GLONASS is the worst, standard deviation (STD) is 3.10 ns, GPS, QZSS, BDS-3 and BDS-2 are 0.65 ns, 0.75 ns, and 1.72 ns, respectively. For signal-in-space ranging errors (SISRE), the SISRE results of GPS and Galileo systems are the best, fluctuating in the range of 0 m–2 m, followed by QZSS, BDS-3, Galileo, and BDS-2. (3) GPS, BDS, GLONASS, Galileo, GPS/QZSS, and BDS/QZSS were used for positioning experiments. In static PPP, the convergence time and positioning accuracy of GPS show the best performance. The positioning accuracy of GPS/QZSS and BDS/QZSS is improved compared with that of GPS and BDS. In terms of kinematic PPP, the convergence time and positioning accuracy of GPS/QZSS and BDS/QZSS are improved compared with that of GPS and BDS. In addition to GLONASS and Galileo systems, the other combinations outperformed 3 cm, 3 cm, and 5 cm in the east, north, and up directions.

**Keywords:** C/N0; pseudorange multipath; broadcast ephemeris clock offset accuracy; broadcast ephemeris orbit accuracy; SISRE; PPP

**Citation:** Li, M.; Huang, G.; Wang, L.; Xie, W.; Yue, F. Performance of Multi-GNSS in the Asia-Pacific Region: Signal Quality, Broadcast Ephemeris and Precise Point Positioning (PPP). *Remote Sens.* **2022**, *14*, 3028. https://doi.org/10.3390/rs14133028

Academic Editors: Chuang Shi, Shengfeng Gu, Yidong Lou and Xiaopeng Gong

Received: 16 May 2022
Accepted: 22 June 2022
Published: 24 June 2022

**Publisher's Note:** MDPI stays neutral with regard to jurisdictional claims in published maps and institutional affiliations.

**Copyright:** © 2022 by the authors. Licensee MDPI, Basel, Switzerland. This article is an open access article distributed under the terms and conditions of the Creative Commons Attribution (CC BY) license (https://creativecommons.org/licenses/by/4.0/).

## 1. Introduction

The accurate positioning, navigation, and timing (PNT) services can be provided by the Global Navigation Satellite System (GNSS), and it has been widely used in many areas such as agriculture [1], weather monitoring [2], time and frequency transfer [3], and disaster monitoring [4]. As of the end of April 2022, there are about 120 GNSS satellites, including GPS, GLONASS, Galileo, BDS-2, BDS-3, and QZSS systems, which can provide PNT services to users around the world [5–7]. The available satellites, frequency, and PRN for each system are listed in Table 1. For GNSS, the observation data quality on the receiver side directly determines the result of precise data processing at the GNSS server and user end. The positioning performance at the GNSS user side plays a key role in the quality of GNSS service, and the broadcast ephemeris directly affects the performance of real-time navigation and positioning. The stations located in the Asia-Pacific region can receive the signals broadcasted by five navigation satellite systems at the same time. Therefore, it is worthy to investigate and evaluate the performance of GNSS observation data quality, broadcast ephemeris, and positioning performance in the Asia-Pacific region.

**Table 1.** Available satellites of each system at the end of April 2022.

| System | Satellite Types | Signals | Number of Satellites | PRN |
|---|---|---|---|---|
| GPS | IIR | L1C/A, L1/L2 | 1 | G13 G20 G22 G16 G21 G19 G02 G17 G31 G12 G15 G25 G07 |
| | IIF | L1C/A, L2C, L5 | 11 | G01 G24 G27 G30 G06 G09 G03 G26 G08 G10 G32 |
| | III | L1C/A, L2C, L5 | 5 | G04 G11 G14 G18 G23 |
| BDS-2 | GEO | B1I, B2I, B3I | 5 | C01 C02 C03 C04 C05 |
| | IGSO | B1I, B2I, B3I | 7 | C06 C07 C08 C09 C10 C13 C16 |
| | MEO | B1I, B2I, B3I | 3 | C11 C12 C14 |
| BDS-3 | MEO | B1I, B3I, B1C, B2a, B2b | 24 | C19 C20 C21 C22 C23 C24 C25 C26 C27 C28 C29 C30 C32 C33 C34 C35 C36 C37 C41 C42 C43 C44 C45 C46 |
| | IGSO | B1I, B3I, B1C, B2a, B2b | 3 | C38 C39 C40 |
| GLONASS | GLONASS-M | G1C, G2C | 20 | R01 R02 R03 R04 R05 R06 R07 R08 R10 R12 R13 R14 R15 R16 R18 R19 R20 R22 R23 R24 |
| | GLONASS-M+ | G1C, G2C, L3 | 2 | R17 R21 |
| | GLONASS-K1 | G1C, G2C, L3 | 2 | R09 R11 |
| Galileo | IOV | E1, E5a/b/ab | 4 | E11 E12 E19 E20 |
| | FOC | E1, E5a/b/ab, E6 | 22 | E01 E02 E03 E04 E05 E07 E08 E09 E10 E13 E14 E15 E18 E22 E24 E26 E27 E30 E31 E33 E34 E36 |
| QZSS | GEO | L1C/A, L1C, L2C, L5 | 1 | J07 |
| | QZO | L1C/A, L1C, L2C, L5 | 3 | J01 J02 J03 |

For the observation data quality of GNSS: the BDS-3 satellites signals were evaluated by Zhang et al.; the B1I/B2I/B3I signals of the BDS-2 satellites, L1/L2/L5 of the GPS Block IIF satellites, and E1/E5a/E5b of the Galileo satellites are also evaluated for comparison in their study [8]. The observation data quality of the BDS-3 signal was studied and analyzed by Yang et al. [9]. The signal quality of BDS-2 and BDS-3 satellites was evaluated by Yan et al. [10]. The observation data quality and positioning performance of BDS/QZSS satellites were studied by Bu et al. [11]. The data quality of BDS/GPS/GLONASS satellites was investigated by Zuo et al. [12]. The signal quality of Galileo/BDS/GPS satellites was evaluated by Tian et al. [13]. It can be seen that previous studies about signal quality mainly focus on single and dual systems; the investigation and comparison of observation data quality among five GNSS systems in Asia-Pacific regions are limited and still needs further study.

In the context of GNSS broadcast ephemeris accuracy assessment, the GPS/GLONASS/BDS/Galileo/QZSS broadcast ephemeris was evaluated by Montenbruck et al. [14], but the BDS-3 satellite constellation had not been built at that time. After the BDS-3 satellites were launched, their broadcast ephemeris orbit and clock offset accuracy were evaluated by many scholars [15,16], and the comparison of broadcast ephemeris orbit and clock offset accuracy between BDS-2 and BDS-3 was also conducted [17]. However, with the modernization of GNSS satellites, the performance of broadcast ephemeris orbit and clock offset accuracy of five GNSS systems satellites is still unknown and unrevealed, it is urgent to conduct a comprehensive evaluation and comparison of the broadcast ephemeris performance for five GNSS systems satellites.

In terms of multi-GNSS positioning, the contribution of QZSS to the single-frequency PPP of GPS/BDS/GLONASS/Galileo satellites was studied by Hong et al. [18]. The positioning performance of BDS/QZSS in the Asia-Pacific region was evaluated by Bu et al. [11]. The precise positioning performance of QZSS and GPS in the Asia-Pacific region was investigated by Li [19]. The positioning performance of BDS-2/BDS-3 in the Asia-Pacific region was analyzed by Cao et al. [20]. It can be seen that the previous studies about PPP in the Asia-Pacific region are mainly focused on single or dual systems, while the PPP performance of multi-GNSS in the Asia-Pacific region is still limited.

In this contribution, multi-GNSS (GPS/GLONASS/Galileo/BDS-2/BDS-3/QZSS) observation data quality, broadcast ephemeris orbit, and clock offset performance and PPP performance in the Asia-Pacific region are investigated in detail. Based on the observation data from 10 Asia-Pacific MGEX stations, multi-GNSS broadcast ephemeris and precise satellite orbit and clock offset products, the observation data quality, broadcast ephemeris orbit, clock offset performance, and PPP performance in the Asia-Pacific region are evaluated and compared from DOY 283 to 289, in 2021. This paper is organized as follows: after this introduction, the observation data quality of five GNSS systems from 10 stations located in the Asia-Pacific region are investigated and analyzed in terms of carrier-to-noise-density ratio(C/N0) and pseudorange multipath in Section 2. The accuracy of the broadcast ephemeris orbit and clock offset and signal-in-space ranging errors (SISRE) of the five systems is investigated, evaluated, and compared in Section 3. The static and kinematic PPP performance of five GNSS systems in the Asia-Pacific region is evaluated and compared from convergence time and positioning accuracy in Section 4. Finally, the conclusions are presented in Section 5.

## 2. Data Quality

The observation data of 10 stations (CEDU, DARW, JFNG, MIZU, NNOR, PIMO, SIN1, USUD, YARR) from the MGEX network located in the Asia-Pacific region from day of year (DOY) 283 to 289 in 2021 are applied. The distribution of these stations is shown in Figure 1, and the latitude, longitude, receiver type, and antenna type of each station are listed in Table 2. These 10 stations are evenly distributed in different latitudes and longitudes around the Asia-Pacific region, and all frequencies of GPS, BDS-2, BDS-3, GLONASS, Galileo, and QZSS satellites can be received by these stations, which can better reflect the observation data quality and positioning performance of multi-GNSS in the Asia-Pacific region. The observation data quality was studied in the Asia-Pacific region in terms of two indicators: C/N0 and pseudorange multipath, in which the C/N0 can reflect the ability of the signal strength from satellite to receiver, whereas the impact of the satellite signal due to ground interference on the receiver can be reflected in pseudorange multipath. The C/N0 and pseudorange multipath are important indicators in the observation data quality assessment, and their performance directly affects the performance of PNT services.

**Table 2.** Receiver information.

| | Longitude | Latitude | Receiver Type | Antenna Type |
|---|---|---|---|---|
| CEDU | 133.81 | −30.13 | SEPT POLARX5 | AOAD/M_T |
| DARW | 131.13 | −11.16 | SEPT POLARX5 | JAVRINGANT_DM |
| JFNG | 114.49 | 30.52 | TRIMBLE ALLOY | TRM59800.00 |
| LAUT | 177.45 | −16.39 | SEPT POLARX5 | JAVRINGANT_DM |
| MIZU | 141.13 | 39.14 | SEPT POLARX5 | SEPCHOKE_B3E6 |
| NNOR | 116.19 | −30.95 | SEPT POLARX5TR | SEPCHOKE_B3E6 |

**Table 2.** *Cont.*

| | Longitude | Latitude | Receiver Type | Antenna Type |
|---|---|---|---|---|
| PIMO | 121.08 | 14.64 | JAVAD TRE_G3TH DELTA4.1.01 | ASH701945C_M |
| SIN1 | 103.68 | 1.34 | TRIMBLE NETR9 | LEIAR25.R3 |
| USUD | 138.36 | 36.13 | SEPT POLARX5 | AOAD/M_T |
| YARR | 115.35 | −28.95 | SEPT POLARX5 | LEIAT504 |
| CEDU | 133.81 | −30.13 | SEPT POLARX5 | AOAD/M_T |
| DARW | 131.13 | −11.16 | SEPT POLARX5 | JAVRINGANT_DM |

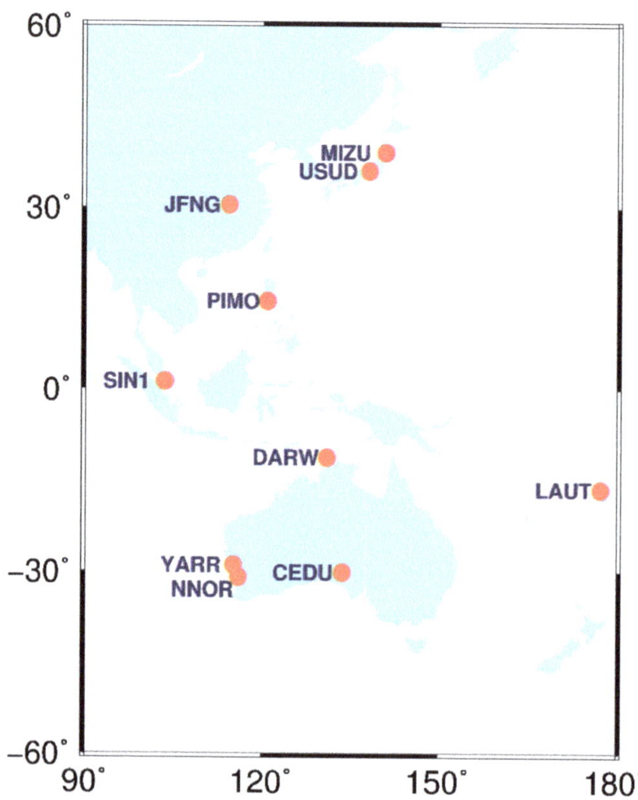


**Figure 1.** Distribution of selected stations.

### *2.1. Carrier-to-Noise-Density Ratio*

The C/N0 is the ratio of the carrier signal to noise, which can reflect the signal strength of GNSS observations on the receiver side. The larger the C/N0, the smaller the noise and the better the signal quality, and vice versa. The C/N0 of every satellite at each epoch can be directly obtained from the observation files. In this study, the relationship between the C/N0 and the elevation of the GNSS satellite signal is investigated and analyzed. When obtaining the C/N0, the sampling interval of the observation data is set as 30 s and the elevation mask is set to 0°. All C/N0 values within 5 degrees of elevation angle are grouped into one group, and then the average of C/N0 within each group of elevation angle is calculated [21].

The average C/N0 corresponding to elevation for BDS-2 MEO, BDS-2 IGSO, BDS-3 MEO, BDS-3 IGSO, Galileo, GPS, GLONASS, and QZSS satellite is calculated, and one typical satellite of each system was shown in Figure 2, respectively. For the BDS-2 MEO satellite, the C/N0 of the B1I signal for the C12 satellite is slightly worse than that of B3I and B2I; although the signal strength is different, the C/N0 variation of these three frequencies shows the same variation trend. The C/N0 of B1I, B3I, and B2I signals have comparable performance for all elevation angles in terms of BDS-2 IGSO satellites. If these three signals can be selected for users, the difference in C/N0 does not need to be considered. In terms of the BDS-3 satellites, the C/N0 of the B2b signal is worse than that of B1C, B1I, B2a, and B3I whether for the IGSO or MEO satellites. It can be clearly seen that the C/N0 of BDS-3 satellites is better than that of BDS-2, which may be that the Binary Offset carrier (BOC) and Quadrature Multiplexed Composite Binary Offset Carrier (QMBOC) signal design is applied to BDS-3 satellites. For MEO satellites, the C/N0 of BDS-3 is higher than that of BDS-2 with 1–2 dB-Hz; while these values are 2–3 dB-Hz for IGSO satellites. The C/N0 of the Galileo E5 signal shows the best performance among the five Galileo frequency bands. The C/N0 of the L2 Z-tracking signal channel for GPS satellites is poorer than that of other frequencies. Moreover, the C/N0 for the L1 C/A signal is slightly poorer than that of L1C, L2C, and L5 I + Q. The C/N0 of GLONASS G1 frequency is better than that of G2, which can be attributed to the lower frequency value of the G2 signal. The C/N0 values for five QZSS frequencies present a similar performance at the different elevation angles. In terms of the five systems: the GPS and Galileo satellites show the best performance, and the C/N0 value can reach 55 dB-Hz when the elevation is nearly 90 degrees, both the BDS and GLONASS are poorer than that of GPS and Galileo, and the QZSS presents the worst performance among five GNSS systems.

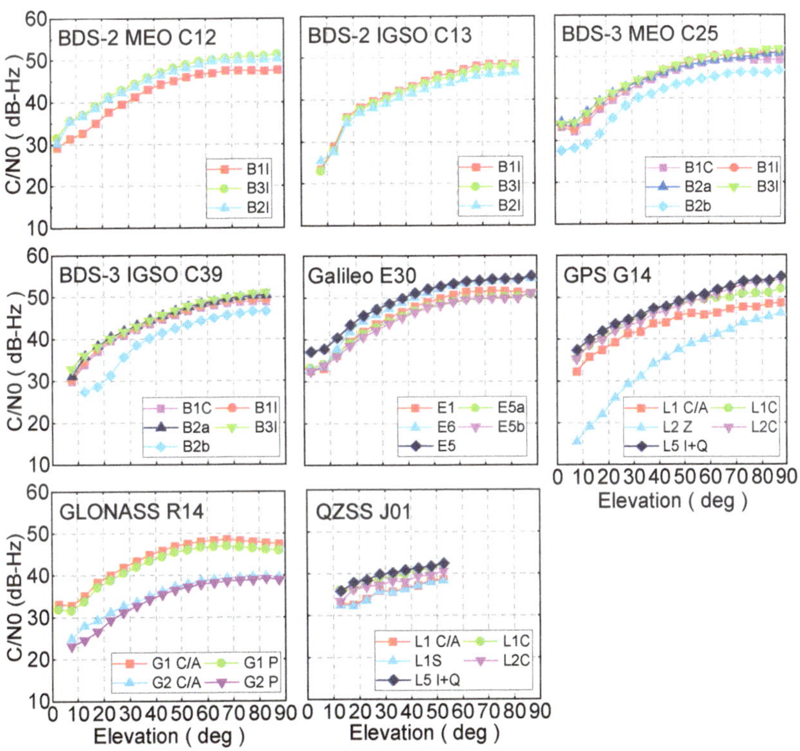


**Figure 2.** The average C/N0 versus elevation angle of BDS-2/ BDS-3/GPS/QZSS/GLONASS/Galileo from DOY 283 to 289, 2021.

### 2.2. Pseudorange Multipath

The combination of the pseudorange and carrier phase observation is used to calculate pseudorange multipath, which can eliminate the effect of tropospheric and ionospheric delays [22], and it can be expressed as follows:

$$MP_k = P_k - L_i - \beta(L_i - L_j) = P_k + \alpha L_i + \beta L_j + \varepsilon \cdot \alpha \tag{1}$$

$$\alpha = f_1^2 / f_2^2 \tag{2}$$

$$\beta = -\frac{\left(f_j^2 + f_k^2\right) f_j^2}{\left(f_i^2 + f_j^2\right) f_k^2} \tag{3}$$

where $k$, $i$, $j$ are frequency, $MP_k$ denotes the pseudorange multipath; $P_k$ represents the pseudorange observation; $L_i$ and $L_j$ are the carrier phase observation on frequency $i$ and $j$, respectively. $f_i$ and $f_j$ are frequencies; $\varepsilon$ is the noise.

The characteristics of pesudorange multipath for GPS, GLONASS, BDS-2, BDS-3, Galileo, and QZSS satellites are investigated in this study using 10 stations from the Asia-Pacific region. The pesudorange multipath values versus elevation angle for the BDS-2 C11, C16, BDS-3 C30, and C39 on DOY 283 for the JFNG station was shown in Figure 3. It can be seen that there is an opposite relationship between pseudorange multipath and elevation angle, the larger the pseudorange multipath, the smaller the elevation angle, and vice versa. The pseudorange multipath is significantly large when the elevation is extremely small in some cases, which may be caused by the observation noise. For the BDS-2 satellites, the pseudorange multipath of the B3I signal is better than those of B1I and B2I. In terms of the BDS-3 satellites, the B1I frequency band of the MEO satellite presents the best performance, its pseudorange multipath is the smallest. Overall, the BDS-3 pseudorange multipath is around 0.28 m, while it is about 0.3 m for the BDS-2 satellites. The time series of pseudorange multipath with respect to elevation on the JFNG station for GPS, GLONASS, Galileo, and QZSS is shown in Figure 4. Similar to BDS-2 and BDS-3 satellites, the pseudorange multipath shows an opposite relationship with the elevation angle. Among them, the pseudorange multipath value of Galileo is the smallest, and it is around 0.2 m, the GPS and QZSS systems show comparable performance, and its value is around 0.3 m, while the value is about 0.4 m for GLONASS satellites. Since the pseudorange multipath effect is an important error source at the receiver side, it has a negative impact on GNSS precise data processing. In GNSS precise data processing, the following measures can be adopted to eliminate or weaken the effect of it: Firstly, reducing or eliminating the weight of observation with low elevation; secondly, modeling the pseudorange multipath errors.

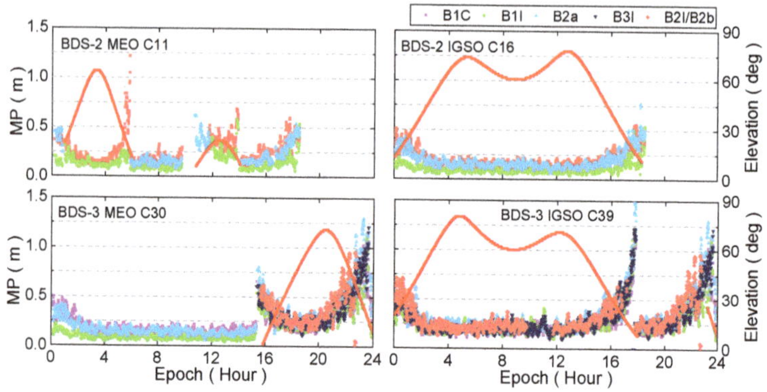


**Figure 3.** The MP series of BDS-2 and BDS-3 for DOY 283 on JFNG station. (The red wave indicates the sequence of elevation angle).

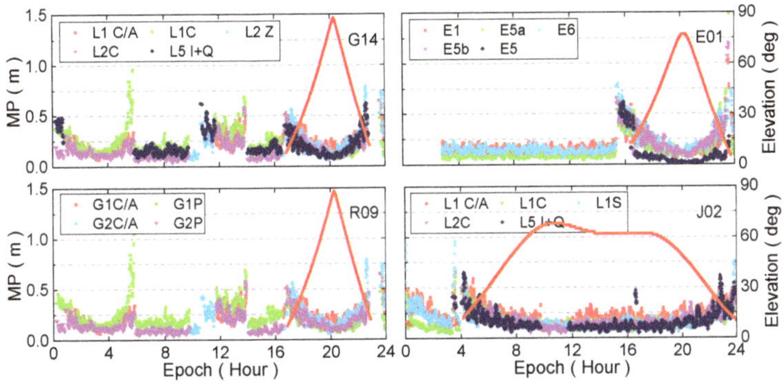


**Figure 4.** The MP series of GPS/QZSS/GLONASS/Galileo for DOY 283 on JFNG station. (The red wave indicates the sequence of elevation angle).

## 3. Broadcast Ephemeris Performance

### *3.1. Broadcast Ephemeris Clock Offset Performance*

The broadcast ephemeris clock offset accuracy of GPS, BDS-2, BDS-3, GLONASS, Galileo, and QZSS satellites from DOY 283 to 289 in 2021 are depicted in Figure 5, and the mean clock offset accuracy is listed in Table 3. The clock offset accuracy assessment method is referred to Huang et al. [23]. For GPS, the clock offset accuracy of the G08 satellite is 2.57 ns, which shows the poorest performance among all GPS satellites since the cesium atomic clock was installed on it. Furthermore, except for the G03, G17, G28, and G29 satellite clock, the clock offset accuracy of other satellites is better than 1 ns. The mean accuracy is 0.65 ns. The broadcast ephemeris clock offset accuracy is 0.98 ns, 1.72 ns, and 2.11 ns for BDS-2 GEO, IGSO, and MEO satellites, respectively. The clock offset accuracy of GEO satellites outperforms that of IGSO and MEO satellites, the reason is that the stations applied to estimate broadcast ephemeris clock offset are mainly located in China. Compared to the IGSO and MEO satellites, the observation arc of GEO satellites is longer, and the data used for predicting clock offset are more, resulting in the clock offset accuracy being higher, whereas the observation data of MEO satellites are few, leading to inferior clock offset accuracy. Moreover, the frequency stability of BDS-2 onboard satellite clocks is poorer, which has a negative impact on broadcast ephemeris clock offset accuracy. The broadcast ephemeris clock offset accuracy of the BDS-3 satellites is about 1 ns apart from the C38, C39, and C40 satellites, the operation period of these three satellites is shorter, and the stations can receive the signal of these three satellites are few. In a word, the mean accuracy is 1.04 ns, its accuracy is improved compared to that of BDS-2 satellites, which can be attributed to the following reasons: Firstly, the improved rubidium atomic clocks and high-performance Passive Hydrogen Masers (PHM) are equipped on BDS-3 satellites, the frequency stability is extremely improved compared to BDS-2 satellites; secondly, since the Inter-Satellite Link (ISL) technology is employed to estimate BDS-3 broadcast ephemeris clock offset [24], the broadcast ephemeris clock offset accuracy can be significantly improved.

**Table 3.** Clock offset accuracy of each system (units: ns).

| System | Clock Offset STD |
|---|---|
| GPS | 0.65 |
| BDS-2 | 1.72 |
| BDS-3 | 1.04 |
| GLONASS | 3.10 |
| Galileo | 0.61 |
| QZSS | 0.75 |

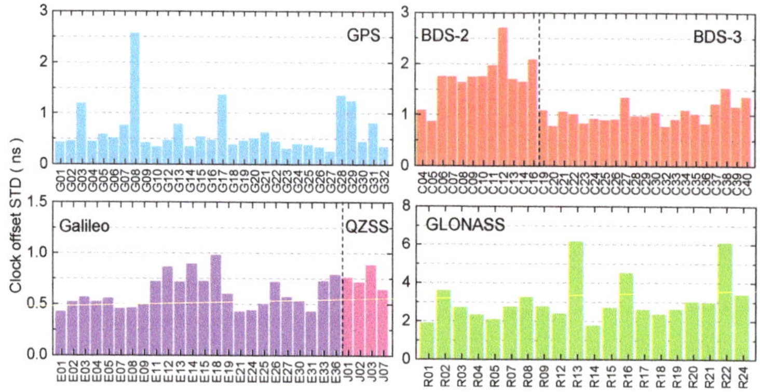


**Figure 5.** Clock offset accuracy of GPS/BDS-2/BDS-3/Galileo/GLONASS/QZSS satellites from broadcast ephemeris.

The cesium atomic clocks are installed on GLONASS satellites, previous studies have demonstrated that the performance of cesium atomic clocks is poorer than that of rubidium atomic clocks and PHM [25]. Compared to other systems, the broadcast ephemeris clock offset accuracy of GLONASS is worse, and the broadcast ephemeris clock offset accuracy of R13, R16, and R22 satellites is poorer than 4 ns. The mean clock offset accuracy is 3.10 ns. In terms of Galileo satellites, the broadcast ephemeris clock offset accuracy of each satellite is better than 1 ns, and the mean is 0.61 ns, which shows superior performance; the reason may be that the high-precision rubidium atomic clocks and PHM are employed on Galileo satellites [26]. Moreover, the update frequency of Galileo satellites is high, it is updated every 10 min, which also guarantees the high-precision performance of broadcast ephemeris clock offset. Each QZSS satellite clock offset accuracy is better than 1 ns, and the mean is 0.75 ns, the atomic clocks equipped on QZSS satellites are the same as that of GPS, while the clock offset accuracy is poorer than that of GPS, which may be related to the satellite orbit type. The QZSS broadcast ephemeris clock offset difference from DOY 283 to 289 is presented in Figure 6. It can be clearly seen that the clock offset difference time series is rather stable, and the value is between −2 ns and 2 ns, the mean is 0.75 ns, 0.71 ns, 0.88 ns, and 1.64 ns for J01, J02, J03, and J07 satellites, respectively.

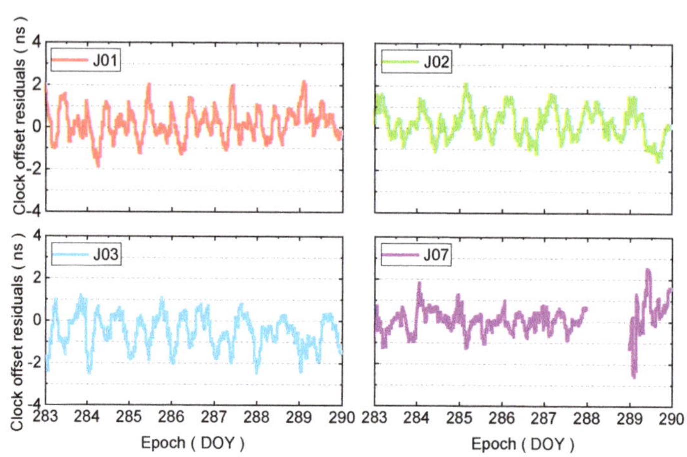


**Figure 6.** Clock offset difference time series of QZSS satellites from broadcast ephemeris.

### 3.2. Broadcast Ephemeris Orbit Performance

The broadcast ephemeris orbit accuracy of GPS, BDS-2, BDS-3, GLONASS, Galileo, and QZSS satellites from DOY 283 to 289 in 2021 are presented in Figure 7, and the mean orbit accuracy is listed in Table 4. For GPS satellites, the radial accuracy of G04, G14, G18, and G23 satellite orbit is worse than that of other satellites, which may be that these four satellites belong to Block III, and the operation period is shorter, and their broadcast ephemeris orbit accuracy still needs to be improved; the orbit accuracy in the radial direction of other satellites is better than 0.25 m. For the along and cross components, the orbit accuracy is between 0.25 m and 1.5 m, and 0.25 m and 1 m, respectively. The mean 3D RMS is 0.60 m. The orbit accuracy is significantly poor for BDS-2 satellites, especially for GEO satellites, which mainly be since the GEO satellites are static relative to the ground stations, resulting in a strong correlation among the observations at different epochs. Compared to BDS-2 satellites, the broadcast ephemeris orbit accuracy of BDS-3 satellites is considerably improved, the satellite orbit accuracy in radial, along, cross, and 3D RMS is 0.11 m, 0.25 m, 0.25 m, and 0.21 m, respectively, and the improvements are 94.24%, 81.06%, 90.23%, 89.29%, respectively. The reasons can be attributed to the following: Firstly, the inter-satellites link technology is employed to determine the satellite orbit. Secondly, the number of BDS-3 satellites is much more than that of BDS-2, and the satellite type is mainly MEO satellites, the redundancy of observation data has been improved. The satellite orbit accuracy after the PRN 38 is slightly inferior, and the operation period of these satellites is shorter, the stations that can receive the signal of these satellites are few. With the increase of stations, satellite orbit accuracy can be improved in the near future.

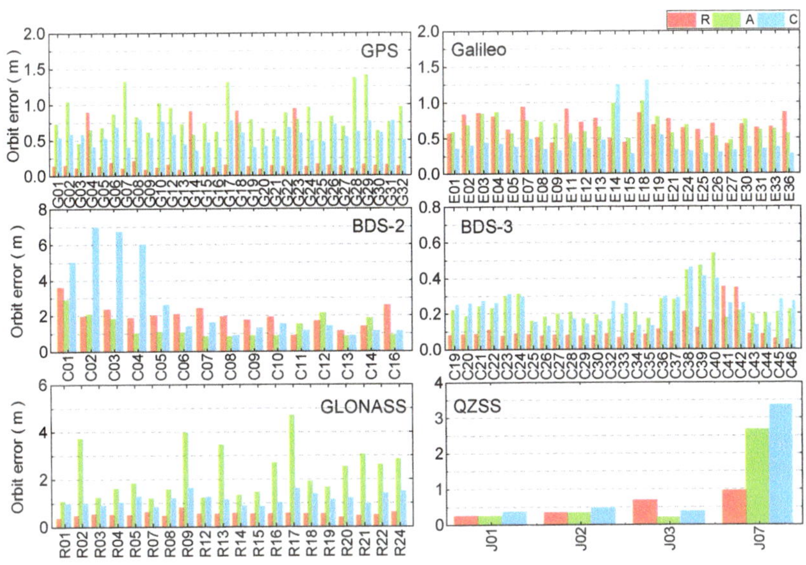


Figure 7. Broadcast orbit accuracy of GNSS five systems.

**Table 4.** Average accuracy of DOY 283–289 broadcast orbit for Multi-GNSS satellites (unit: m).

| System | Radial | Along-Track | Cross-Track | 3D RMS |
|---|---|---|---|---|
| GPS | 0.24 | 0.84 | 0.57 | 0.60 |
| BDS-2 | 1.91 | 1.32 | 2.56 | 1.99 |
| BDS-3 | 0.11 | 0.25 | 0.25 | 0.21 |
| GLONASS | 0.46 | 1.92 | 0.99 | 1.27 |
| Galileo | 0.69 | 0.68 | 0.46 | 0.62 |
| QZSS | 0.56 | 0.80 | 1.06 | 0.83 |

The satellite orbit accuracy in radial and cross is small for GLONASS satellites, most satellites are better than 1.5 m, whereas the broadcast ephemeris orbit accuracy in along component shows poorer performance, the mean orbit accuracy is 0.46 m, 0.99 m, 1.92 m, and 1.27 m for radial, cross, along, and 3D direction, respectively. The orbit difference of each Galileo satellite is small, and the orbit accuracy for most satellites is better than 1.0 m in along, cross, and radial directions, and the mean 3D RMS is 0.62 m, which is comparable to that of GPS. The orbit accuracy of QZSS satellites is similar to that of BDS-2, the orbit accuracy of three IGSO satellites is tremendously better than that of GEO satellites. The mean 3D RMS is 0.83 m, which is poorer than that of BDS-3, GPS, and Galileo satellites.

### *3.3. Signal-in-Space Ranging Errors*

To comprehensively assess the accuracy of GNSS broadcast ephemeris orbits and clock offset, the SISRE of all systems was investigated [27]. The SISRE for GPS/BDS-2/BDS-3/GLONASS/Galileo/QZSS satellites from 283 to 289 days in 2021 is given in Figure 8. The mean and RMS of SISRE are given in Table 5. It can be found that the SISRE of GPS and Galileo show the best performance among the five systems, and its fluctuation is between 0 m and 2 m. However, the SISRE of Galileo is larger at some epochs. The rank of SISRE from best to poorest is QZSS, BDS-3, GLONASS, and BDS-2. For the BDS MEO satellites, the SISRE value of the BDS-2 MEO satellites (C11, C12) and the BDS-3 MEO satellites (C25, C30) are from 0 m to 7 m and 0 m to 4 m, respectively. In terms of BDS IGSO satellites, the fluctuation of BDS-2 IGSO satellites (C13, C16) and BDS-3 IGSO satellites (C39, C40) are between 0 m and 6 m, and 0 m and 4 m, respectively. The SISRE of BDS-3 is better than BDS-2 for both MEO and IGSO satellites. Since SISRE can reflect the combined error of orbit and clock offset, and the accuracy of broadcast ephemeris orbit and clock offset are calculated using the observations from ground stations, there is a certain relationship between the observation data quality and SISRE. By comparing the observation data quality and SISRE, it can be found that when the observation data quality is better, the broadcast ephemeris SISRE is also better, and vice versa.

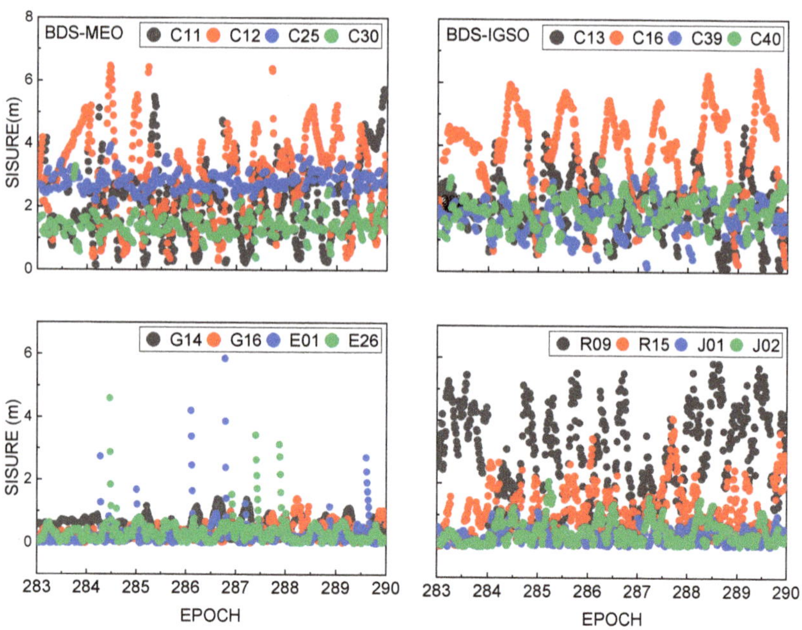


Figure 8. SISIRE of GPS/BDS-2/BDS-3/GLONASS/Galileo/QZSS.

**Table 5.** The average accuracy and RMS of DOY 283–289 SISRE for Multi-GNSS satellites (unit: m).

| System | Average | RMS |
|---|---|---|
| GPS | 0.62 | 0.21 |
| BDS-2(MEO) | 2.19 | 0.65 |
| BDS-2(IGSO) | 3.73 | 0.90 |
| BDS-3(MEO) | 1.97 | 0.71 |
| BDS-3(IGSO) | 2.01 | 0.35 |
| Galileo | 0.62 | 0.79 |
| GLONASS | 2.48 | 1.04 |
| QZSS | 0.98 | 1.18 |

## 4. PPP Accuracy Evaluation

To investigate the positioning performance of multi-GNSS in the Asia-Pacific region, 10 MGEX stations were selected to conduct static and kinematic PPP experiments. For the data processing strategies, the sampling interval of observation data is 30 s, and the period is from DOY 283 to 289 in 2021. At present, several MGEX analysis centers can provide precise satellite orbit and clock offset products for the five systems, and the consistency between WUM orbit and clock offset products and other MGEX analysis centers is about 3–10 cm and 0.1–0.3 ns, respectively, showing better consistency with the products from other analysis centers. The positioning performance in the Asia-Pacific region can be reflected using the WUM orbit and clock offset products [28,29]. Therefore, the multi-GNSS final satellite orbit, clock offset, and earth rotation parameter (ERP) products generated from Wuhan University are applied in this study [30]. The satellite antenna phase center variation (PCV) and phase center deviation (PCO) values are used from igs14.atx [31]. The dual-frequency ionosphere-free is employed to eliminate the effect of the first-order ionosphere and the higher-order ionosphere is ignored [32]. The zenith hydrostatic delay of the troposphere is corrected using the Saastamoinen model [33], while the zenith wet delay is estimated as the parameter. The carrier phase ambiguities are estimated as float solutions [34]. The station coordinates of the static PPP are estimated as a constant, while it is estimated as white noise in the kinematic model. The receiver clock offset is estimated as white noise. In addition, the relativistic effects, satellite antenna phase wind-up, and station tides are weakened or eliminated using existing models [35–37]. To compare the positioning performance differences between single-system and multi-GNSS combinations in the Asia-Pacific region, six mode combinations in the static and kinematic PPP experiments were conducted, which is: GPS(G), GPS/QZSS combination (GJ), BDS (C), BDS/QZSS combination (CJ), GLONASS (R), and Galileo (E), respectively. The convergence time and positioning accuracy are used to evaluate the positioning performance. The convergence time is that the current epoch with 20 consecutive epochs is better than 10 cm, and the positioning accuracy is the RMS of the positioning error after convergence [38]. It is noted that the positioning error is the positioning difference between the PPP solutions and IGS weekly solution [39].

### *4.1. Static PPP Performance*

The mean convergence time for static PPP of six combinations for each station in the east, north, and up directions is presented in Figure 9, and the mean convergence time of each combination is listed in Table 6. It can be seen from Figure 9 and Table 6 that the convergence time of GPS is the shortest among six combinations, which are 6.01 min, 5.53 min, and 16.52 min for the east, north, and up directions, respectively. For the east and north directions, the convergence time of Galileo is faster than that of BDS, and the GLONASS show the longest convergence time, which may be that all selected stations are located in the region with low latitudes, and better positioning performance can be achieved at high latitudes than low latitudes for GLONASS [38]. In terms of up component, the convergence time of Galileo is shorter than that of GLONASS, while the BDS presents the poorest convergence performance in the up component among the six combinations, and its convergence time is 32.17 min. Furthermore, compared to the GPS-only solution, the convergence time of the GPS/QZSS combination can be shorted, and the improvements are 10.37%, 0.90%, and 1.15% in the east, north, and up directions, respectively. While compared to the BDS-only solution, the BDS/QZSS solutions only short the convergence time in the up direction, the improvement is 1.65%. The reason may be that the number of BDS satellites in Asia-Pacific is larger than 20, when conducting PPP, by adding QZSS satellites, the improvements in convergence time for static PPP are limited.

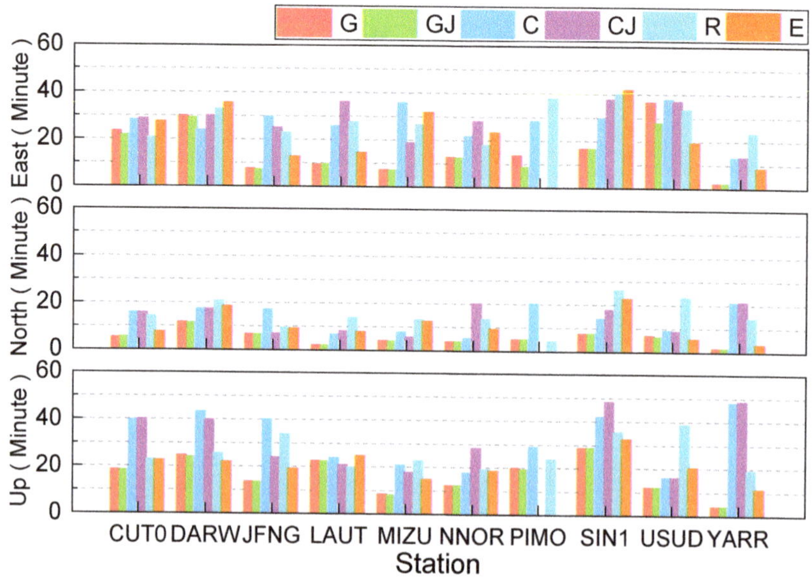


**Figure 9.** Mean convergence time for static PPP in each station.

**Table 6.** Convergence time and positioning accuracy of different combination static PPP.

| System | Convergence Time (min) | | | Positioning Accuracy (cm) | | |
|---|---|---|---|---|---|---|
| | E | N | U | E | N | U |
| G | 16.01 | 5.53 | 16.52 | 1.09 | 0.78 | 1.68 |
| GJ | 14.35 | 5.48 | 16.33 | 1.08 | 0.77 | 1.66 |
| C | 27.47 | 13.58 | 32.17 | 1.40 | 1.03 | 2.08 |
| CJ | 28.29 | 13.59 | 31.64 | 1.38 | 1.01 | 2.04 |
| R | 28.44 | 15.44 | 26.26 | 1.93 | 1.32 | 3.07 |
| E | 23.93 | 10.78 | 20.79 | 2.24 | 1.12 | 2.65 |

The mean positioning accuracy for static PPP of six combinations for each station in the east, north, and up directions are presented in Figure 10, and the mean positioning accuracy of each combination is listed in Table 6. It can be clearly seen that the positioning accuracy is better than 3 cm in the east direction except for PIMO and SIN1 stations. Apart from the USUD station, the positioning accuracy of other stations is better than 2 cm in the north component. The positioning accuracy is outperformed 5 cm in the up direction except for YARR station. The positioning accuracy of GPS is the best, and it is 1.09 cm, 0.78 cm, and 1.68 cm in east, north, and up directions, respectively. For three components, the positioning accuracy using BDS is better than that of GLONASS, while Galileo shows the worst performance. Compared to the GPS-only solution, the positioning accuracy of GPS/QZSS solutions can be improved, and the improvement is 0.92%, 1.28%, and 1.19% in the east, north, and up directions, respectively. The few improvements may be caused by the limited number of QZSS satellites. The improvement of the BDS/QZSS solution in terms of positioning accuracy is 1.43%, 1.94%, and 1.92% in east, north, and up directions compared to the single BDS-only solution. Totally, the positioning accuracy of six combinations is better than 3 cm, 2 cm, and 4 cm in the east, north, and up directions, respectively.

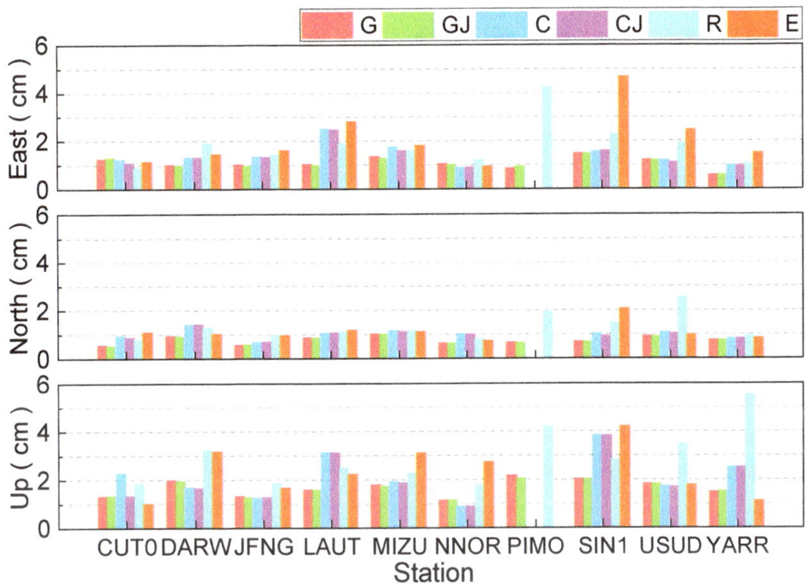


**Figure 10.** Positioning accuracy of static PPP solution in each station.

## 4.2. *Kinematic PPP Performance*

The mean convergence time for kinematic PPP of six combinations for each station in the east, north, and up directions is presented in Figure 11, and the mean convergence time of each combination is listed in Table 7. It can be seen that the convergence time of the Galileo solution is the shortest in the east direction, it is 23.82 min, while the shortest convergence time is the GPS-only solution in the north and up directions, with convergence times being 8.49 min and 24.4 min, respectively. For the east direction, the convergence time of the GPS-only solution is faster than that of the GLONASS-only solution, and the BDS-only solution is the longest, nearly one hour is still needed to obtain the centimeter-level position accuracy in kinematic PPP mode, whereas it is about 20 min for the GPS/QZSS solutions. Compared to the GPS-only and BDS-only solutions, the improvement of convergence time for the GPS/QZSS and BDS/QZSS solutions is 24.82%, 7.66%, 10.90%, and 11.06%, 19.94%, and 6.66% in the east, north, and up components, respectively. Compared to the static PPP, the improvement rate of convergence time for the GPS/QZSS and BDS/QZSS solutions is larger, which may be that the increased number of satellites and better geometry distribution of satellites are beneficial to the convergence of kinematic PPP.

**Table 7.** Convergence time and positioning accuracy of different combination kinematic PPP.

| System | Convergence Time (min) | | | Positioning Accuracy (cm) | | |
|---|---|---|---|---|---|---|
| | E | N | U | E | N | U |
| G | 24.78 | 8.49 | 24.40 | 2.80 | 2.03 | 4.65 |
| GJ | 18.63 | 7.84 | 21.74 | 2.43 | 1.91 | 4.15 |
| C | 55.62 | 31.09 | 45.37 | 1.77 | 1.59 | 4.69 |
| CJ | 49.47 | 24.89 | 42.35 | 1.76 | 1.57 | 4.50 |
| R | 43.39 | 47.68 | 50.98 | 6.32 | 5.59 | 14.87 |
| E | 23.82 | 17.80 | 34.02 | 6.36 | 5.68 | 8.79 |

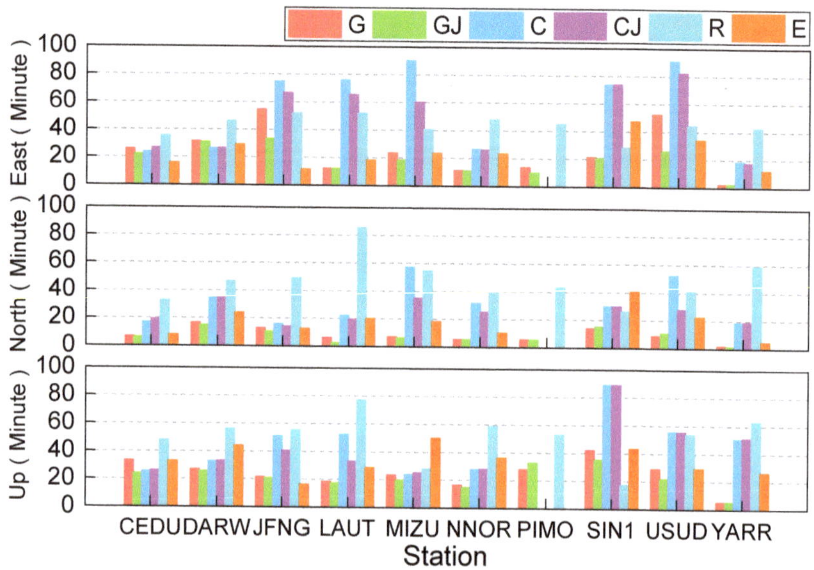


**Figure 11.** Mean convergence time for kinematic PPP in each station.

The kinematic PPP mean positioning accuracy of six combinations for each station in the east, north, and up directions are presented in Figure 12, and the mean positioning accuracy of each combination is listed in Table 7. One can see that the GPS-only, GPS/QZSS, BDS-only, and BDS/QZSS solutions show similar positioning accuracy in three directions except for the LAUT and MIZU station, which is better than 3 cm in the three directions. The positioning performance of GLONASS and Galileo in the Asia-Pacific region is relatively poorer. The positioning accuracy of BDS presents the best performance in the east and north component, and the positioning accuracy is 1.77 cm and 1.59 cm, respectively. This is due to the launch of BDS-3 satellites, the number of BDS satellites is more than other satellite systems in the Asia-Pacific region. The positioning accuracy in the up direction of GPS shows the best performance, and it is 4.65 cm. For the east and north directions, the positioning accuracy of GPS is better than that of Galileo, whereas the GLONASS is the worst. In terms of the up direction, the positioning accuracy of BDS is better than that of Galileo, while the GLONASS is still the worst. Moreover, compared to the GPS-only solution and BDS-only solutions, the positioning accuracy of the GPS/QZSS and BDS/QZSS solutions can be improved from 2.80 cm, 2.03 cm and 4.65 cm to 2.43 cm, 1.91 cm and 4.15 cm, with the improvement being 13.21%, 5.91%, and 10.75%, from 1.77 cm, 1.59 cm, and 4.69 cm to 1.76 cm, 1.57 cm, and 4.50 cm for the east, north, and up directions, with the improvement being 0.56%, 1.26%, and 4.05%, respectively. Except for the GLONASS-only and Galileo-only solutions, the positioning accuracy of 3 cm, 3 cm, and 5 cm in the east, north, and up components for kinematic PPP can be achieved.

The consistency of performance improvement among systems based on multi-GNSS data quality, broadcast ephemeris accuracy, and precision positioning performance is analyzed. It can be found that the observation data quality of GPS is improved by 28.11%, and the improvement for SISRE, static PPP, and kinematic PPP is 75%, 44.31%, and 69.24% compared to GLONASS, respectively. Compared to GLONASS, the improvement of data quality, SISRE, static PPP, and kinematic PPP of Galileo is improved by 32.71%, 75%, 5.52%, and 28.37%, respectively. The improvement of BDS-3 is 33% and 32.78% compared to BDS-2 in terms of observation data quality and SISRE, respectively. These results show that there is a consistency between data quality, SISRE, and PPP.

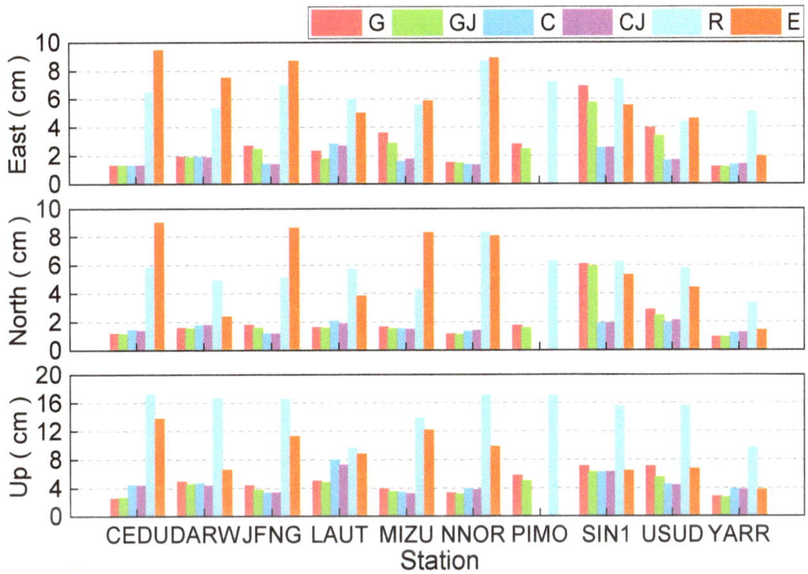


**Figure 12.** Positioning accuracy of kinematic PPP for each station.

## 5. Conclusions

This paper focused on the performance of multi-GNSS in the Asia-Pacific region. One-week period observation data from 10 MGEX stations, precise satellite orbit and clock offset products, and broadcast ephemeris were used to investigate the performance of C/N0, pseudorange multipath, broadcast ephemeris clock offset accuracy, broadcast ephemeris orbit accuracy, SISRE, static PPP, and kinematic PPP in the Asia-Pacific region, and the conclusions are as follows:

(1) The C/N0 and pseudorange multipath among GPS, BDS-2, BDS-3, QZSS, GLONASS, and Galileo were investigated and compared; the C/N0 of GPS and Galileo shows the best performance, and the C/N0 for BDS and GLONASS are worse, and the C/N0 of QZSS is the poorest. The pseudorange multipath error of Galileo has the smallest, about 0.2 m, GPS and QZSS are comparable at about 0.3 m, and GLONASS is around 0.4 m.

(2) The broadcast ephemeris orbit and broadcast ephemeris clock offset of GPS, BDS-2, BDS-3, QZSS, GLONASS, and Galileo were analyzed, and the results show that the broadcast ephemeris clock offset accuracy of Galileo is the best, followed by GPS, QZSS, BDS-3, BDS-2, and GLONASS, respectively. For the broadcast ephemeris orbit, the accuracy of BDS-3 is the best, followed by GPS, Galileo, QZSS, GLONASS, and BDS-2, respectively. The SISRE of BDS-3 is better than BDS-2 for both MEO and IGSO satellites, and by comparing the observation data quality and SISRE, it can be found that when the observation data quality is better, the broadcast ephemeris SISRE is also better, and vice versa, indicating that the magnitude of SISRE has a certain correlation with the observation data quality results.

(3) The static PPP and kinematic PPP performance were analyzed in terms of GPS, GPS/QZSS combination, BDS, BDS/QZSS combination, GLONASS, and Galileo. For static PPP, the positioning accuracy of GPS and GLONASS present the best and worst performance, respectively. In terms of kinematic PPP, the positioning accuracy in the east and north direction of GPS is better than that of Galileo. For up direction, the positioning accuracy of BDS is better than that of Galileo, the GLONASS shows the worst positioning performance in the east, north, and up component. Compared to the GPS-only and BDS-only solutions, the positioning accuracy of GPS/QZSS and BDS/QZSS solutions is improved, and the improvement rate of kinematic is larger than static PPP.

**Author Contributions:** M.L. and G.H. conceived and designed these experiments and wrote the paper; M.L., G.H., L.W. and W.X. performed the experiments, analyzed the data, drew pictures, and wrote the paper; F.Y. reviewed the paper. All authors have read and agreed to the published version of the manuscript.

**Funding:** This work was supported by the Programs of the National Natural Science Foundation of China (42127802), the Key R&D Program of Shaanxi Province (2022ZDLSF07-12), the Special Fund for Basic Scientific Research of Central Colleges (Grant No. CHD300102269305, CHD300102268305, Chang'an University).

**Data Availability Statement:** The observation data and precise products used in the research are available on the FTP of Wuhan University (ftp://igs.gnsswhu.cn/pub/ accessed from 10 to 16 October 2021).

**Acknowledgments:** The IGS and WHU are greatly acknowledged for providing the multi-GNSS tracking data, SINEX coordinates, and satellite orbit and clock offset products. We would also like to thank Zhongyang Zhao, an employee of Xi'an Honor Device Co., Ltd. for his suggestions on this paper.

**Conflicts of Interest:** The authors declare no conflict of interest.

## References

1. Guo, J.; Li, X.; Li, Z.; Hu, L.; Yang, G.; Zhao, C.; Fairbairn, D.; Watson, D.; Ge, M. Multi-GNSS precise point positioning for precision agriculture. *Precis. Agric.* **2018**, *19*, 895–911. [CrossRef]
2. Lu, C.; Li, X.; Nilsson, T.; Ning, T.; Heinkelmann, R.; Ge, M.; Glaser, S.; Schuh, H. Real-time retrieval of precipitable water vapor from GPS and BeiDou observations. *J. Geod.* **2015**, *89*, 843–856. [CrossRef]
3. Defraigne, P.; Aerts, W.; Pottiaux, E. Monitoring of UTC(k)'s using PPP and IGS real-time products. *GPS Solut.* **2015**, *19*, 165–172. [CrossRef]
4. Wright, T.J.; Houlié, N.; Hildyard, M.; Iwabuchi, T. Real-time, reliable magnitudes for large earthquakes from 1 Hz GPS precise point positioning: The 2011 Tohoku-Oki (Japan) earthquake. *Geophys. Res. Lett.* **2012**, *39*, L12302. [CrossRef]
5. Yang, Y.; Li, J.; Xu, J.; Tang, J.; Guo, H.; He, H. Contribution of the Compass satellite navigation system to global PNT users. *Chin. Sci. Bull.* **2011**, *56*, 2813–2819. [CrossRef]
6. Yang, Y.; Mao, Y.; Sun, B. Basic performance and future developments of BeiDou global navigation satellite system. *Satell. Navig.* **2020**, *1*, 1. [CrossRef]
7. Hein, G. Status, perspectives and trends of satellite navigation. *Satell. Navig.* **2020**, *1*, 22. [CrossRef] [PubMed]
8. Zhang, X.; Wu, M.; Liu, W.; Li, X.; Yu, S.; Lu, C.; Wickert, J. Initial assessment of the COMPASS/BeiDou-3: New-generation navigation signals. *J. Geod.* **2017**, *91*, 1225–1240. [CrossRef]
9. Yang, Y.; Xu, Y.; Li, J.; Yang, C. Progress and performance evaluation of BeiDou global navigation satellite system: Data analysis based on BDS-3 demonstration. *Sci. China Earth Sci.* **2018**, *61*, 614–624. [CrossRef]
10. Yan, X.; Huang, G.; Zhang, Q.; Liu, C.; Wang, L.; Qin, Z. Early analysis of precise orbit and clock offset determination for the satellites of the global BeiDou-3 system. *Adv. Space Res.* **2018**, *63*, 1270–1279. [CrossRef]
11. Bu, J.; Zuo, X.; Li, X.; Chang, J.; Zhang, X. Evaluation and analysis on positioning performance of BDS/QZSS satellite navigation systems in Asian-Pacific region. *Adv. Space Res.* **2018**, *63*, 2189–2211. [CrossRef]
12. Zuo, X.; Bu, J.; Li, X.; Chang, J.; Li, X. The quality analysis of GNSS satellite positioning data. *Cluster Comput.* **2019**, *22*, S6693–S6708. [CrossRef]
13. Tian, Y.; Sui, L.; Xiao, G.; Zhao, D.; Tian, Y. Analysis of Galileo/BDS/GPS signals and RTK performance. *GPS Solut.* **2019**, *23*, 37. [CrossRef]
14. Montenbruck, O.; Steigenberger, P.; Hauschild, A. Broadcast versus precise ephemerides a multi-GNSS perspective. *GPS Solut.* **2015**, *19*, 321–333. [CrossRef]
15. Lv, Y.; Geng, T.; Zhao, Q.; Xie, X.; Zhou, R. Initial assessment of BDS-3 preliminary system signal-in-space range error. *GPS Solut.* **2020**, *24*, 16. [CrossRef]
16. Shi, J.; Ouyang, C.; Huang, Y.; Peng, W. Assessment of BDS-3 global positioning service: Ephemeris, SPP, PPP, RTK, and new signal. *GPS Solut.* **2020**, *24*, 81. [CrossRef]
17. Jiao, G.; Song, S.; Liu, Y.; Su, K.; Cheng, N.; Wang, S. Anaysis and Assessment of BDS-2 and BDS-3 Broadcast Ephemeris: Accuracy, the Datum of Broadcast Clocks and Its Impact on Single point positioning. *Remote Sens.* **2020**, *12*, 2081. [CrossRef]
18. Hong, J.; Tu, R.; Zhang, R.; Fan, L.; Zhang, P.; Han, J. Contribution analysis of QZSS to single-frequency PPP of GPS/BDS/GLONASS/Galileo. *Adv. Space Res.* **2020**, *63*, 1803–1807. [CrossRef]
19. Li, X.; Pan, L.; Yu, W. Assessment and Analysis of the Four-Satellite QZSS Precise Point Positioning and the Integrated Data Processing with GPS. *IEEE Access* **2021**, *9*, 116376–116394. [CrossRef]
20. Cao, X.; Shen, F.; Zhang, S.; Li, J. Satellite availability and positioning performance of uncombined precise point positioning using BeiDou-2 and BeiDou-3 multi-frequency signals. *Adv. Space Res.* **2021**, *67*, 1303–1306. [CrossRef]
21. Bilich, A.; Larson, K. Mapping the GPS multipath environment using the signal-to-noise ratio (SNR). *Radio Sci.* **2007**, *42*, RS6003. [CrossRef]

22. Wu, X.; Zhou, J.; Wang, G.; HU, X.; CAO, Y. Multipath error detection and correction for GEO/IGSO satellites. *Sci. China-Phys. Mech. Astron.* **2012**, *55*, 1297–1306. [CrossRef]
23. Huang, G.; Cui, B.; Zhang, Q.; Fu, W.; Li, P. An improved predicted model for BDS ultra-rapid satellite clock offsets. *Remote Sens.* **2018**, *10*, 60. [CrossRef]
24. Yang, Y.; Yang, Y.; Hu, X.; Tang, C.; Guo, R.; Zhou, S.; Xu, J.; Pan, J.; Su, M. BeiDou-3 broadcast clock estimation by integration of observations of regional tracking stations and inter-satellite links. *GPS Solut.* **2021**, *25*, 57. [CrossRef]
25. Xie, W.; Huang, G.; Wang, L.; Li, P.; Cui, B.; Wang, H.; Cao, Y. Long-term performance detection and evaluation of GLONASS onboard satellite clocks. *Measurement* **2021**, *175*, 109091. [CrossRef]
26. Huang, G.; Cui, B.; Xu, Y.; Zhang, Q. Characteristics and performance evaluation of Galileo on-orbit satellites atomic clocks during 2014–2017. *Adv. Space Res.* **2019**, *63*, 2899–2911. [CrossRef]
27. Peter, S.; Oliver, M. Consistency of MGEX Orbit and Clock Products. *Engineering* **2020**, *6*, 898–903.
28. Li, X.; Zhu, Y.; Zheng, K.; Yuan, Y.; Liu, G.; Xiong, Y. Precise Orbit and Clock Products of Galileo, BDS and QZSS from MGEX Since 2018: Comparison and PPP Validation. *Remote Sens.* **2020**, *12*, 1415. [CrossRef]
29. Montenbruck, O.; Steigenberger, P.; Prange, L.; Deng, Z.; Zhao, Q.; Perosanz, F.; Romero, I.; Noll, C.; Stürze, A.; Weber, G.; et al. The Multi-GNSS Experiment (MGEX) of the International GNSS Service (IGS)—Achievements, Prospects and Challenges. *Adv. Space Res.* **2017**, *59*, 1671–1697. [CrossRef]
30. Guo, J.; Xu, X.; Zhao, Q.; Liu, J. Precise orbit determination for quadconstellation satellites at Wuhan University: Strategy, result validation, and comparison. *J. Geod.* **2016**, *90*, 143–159. [CrossRef]
31. Zhou, R.; Hu, Z.; Zhao, Q.; Cai, H.; Liu, X.; Liu, C.; Wang, G.; Kan, H.; Chen, L. Consistency Analysis of the GNSS Antenna Phase Center Correction Models. *Remote Sens.* **2022**, *14*, 540. [CrossRef]
32. Su, K.; Jin, S. Three Dual-Frequency Precise Point Positioning Models for the Ionospheric Modeling and Satellite Pseudorange Observable-Specific Signal Bias Estimation. *Remote Sens.* **2021**, *13*, 5093. [CrossRef]
33. Saastamoinen, J. Contributions to the theory of atmospheric refraction—part II. Refraction corrections in satellite geodesy. *Bull. Géod.* **1973**, *47*, 13–34. [CrossRef]
34. Kiliszek, D.; Kroszczyński, K.; Araszkiewicz, A. Analysis of Different Weighting Functions of Observations for GPS and Galileo Precise Point Positioning Performance. *Remote Sens.* **2022**, *14*, 2223. [CrossRef]
35. Kouba, J.; Héroux, P. Precise point positioning using IGS orbit and clock products. *GPS Solut.* **2001**, *5*, 12–28. [CrossRef]
36. Wang, J.; Huang, G.; Yang, Y.; Zhang, Q.; Gao, Y.; Zhou, P. Mitigation of Short-Term Temporal Variations of Receiver Code Bias to Achieve Increased Success Rate of Ambiguity Resolution in PPP. *Remote Sens.* **2020**, *12*, 796. [CrossRef]
37. Wang, J.; Huang, G.; Zhang, Q.; Gao, Y.; Gao, Y.; Luo, Y. GPS/BDS-2/Galileo Precise Point Positioning Ambiguity Resolution Based on the Uncombined Model. *Remote Sens.* **2020**, *12*, 1853. [CrossRef]
38. Zhou, F.; Cao, X.; Ge, Y.; Li, W. Assessment of the positioning performance and tropospheric delay retrieval with precise point positioning using products from different analysis centers. *GPS Solut.* **2020**, *24*, 12. [CrossRef]
39. Wang, J.; Zhang, Q.; Huang, G. Estimation of fractional cycle bias for GPS/ BDS-2/ Galileo based on international GNSS monitoring and assessment system observations using the uncombined PPP model. *Satell. Navig.* **2021**, *2*, 9. [CrossRef]

*remote sensing*

MDPI

*Article*

# Satellite Clock Batch Estimation Accuracy Analysis and Its Impacts on PPP

**Menghao Li, Weiquan Huang *, Hui Li, Renlong Wang and Peng Cui**

College of Intelligent Systems Science and Engineering, Harbin Engineering University, Harbin 150001, China
* Correspondence: huangweiquan@hrbeu.edu.cn; Tel.: +86-153-0366-9953

**Abstract:** The ultra-rapid satellite clock product based on the satellite clock batch estimation is commonly used for high-precision and reliable precise point positioning (PPP) services. In order to clarify the effect of different ranging errors on the satellite clock batch estimation accuracy, the source of the satellite clock bias induced by the batch observation model is classified into the initial clock bias (ICB) and time-dependent bias (TDB). In addition to the effect of the ICB and TDB, the analytic relationship between the observation redundancy and the satellite clock batch estimation accuracy are derived and verified. The suitable number of stations is suggested to be 40 for the satellite clock batch estimation to achieve the counterbalance between the efficiency and saturable accuracy. For the PPP based on the batch-estimated satellite clock, the impacts of the ICB and TDB on PPP are clarified. The satellite clock batch estimation and PPP experiments are carried out to investigate the impacts of the ICB and TDB on the satellite clock batch estimation accuracy and the PPP performance. The ICB causes a significant bias for the batch-estimated satellite clock. The TDB is impacted by the assimilation ability of the batch-estimated satellite clock to the satellite orbit error. The convergence time and the positioning accuracy after the convergence of PPP are primarily affected by the ICB and TDB, respectively.

**Keywords:** satellite clock batch estimation; ICB; TDB; saturable accuracy; PPP

**Citation:** Li, M.; Huang, W.; Li, H.; Wang, R.; Cui, P. Satellite Clock Batch Estimation Accuracy Analysis and Its Impacts on PPP. *Remote Sens.* **2022**, *14*, 3932. https://doi.org/10.3390/rs14163932

Academic Editors: Shengfeng Gu, Xiaopeng Gong, Yidong Lou and Chuang Shi

Received: 27 June 2022
Accepted: 11 August 2022
Published: 13 August 2022

**Publisher's Note:** MDPI stays neutral with regard to jurisdictional claims in published maps and institutional affiliations.

**Copyright:** © 2022 by the authors. Licensee MDPI, Basel, Switzerland. This article is an open access article distributed under the terms and conditions of the Creative Commons Attribution (CC BY) license (https://creativecommons.org/licenses/by/4.0/).

## 1. Introduction

The precise satellite clock product is indispensable to obtain high-precision and reliable precise point positioning (PPP) services [1], which can be divided into the real-time, ultra-rapid, rapid and final satellite clock products [2,3]. Compared with the other satellite clock products, the ultra-rapid satellite clock product offers better performance in terms of stability and latency [4,5]. Therefore, the ultra-rapid satellite clock product can be beneficial for the PPP with the relatively low real-time requirement [6]. However, the accuracy of the ultra-rapid satellite clock product needs to be improved to meet the centimeter-level PPP application.

The ultra-rapid satellite clock product typically includes the observation and prediction sessions [7]. The observation session is batch estimated based on the observed receiver–satellite range [8]. In the prediction session, the satellite clock is obtained by the fitting and extrapolation based on the observation session and an accurate satellite clock prediction model [9]. It is noted that the satellite clock batch estimation accuracy determines the quality of the ultra-rapid satellite clock product in the observation and prediction sessions. For the PPP application based on the ultra-rapid satellite clock product, the PPP performance is affected by the satellite clock batch estimation accuracy [3]. Therefore, clarifying the error propagation of the satellite clock batch estimation and its effect on PPP can improve the accuracy of the ultra-rapid satellite clock product and the performance for the PPP application.

The batch observation model needs to be constructed for the satellite clock batch estimation. The traditional observation models for the satellite clock estimation generally include undifferenced, epoch-differenced and mixed-differenced models [10,11]. The undifferenced and mixed-differenced observation models for the satellite clock estimation are analytically equivalent [10], whilst the epoch-differenced model suffers from the quality of the initial satellite clock, resulting in significant satellite clock biases [11]. For the generation of the ultra-rapid satellite clock product in the observation session, the undifferenced observation model is extensively applied due to the lower observation noise and smaller satellite clock biases than the others. Given the batch observation model, the batch-estimated satellite clock assimilates the ranging errors due to the model strength. Hence, the satellite clock bias is introduced so that the accuracy of the batch-estimated satellite clock is affected. The ranging errors absorbed by the batch-estimated satellite clock mainly include the code hardware delay for satellites and the fixed satellite orbit error. Considering the characteristic of the hardware delay over time, the time-invariant and time-variant hardware delays affect the satellite clock estimation accuracy [10]. For the satellite clock batch estimation within observation arcs, the time-variant hardware delay can be roughly ignored due to its stability [12]. Additionally, the impacts of the satellite orbit error on the real-time satellite clock estimation were already analyzed and verified preliminarily in previous studies [13,14]. The radial and tangential satellite orbit errors can be absorbed to various extents by the estimated satellite clock. This absorption increases as the size of the station network used for the satellite clock estimation decreases, especially for the regional distributed network [13]. In view of the ultra-rapid satellite clock product for the global PPP service, the satellite clock is batch estimated based on the global distributed network [8]. The impacts of the satellite orbit error on the satellite clock batch estimation based on the global distributed network need to be further analyzed. Therefore, the satellite clock batch estimation observation model can be constructed in the observation session by considering the time-invariant hardware delays and the satellite orbit error.

Based on the constructed batch estimation observation model, the satellite clock bias is induced by the assimilated ranging errors and the model accuracy. For the real-time satellite clock estimation based on the undifferenced observation model, the satellite clock biases induced by the model are grouped into the initial clock bias and time-dependent bias [10,15]. The initial clock bias is the satellite clock error at the initial epoch, which depends on the accuracy of the initial satellite clock and the bias introduced by the pseudorange observation [16,17]. The time-dependent bias is determined by phase observation [10]. This classification can be used as the reference for the satellite clock batch estimation. However, for the satellite clock batch estimation, the initial clock bias was predetermined with relatively high precision. The impacts of the initial satellite clock accuracy on the initial clock bias can be ignored, which will not be further analyzed in this contribution. Moreover, the accuracy of the fixed satellite orbit product is not precise enough. The satellite orbit error needs to be considered in the classification of the batch-estimated satellite clock bias. Therefore, referring to the typical classification of the satellite clock bias in [10], we clarify the source of the batch-estimated satellite clock biases to analyze the impacts of these biases on the satellite clock batch estimation accuracy.

In addition to satellite clock biases, observation redundancy is another critical factor affecting the satellite clock batch estimation accuracy. Generally, the parameter estimation accuracy depends on the observation redundancy [18,19]. Since the satellite clock estimation accuracy becomes approximately saturable when the observation redundancy has adequately increased. The benefit of increasing more observation redundancy is negligible for the accuracy improvement of the satellite clock estimation [20]. Furthermore, the excessive observation redundancy will burden the efficiency and may increase the update delay, which is not conducive for the generation of the ultra-rapid satellite clock product with low latency [4,5]. The lower latency for the batch-estimated satellite clock brings the better performance for the satellite clock prediction and the PPP based on the predicted satellite clock [21,22]. The update interval and the update rate simultaneously determine the latency of the batch-estimated satellite clock, which can be improved by the efficient sliding satellite clock batch estimation with a short window and the satellite clock prediction with a short period [23,24]. In order to achieve the counterbalance between efficiency and estimation accuracy, many studies have shown the number of stations suitable for the real-time satellite clock estimation [11,20]. Such a number of stations may not be suitable for satellite clock batch estimation because the observation redundancy has already been improved for the batch estimation model by combining observations within observation arcs. In addition, the variations in the number of stations may affect the station distribution. Since the ultra-rapid satellite clock product for the global PPP users is based on the global distributed network, the impacts of the baseline distance among stations on the satellite clock batch estimation can be ignored and will not be considered in this contribution [25]. Therefore, the number of stations suitable for the satellite clock batch estimation needs to be determined to achieve the counterbalance between the efficiency and the batch estimation accuracy.

For the PPP based on the estimated satellite clock, the impacts of the estimated satellite clock on PPP have already been studied in many studies [10,26]. The absolute accuracy, latency and sampling interval of the satellite clock product are the main factors affecting PPP performance, including the convergence time and the positioning accuracy after convergence [21,26]. Furthermore, for the PPP based on the real-time satellite clock, the impacts of the classified satellite clock biases on PPP are analyzed [10]. Two such types of satellite clock biases, i.e., the initial clock bias and the time-dependent bias, affect the pseudorange and phase observations in the typical PPP observation model, which determine the convergence time and the positioning accuracy after convergence, respectively [15,27]. However, few studies have performed the comprehensive analysis for the impacts of the batch-estimated satellite clock on PPP. Therefore, we clarify the error propagation of the batch-estimated satellite clock biases in the PPP observation model and analyze their impacts on PPP.

We begin by constructing the batch observation model for the satellite clock batch estimation and clarify the error propagation of the satellite clock batch estimation. Then, the batch-estimated satellite clock biases are classified according to their impacts on the satellite clock batch estimation accuracy. Furthermore, we derive the analytic relationship between the observation redundancy and the satellite clock batch estimation accuracy to determine the number of stations suitable for satellite clock batch estimation. Moreover, we clarify the error propagation of the batch-estimated satellite clock biases in the PPP observation model. In the experimental verification, we present the data collection and validation strategies for the satellite clock batch estimation and PPP. The effect analysis of the satellite clock batch estimation and the PPP based on the batch-estimated satellite clock are verified. Finally, we give the discussion and summarize the main conclusions of this study.

## 2. Methodology

We first introduce the typical observation model of the satellite clock batch estimation. The source of the satellite clock bias induced by the batch observation model is clarified and classified. Then, based on the constructed batch observation model, the observation redundancy suitable for the satellite clock batch estimation is determined by deriving the covariance–variance matrix. Finally, the impacts of the classified satellite clock biases from the batch estimation on PPP are illustrated.

### 2.1. Sources of Batch-Estimated Satellite Clock Biases

The undifferenced observation model based on the batch estimation is adopted due to the lower observation noise and the smaller satellite clock biases than other traditional satellite clock estimation models [10]. Limited by the model strength, the batch observation model causes satellite clock biases in the batch-estimated satellite clock [10,28]. According to the impacts of the resulting satellite clock biases on the satellite clock batch estimation accuracy, they are categorized to clarify the error propagation in the satellite clock batch estimation.

The dual-frequency ionosphere-free (IF) combination is widely used for the undifferenced observation model to eliminate the higher-order ionospheric delay. The tropospheric delay is usually corrected for its dry component with the a priori model, and the wet component is estimated with the zenith tropospheric delay and its corresponding mapping function [29]. The station coordinate and the satellite orbit are usually fixed in the IF observation model [30]. It is assumed that the station coordinate is precisely known, which is normally fixed to the PPP weekly solution, while the error exists in the fixed ultra-rapid satellite orbit product due to the different processing strategies for the satellite orbit determination such as the satellite attitude model and the solar radiation pressure model [31,32]. For the batch observation model, all the observations and parameters in the observation session are constructed, which can be expressed as

$$\begin{bmatrix} \boldsymbol{v}^s_{P,IF,r} \\ \boldsymbol{v}^s_{L,IF,r} \end{bmatrix} = \begin{bmatrix} \boldsymbol{G}_r & \boldsymbol{G}^s & \boldsymbol{G}_{ZTD} & \boldsymbol{G}_r & \boldsymbol{G}^s & \boldsymbol{0} & \boldsymbol{0} & \boldsymbol{0} \\ \boldsymbol{G}_r & \boldsymbol{G}^s & \boldsymbol{G}_{ZTD} & \boldsymbol{G}_r & \boldsymbol{G}^s & \boldsymbol{G}_r & \boldsymbol{G}^s & \boldsymbol{G}_N \end{bmatrix} \begin{bmatrix} \delta \boldsymbol{t}_r \\ \delta \boldsymbol{t}^s \\ \boldsymbol{ZTD}_r \\ \boldsymbol{b}_r \\ \boldsymbol{b}^s \\ \boldsymbol{B}_r \\ \boldsymbol{B}^s \\ \boldsymbol{N}^s_r \end{bmatrix} - \begin{bmatrix} \boldsymbol{l}^s_{P,IF,r} + \boldsymbol{G}^s \cdot \boldsymbol{\mu}^s_r \delta \boldsymbol{x}^s \\ \boldsymbol{l}^s_{L,IF,r} + \boldsymbol{G}^s \cdot \boldsymbol{\mu}^s_r \delta \boldsymbol{x}^s \end{bmatrix} \quad (1)$$

where $\boldsymbol{G}_r = \boldsymbol{I}_m \otimes \boldsymbol{e}_n$, $\boldsymbol{G}^s = -\boldsymbol{e}_m \otimes \boldsymbol{I}_n$, $\boldsymbol{G}_{ZTD} = \boldsymbol{I}_m \otimes \boldsymbol{M}_n$, $\boldsymbol{G}_N = \boldsymbol{I}_m \otimes \boldsymbol{e}_n$, $\boldsymbol{G}$ is the designed matrix for the satellite clock batch estimation model, $s$ is the satellite, $r$ is the receiver, $\boldsymbol{I}_i$ is an $i \times i$ identity matrix, $\boldsymbol{e}_i$ is an $i \times 1$ vector with all elements equal to 1, $\boldsymbol{M}_i$ is an $i \times 1$ vector containing the tropospheric mapping function for each receiver, $m$ is the number of the observations, $n$ is the number of the observed satellites at different stations, $\otimes$ is the Kronecker product operation [33], $P$ is the pseudorange observation, $L$ is the phase observation, $IF$ is the IF combination, $\boldsymbol{v}$ is the vector of the a posterior observation residual for each observation, $\boldsymbol{\mu}^s_r$ is the unit vector from the satellite to the receiver, $\delta \boldsymbol{x}^s$ is the vector of the satellite orbit error for each satellite, $\delta \boldsymbol{t}_r$ is the vector of the receiver clock for each receiver, $\delta \boldsymbol{t}^s$ is the vector of the satellite clock for each satellite, $\boldsymbol{ZTD}$ is the vector of the zenith tropospheric delay for each receiver, $\boldsymbol{N}$ is the vector of the ambiguity for each satellite and each receiver, $\boldsymbol{b}_r$ is the vector of the code hardware delay for each receiver, $\boldsymbol{b}^s$ is the vector of the code hardware delay for each satellite, $\boldsymbol{B}_r$ is the vector of the phase hardware delay for each receiver, $\boldsymbol{B}^s$ is the vector of the phase hardware delay for each satellite and $\boldsymbol{l}$ is the vector of the a priori observation residual for each observation. In addition to the ranging errors in (1), the satellite and receiver antenna phase center offsets and variations, relativity, tidal loadings and phase windup should be modeled and precisely corrected in the a priori observation residual [30].

It can be seen from (1) that all the unknown parameters except the zenith tropospheric delay are linearly dependent. The satellite clock, receiver clock, zenith tropospheric delay and ambiguity parameters are the parameters to be estimated. Hence, other unknown parameters are absorbed by the estimated parameters or remain in the observation residual. Such absorption should satisfy the requirement of minimizing the weighted sum of the squares of residuals. According to the correlation between the parameters, and the consistency of the batch observation equation, $\delta \boldsymbol{x}^s$, $\boldsymbol{b}_r$, $\boldsymbol{b}^s$, $\boldsymbol{B}^s$ and $\boldsymbol{B}_r$ are absorbed by $\delta \boldsymbol{t}^s$, $\delta \boldsymbol{t}_r$, $\delta \boldsymbol{t}^s$, $\boldsymbol{N}^s_r$ and $\boldsymbol{N}^s_r$, respectively.

We focus on the ranging errors absorbed by the batch-estimated satellite clock, i.e., the code hardware delay for satellites and the satellite orbit error. Since the hardware delay can be treated as the constant within the observation session, the absorbed code hardware delay for satellites causes the time-invariant bias dependent with satellites. The time-invariant bias of the batch-estimated satellite clock will not affect the standard deviation (STD). For the satellite orbit error, not all of the satellite orbit errors are absorbed by the batch-estimated satellite clock. The assimilation ability depends on the station network distribution used for the satellite clock batch estimation [13]. The pseudorange and phase observations of the global distributed network are used for generating the worldwide satellite clock product. Since the satellite can be approximately regarded always directly above the global distributed network, the projection directions of the satellite orbit radial and tangential errors for each satellite in the signal propagation direction are identical and opposite, respectively. It is well known that the angles of the satellite parallax for the global distributed stations are different, resulting in different projection amounts of the satellite orbit errors. Therefore, the common satellite orbit radial errors observed by different stations in the signal propagation direction can be absorbed by the batch-estimated satellite clock. On the contrary, the satellite orbit tangential error cannot be absorbed due to the opposite projection direction. Therefore, the common satellite orbit error can be absorbed by the batch-estimated satellite clock epoch-wise, which belongs to the time-variant bias and affects the STD of the satellite clock batch estimation. Furthermore, the satellite orbit error not absorbed by the batch-estimated satellite clock is absorbed by the observation residual. The STD of the satellite clock batch estimation is still affected.

It can be seen from (1) that the rank of the coefficient matrix presents deficiency due to the linear dependency between the satellite clock and receiver clock. The clock constraint should be introduced to avoid the rank deficient [34,35]. Therefore, the batch-estimated receiver and satellite clocks are aligned as the clock relative to the reference clock. The reference clock strategies include selecting one satellite clock as zero, one receiver clock as zero and the mean value of satellite clocks as zero, which are equivalent to each other [36]. In order to ensure the stability of the satellite clock batch estimation, we select the zero-mean condition as the reference clock [37]. Inevitably, the batch-estimated satellite clock contains the timescale difference caused by the clock bias of the selected reference clock [15]. The timescale difference belongs to the time-invariant bias independent with satellites, which does not affect the satellite clock batch estimation accuracy when the accuracy of the reference clock is better than $10^{-6}$ s [15].

In summary, the reparametrized satellite clock batch estimation model due to the parameter assimilation can be redefined as

$$\begin{bmatrix} \boldsymbol{v}_{P,IF,r}^{s} \\ \boldsymbol{v}_{L,IF,r}^{s} \end{bmatrix} = \begin{bmatrix} \boldsymbol{G}_r & \boldsymbol{G}^s & G_{ZTD} & \boldsymbol{0} \\ \boldsymbol{G}_r & \boldsymbol{G}^s & G_{ZTD} & G_N \end{bmatrix} \begin{bmatrix} \delta\bar{\boldsymbol{t}}_r \\ \delta\bar{\boldsymbol{t}}^s \\ \boldsymbol{ZTD}_r \\ \overline{\boldsymbol{N}}_r^s \end{bmatrix} - \begin{bmatrix} \bar{\boldsymbol{l}}_{P,IF,r}^{s} \\ \bar{\boldsymbol{l}}_{L,IF,r}^{s} \end{bmatrix} \quad (2)$$

where $\delta\bar{\boldsymbol{t}}_r$, $\delta\bar{\boldsymbol{t}}^s$, $\overline{\boldsymbol{N}}_r^s$, $\bar{\boldsymbol{l}}_{P,IF,r}^{s}$ and $\bar{\boldsymbol{l}}_{L,IF,r}^{s}$ are the reparametrized receiver clock, satellite clock, ambiguity, a priori pseudorange observation residual and a priori phase observation residual, respectively. They can be rewritten as

$$\begin{cases} \delta\bar{\boldsymbol{t}}_r = \delta\boldsymbol{t}_r + \boldsymbol{b}_r \\ \delta\bar{\boldsymbol{t}}^s = \delta\boldsymbol{t}^s + \alpha\boldsymbol{\mu}_r^s\delta\boldsymbol{x}^{s,R} + \beta\boldsymbol{\mu}_r^s\delta\boldsymbol{x}^{s,T} + \boldsymbol{b}^s + \boldsymbol{D}^s \\ \overline{\boldsymbol{N}}_r^s = \boldsymbol{N}_r^s + \boldsymbol{B}_r - \boldsymbol{B}^s + \boldsymbol{b}^s - \boldsymbol{b}_r \\ \bar{\boldsymbol{l}}_{P,IF,r}^{s} = \boldsymbol{l}_{P,IF,r}^{s} - \boldsymbol{G}^s\cdot(1-\alpha)\boldsymbol{\mu}_r^s\delta\boldsymbol{x}^{s,R} - \boldsymbol{G}^s\cdot(1-\beta)\boldsymbol{\mu}_r^s\delta\boldsymbol{x}^{s,t} \\ \bar{\boldsymbol{l}}_{L,IF,r}^{s} = \boldsymbol{l}_{L,IF,r}^{s} - \boldsymbol{G}^s\cdot(1-\alpha)\boldsymbol{\mu}_r^s\delta\boldsymbol{x}^{s,R} - \boldsymbol{G}^s\cdot(1-\beta)\boldsymbol{\mu}_r^s\delta\boldsymbol{x}^{s,T} \end{cases} \quad (3)$$

where $\boldsymbol{D}^s$ is the timescale difference, $\delta\boldsymbol{x}^{s,R}$ and $\delta\boldsymbol{x}^{s,T}$ are the satellite orbit radial and tangential errors, $\alpha$ and $\beta$ are the assimilation proportion of the batch-estimated satellite clock to the satellite orbit radial and tangential errors. The absorption ability can be calculated by using the range of the satellite parallax angle during satellite regular motion [13]. Taking GPS as an example, the estimated satellite clock can absorb 97.1% of the satellite orbit error in the radial direction at least, and 24.0% in the tangential direction at most. The specific assimilation proportion depends on the station network distribution used for the satellite clock batch estimation.

The impact of the observation noise needs to be accounted for the satellite clock batch estimation. Influenced by the ambiguity in the phase observation, the variation and the absolute bias of the batch-estimated satellite clock are determined by the phase and pseudorange observation, respectively [38]. Furthermore, any biases in the ambiguity or the undetected cycle slips in the data preprocessing will cause the time-invariant bias in the batch-estimated satellite clock. Moreover, for the real-time satellite clock estimation, the time-invariant bias will also be introduced by the initial satellite clock with poor accuracy, which arises from the broadcast clock or the predicted satellite clock [10]. However, the impacts of the initial satellite clock on the satellite clock batch estimation can be ignored. This is because the initial value of the satellite clock can be predetermined with high precision in the batch estimation.

In order to categorically analyze the impacts of the satellite clock biases on the accuracy of the batch-estimated satellite clock, the time-invariant and time-variant bias induced by the batch observation model are denoted as the initial clock bias (ICB) and time-dependent bias (TDB), respectively. The ICB and TDB are both estimated satellite clock biases, which cause the satellite clock error compared with the satellite clock reference product. The only difference between them is that the ICB and TDB contribute to the constant and variable components of the batch-estimated satellite clock errors, respectively. The ICBs caused by the code hardware delay assimilation, the reference clock selection and the pseudorange observation noise will not affect the STD of the batch-estimated satellite clock, but will affect the root mean square (RMS). The TDB induced by the satellite orbit error and the phase observation noise will affect the STD and RMS.

### *2.2. Suitable Observation Redundancy for Satellite Clock Batch Estimation*

In addition to the ICB and TDB, the observation redundancy correlated with the number of stations also affects the satellite clock batch estimation. The accuracy of the batch-estimated satellite clock becomes saturable when the observation redundancy adequately increases. The benefit of increased observation redundancy is negligible for the accuracy improvement of the satellite clock batch estimation [20]. Furthermore, excessive observation redundancy will reduce the efficiency, which needs to be considered in the satellite clock batch estimation with low latency. Therefore, based on the clarified batch observation model, we derive the analytic relationship between the observation and the parameter to be estimated, and simultaneously consider the solution efficiency to determine the saturable accuracy of the batch-estimated satellite clock.

In order to reduce the computation burden of estimating all parameters in the normal equation with large dimensions, the parameter pre-elimination and back-substitution are generally adopted in the batch estimation [39]. The parameters to be estimated are divided into the time-variant and time-invariant parameters to perform the parameter pre-elimination and back-substitution [2]. The satellite clock and receiver clock are denoted as the time-variant parameter and estimated as epoch-wise. The zenith tropospheric delay is denoted as the time-invariant parameter and estimated as piece-wise. The ambiguity is generally regarded as the constant in the absence of cycle slips, which is also denoted as the time-invariant parameter. For the processing strategy of the satellite clock batch estimation, the time-variant parameter is generally set up and pre-eliminated epoch-wise due to its potentially large number. The back-substitution is used to obtain the solution of the time-variant parameter. The observation model can be rewritten as

$$\begin{bmatrix} \boldsymbol{v}^s_{P,IF,r} \\ \boldsymbol{v}^s_{L,IF,r} \end{bmatrix} = \begin{bmatrix} \boldsymbol{G}_r & \boldsymbol{G}^s \\ \boldsymbol{G}_r & \boldsymbol{G}^s \end{bmatrix} \begin{bmatrix} \delta\bar{\boldsymbol{t}}_r \\ \delta\bar{\boldsymbol{t}}^s \end{bmatrix} + \begin{bmatrix} \boldsymbol{G}_{ZTD} & \boldsymbol{0} \\ \boldsymbol{G}_{ZTD} & \boldsymbol{G}_N \end{bmatrix} \begin{bmatrix} \boldsymbol{ZTD}_r \\ \overline{\boldsymbol{N}}^s_r \end{bmatrix} - \begin{bmatrix} \vec{\boldsymbol{l}}^s_{P,IF,r} \\ \vec{\boldsymbol{l}}^s_{L,IF,r} \end{bmatrix} \tag{4}$$

Combining with the pseudorange and phase observations, the observation error is accurately characterized by the corresponding weighting matrix. In order to simply represent the processing of the pre-eliminated parameter and back-substitution, the normal equation of the satellite clock batch estimation can be abbreviated in the matrix form as

$$\boldsymbol{PV} = \boldsymbol{PAX} + \boldsymbol{PBY} - \boldsymbol{PL} \tag{5}$$

where $\boldsymbol{P}$ is the weighting matrix related to the observation error, $\boldsymbol{X}$ and $\boldsymbol{Y}$ are the vectors of the time-variant and time-invariant parameters, $\boldsymbol{A}$ and $\boldsymbol{B}$ are their corresponding coefficient matrices, $\boldsymbol{V}$ and $\boldsymbol{L}$ are the vectors of the a posteriori and the a priori observation residuals.

According to the principle of the least square parameter estimation based on the parameter pre-elimination and back-substitution, the batch-estimated parameters can be expressed as

$$\begin{cases} \hat{X} = N_{XX}^{-1}(U_{XL}L - N_{XY}\hat{Y}) \\ \hat{Y} = (N_{YY} - N_{YX}N_{XX}^{-1}N_{XY})^{-1}\left(U_{YL}L - N_{YX}N_{XX}^{-1}U_{XL}L\right) \end{cases} \tag{6}$$

where $N_{XX} = A^{T}CA$, $N_{XY} = A^{T}CB$, $N_{YX} = B^{T}CA$, $N_{YY} = B^{T}CB$, $U_{XL} = A^{T}C$, $U_{YL} = B^{T}C$, $C = P^{T}P$. The diagonal element in the covariance–variance matrix can reflect the STD of the parameter to be estimated. The satellite clock batch estimation accuracy can be obtained by extracting the diagonal elements of the covariance–variance matrix corresponding to the satellite clocks. This evaluated accuracy does not require the additional satellite clock reference product. The clock covariance–variance matrix of the satellite clock batch estimation derived by (6) can be expressed as

$$\mathrm{COV}(\varepsilon_{\hat{X}}, \varepsilon_{\hat{X}}) = N_{XX}^{-1} + N_{XX}^{-1}N_{XY}(N_{YY} - N_{YX}N_{XX}^{-1}N_{XY})^{-1}N_{YX}N_{XX}^{-1} \tag{7}$$

It can be found that the clock covariance–variance matrix consists of the coefficient matrix and the weighting matrix for the satellite clock batch estimation model. The satellite clock batch estimation accuracy depends on the number of stations and the observation error, which are closely related to the coefficient matrix dimension and the weighting matrix, respectively. Thus, the analytic relationship between the observation and the parameter to be estimated has clearly been derived. However, the number of stations suitable for the satellite clock batch estimation and the corresponding saturable estimation accuracy still cannot be determined directly from the analytic expression. This is because the coefficient matrix dimension is also affected by the number of visible satellites at different stations and different epochs, which needs to be determined experimentally.

The GPS data from the day of the year (DOY) 045 to 051 in 2021 are used for determining the number of stations suitable for the satellite clock batch estimation and the corresponding saturable estimation accuracy. We ignore the impacts of different stations with identical quantities on the satellite clock batch estimation. This is due to the identical full coverage for visible satellites in the observation session and the little difference in the observation quality between different International GNSS Service (IGS) stations used. The efficiency of the satellite clock batch estimation has limitations due to the low latency requirements, which should be controlled within one hour including the consumption of hourly data collection. We count the processing time to reflect the efficiency of satellite clock batch estimation [40]. The satellite clock batch estimation based on the final satellite orbit product provided by IGS is performed on the PowerEdge R7525 Server (AMD EPYC 7F72 24-Core Processor @3.69 GHz). The STD is extracted from (7) based on the gradually increasing number of the global distributed network stations and the elevation-dependent stochastic model, whose consistency with the STD for the external coincidence is verified. The STD for the external coincidence is conducted by calculating the batch-estimated satellite clock error based on the double difference strategy [10]. Moreover, the criteria for gradually increasing stations serve to maximize the observation, which is mainly relative to the observed satellites. The mean values of the STD for all satellites and the processing time of the satellite clock batch estimation with a gradually increasing number of stations are shown in Figure 1.

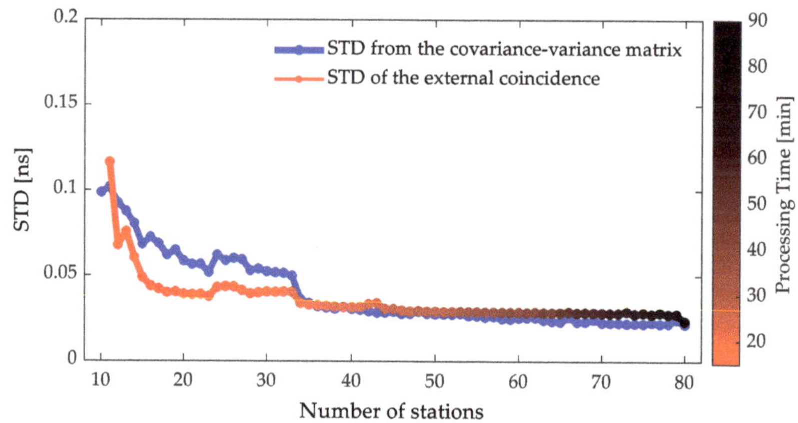


**Figure 1.** STD and processing time with a gradually increasing number of stations for one week.

It can be seen from Figure 1 that the satellite clock batch estimation accuracy is gradually improved as the increasing number of stations. The satellite clock batch estimation accuracy is greatly affected by the observation redundancy for less than 40 stations and ranges from 0.03 to 0.10 ns. However, the GPS satellite clock batch estimation accuracy tends to be stable when the number of stations is more than 40. This indicates that the satellite clock batch estimation accuracy becomes saturable. Considering the counterbalance between the efficiency and estimation accuracy, we suggest that the number of stations suitable for GPS hourly updated satellite clock batch estimation is to be 40.

### 2.3. *Impacts of Batch-Estimated Satellite Clock Biases on PPP*

Based on the suitable observation redundancy for the satellite clock batch estimation, the ICB and TDB of the batch-estimated satellite clock affect the PPP performance. We will clarify the error propagation of these satellite clock biases in the PPP observation model. For the conventional dual-frequency IF combined PPP processing, the consistent satellite clock and orbit products are generally fixed [41]. Since the satellite clock batch estimation is performed based on the ultra-rapid satellite orbit product, the error exists in the fixed satellite clock product for the PPP observation model. Due to the consistency between the satellite clock and orbit product used for PPP, the fixed ultra-rapid satellite orbit product is also fixed in PPP. Thus, the satellite clock and orbit errors cannot be ignored in the PPP observation model. For a typical PPP observation model [42,43], the observations and the parameters are epoch-wise constructed, which can be expressed as

$$\begin{bmatrix} \boldsymbol{v}^s_{P,IF,r} \\ \boldsymbol{v}^s_{L,IF,r} \end{bmatrix} = \begin{bmatrix} \boldsymbol{K}_r & \boldsymbol{K}_r & \boldsymbol{K}_{ZTD} & \boldsymbol{K}_r & \boldsymbol{K}^s & \boldsymbol{0} & \boldsymbol{0} & \boldsymbol{0} \\ \boldsymbol{K}_r & \boldsymbol{K}_r & \boldsymbol{K}_{ZTD} & \boldsymbol{K}_r & \boldsymbol{K}^s & \boldsymbol{K}_r & \boldsymbol{K}^s & \boldsymbol{K}_N \end{bmatrix} \begin{bmatrix} \boldsymbol{\mu}^s_r \delta \boldsymbol{x}_r \\ \delta \boldsymbol{t}_r \\ \boldsymbol{ZTD}_r \\ \boldsymbol{b}_r \\ \boldsymbol{b}^s \\ \boldsymbol{B}_r \\ \boldsymbol{B}^s \\ \boldsymbol{N}^s_r \end{bmatrix} - \begin{bmatrix} \boldsymbol{l}^s_{P,IF,r} + \boldsymbol{K}^s \cdot \boldsymbol{\mu}^s_r \delta \boldsymbol{x}^s - \boldsymbol{K}^s \cdot \delta \bar{\boldsymbol{t}}^s \\ \boldsymbol{l}^s_{L,IF,r} + \boldsymbol{K}^s \cdot \boldsymbol{\mu}^s_r \delta \boldsymbol{x}^s - \boldsymbol{K}^s \cdot \delta \bar{\boldsymbol{t}}^s \end{bmatrix} \tag{8}$$

where $\boldsymbol{K}_r = \boldsymbol{I}_p \otimes \boldsymbol{e}_q$, $\boldsymbol{K}^s = -\boldsymbol{e}_p \otimes \boldsymbol{I}_q$, $\boldsymbol{K}_{ZTD} = \boldsymbol{I}_p \otimes \boldsymbol{M}_q$, $\boldsymbol{K}_N = \boldsymbol{I}_p \otimes \boldsymbol{e}_q$, $\boldsymbol{K}$ is the designed matrix for the PPP observation model, $p$ is the number of the observations, $q$ is the number of observed satellites at different stations and $\delta x_r$ is the increment for the a priori receiver position vector. The receiver position increment, the receiver clock, the zenith tropospheric delay and the ambiguity parameters are the parameters to be estimated in the PPP observation model. Given that the hardware delay is time-invariant, the redefined PPP observation model due to the correlation among the unknown parameters can be expressed as

$$\begin{bmatrix} v^s_{P,IF,r} \\ v^s_{L,IF,r} \end{bmatrix} = \begin{bmatrix} K_r & K_r & K_{ZTD} & 0 \\ K_r & K_r & K_{ZTD} & K_N \end{bmatrix} \begin{bmatrix} \mu^s_r \delta x_r \\ \delta\tilde{t}_r \\ ZTD_r \\ \tilde{N}^s_r \end{bmatrix} - \begin{bmatrix} \tilde{l}^s_{P,IF,r} \\ \tilde{l}^s_{L,IF,r} \end{bmatrix} \tag{9}$$

where $\delta\tilde{t}_r$, $\tilde{N}^s_r$, $\tilde{l}^s_{P,IF,r}$ and $\tilde{l}^s_{L,IF,r}$ are the reparametrized receiver clock, ambiguity, a priori pseudorange observation residual and a priori phase observation residual, respectively. They can be rewritten as

$$\begin{cases} \delta\tilde{t}_r = \delta t_r + b_r + K^s \cdot D^s \\ \tilde{N}^s_r = N^s_r + B_r - b_r + \left[\delta\tilde{t}^s(i_0) - K^s \cdot b^s\right] \\ \tilde{l}^s_{P,IF,r} = l^s_{P,IF,r} + K^s \cdot \mu^s_r \delta x^{s,R} + K^s \cdot \mu^s_r \delta x^{s,T} - K^s \cdot \sum\limits_{j=i_0+1}^{i} \Delta\delta\tilde{t}^s(i) \\ \tilde{l}^s_{L,IF,r} = l^s_{L,IF,r} + K^s \cdot \mu^s_r \delta x^{s,R} + K^s \cdot \mu^s_r \delta x^{s,T} - K^s \cdot \sum\limits_{j=i_0+1}^{i} \Delta\delta\tilde{t}^s(i) \end{cases} \tag{10}$$

where $\delta\tilde{t}^s(i_0)$ is the ICB of the fixed batch-estimated satellite clock at the initial epoch $i_0$, $\Delta\delta\tilde{t}^s(i)$ is the TDB of the fixed batch-estimated satellite clock at the $i$th epoch.

We focus on the impacts of the satellite clock bias for the fixed batch-estimated satellite clock on the PPP observation model. It can be found from (10) that the timescale difference can be absorbed by the receiver clock in the PPP observation model without affecting the PPP positioning accuracy. The ICB caused by the code hardware delay for satellites can be canceled by the code hardware delay for satellites in the PPP observation model if the identical combination of observation is used for PPP and satellite clock batch estimation, i.e., the dual-frequency IF combination. The TDB induced by the absorbed satellite orbit error can mostly be canceled by the fixed satellite orbit error in the PPP observation model. The reason for the cancellation is that the satellite clock and orbit product used for PPP is consistent. Furthermore, the cancellation depends on the assimilation ability of the batch-estimated satellite clock to the satellite orbit error, resulting in incomplete cancellation.

It can be found from (10) that the ICB not caused by the code hardware delay for satellites is absorbed by ambiguity in the PPP observation model. Furthermore, the TDB caused by the unabsorbed satellite orbit error or others is absorbed by the observation residual in the PPP observation model. For the ICB and TDB of the batch-estimated satellite clock caused by these reasons, we investigate their impacts on the PPP positioning performance, including the convergence time and the positioning accuracy after convergence. The convergence time and positioning accuracy after convergence depend on the speed of accurately estimating parameters and the observation noise, respectively. Since the ICB is absorbed by the ambiguity, the estimation of the ambiguity requires a long convergence process and mainly impacts the convergence time of PPP. Furthermore, the inaccurate ambiguity parameter may impact the phase observation, resulting in affecting the positioning accuracy after convergence. The TDB can be treated as impacting the observation noise when estimating the receiver coordinates. The large TDB is equivalent to increasing the observation noise, and mainly impacts the accuracy of the receiver coordinate estimation, i.e., the positioning accuracy after convergence. Meanwhile, the speed of accurately estimating the receiver coordinate, i.e., the convergence time of PPP, may also be affected.

## 3. Results

The satellite clock batch estimation based on the global distributed network is implemented to verify the impacts of the ICB and TDB. Since the long absence of the satellite clock reference product and the fixed satellite orbit product for G11, the satellite clock batch estimation and the evaluation does not include this satellite. The experimental setup and processing strategy is shown in Table 1. The station selection of the global distributed network is performed in the preprocessing of satellite clock batch estimation, which depends on the quality of observations at different stations. One example of the station selection is shown in Figure 2. The number of selected stations is 40, which is the suitable observation redundancy for the satellite clock batch estimation verified in the previous experiments. In order to evaluate the batch-estimated satellite clock and the fixed satellite orbit, the GPS final satellite clock and orbit product provided by IGS are introduced as the reference, respectively.

**Table 1.** Experimental setup and processing strategy.

| Items | Settings |
| --- | --- |
| Time | DOY from 001 to 365 in 2021 |
| Basic observations | Undifferenced pseudorange and phase observations |
| Combination model | Dual-frequency IF combination |
| Observation signal | GPS: L1/L2 |
| Elevation cutoff angle | 5° |
| Sampling rate | 300 s |
| Stochastic model | A priori precision of pseudorange: 0.1 m in unit weighting;<br>A priori precision of phase: 0.001 m in unit weighting;<br>Elevation angle weighting |
| Station network | Global distributed network with 40 stations |
| Station coordinate | Fixed station coordinates provided for IGS |
| Hardware delay | Absorbed |
| Tropospheric delay | Dry components: model correction;<br>Wet components: mapping function and estimated every 2 h |
| Ionospheric delay | Lower order: eliminate by dual-frequency IF combined;<br>Higher order: ignore |
| Receiver clock | Estimated by single point positioning as a priori value,<br>and epoch-wise estimated |
| Satellite clock | Select an initial satellite clock, and epoch-wise estimated |
| Satellite orbit | Fixed by the satellite orbit products |
| Ambiguity | Float solutions and estimated as constant for continuous arc |
| Estimation strategy | Batch least square algorithms |

The TDB due to the satellite orbit error assimilation on satellite clock batch estimation IS clarified by the correlation analysis between the satellite clock error and the satellite orbit error. The batch-estimated satellite clock errors are evaluated to verify the impacts of the ICB and TDB on the satellite clock batch estimation. Furthermore, the convergence time and the positioning accuracy after convergence are analyzed to reveal the impacts of the ICB and TDB on PPP.

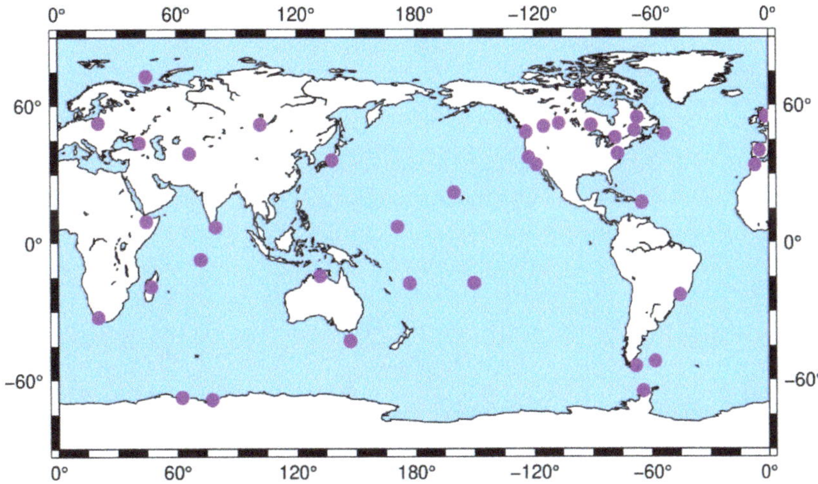


**Figure 2.** One example of station selection with 40 stations for satellite clock batch estimation. The purple circles represent selected stations.

### 3.1. *Effect Analysis of Satellite Clock Batch Estimation*

The bias and dispersion of the batch-estimated satellite clock error can reveal the ICB and TDB, respectively. We use the double difference strategy to extract the batch-estimated satellite clock error. The reference satellite needs be selected in advance. The reference satellite clock strategies typically include selecting one satellite clock and the mean value of satellite clocks for all satellites. In order to preserve the satellite clock error sequences for all satellites, the mean value of all satellite clocks is selected as the reference in this contribution, which can be expressed as

$$\nabla\Delta\delta\bar{\boldsymbol{t}}^{s} = \left[\delta\bar{\boldsymbol{t}}^{s} - \frac{1}{n}\sum_{i=1}^{n}\delta\bar{\boldsymbol{t}}^{s}(i)\right] - \left[\delta\boldsymbol{t}^{s}_{ref} - \frac{1}{n}\sum_{i=1}^{n}\delta\boldsymbol{t}^{s}_{ref}(i)\right] \tag{11}$$

where $\nabla\Delta$ is the double difference operator, $ref$ is the satellite clock reference product provided by IGS, $n$ is the total number of epochs and $i$ is the epoch. In addition, we used the metrics of the RMS and STD to reflect the bias and dispersion for the batch-estimated satellite clock error, respectively.

There are two classifications of biases contributing to the batch-estimated satellite clock, i.e., the ICB and the TDB, in which we will not conduct additional experiments to verify the impacts of the ICB on the satellite clock batch estimation. The ICB is reflected by the RMS of the batch-estimated satellite clock error. The TDB, another classification of the batch-estimated satellite clock, is impacted by the assimilation of the satellite orbit error. The assimilation ability can be indicated by the correlation coefficients between the satellite clock error and the satellite orbit error. The epoch-wise satellite orbit error can be obtained by comparing it with the satellite orbit reference product provided by IGS. The satellite orbit errors in the radial, along-track and cross-track directions can be computed as [44]

$$\begin{bmatrix}\Delta e_R \\ \Delta e_A \\ \Delta e_C\end{bmatrix} = \begin{bmatrix}\frac{\boldsymbol{r}}{|\boldsymbol{r}|} & \frac{\boldsymbol{r}\times\boldsymbol{v}}{|\boldsymbol{r}\times\boldsymbol{v}|}\times\frac{\boldsymbol{r}}{|\boldsymbol{r}|} & \frac{\boldsymbol{r}\times\boldsymbol{v}}{|\boldsymbol{r}\times\boldsymbol{v}|}\end{bmatrix}^{\mathrm{T}} \begin{bmatrix}\Delta x \\ \Delta y \\ \Delta z\end{bmatrix} \tag{12}$$

where $\Delta e_R$, $\Delta e_A$ and $\Delta e_C$ are the satellite orbit errors in the radial, along-track and cross-track directions, respectively. $\Delta x$, $\Delta y$ and $\Delta z$ are the satellite orbit errors in the Earth-centered Earth-fixed (ECEF) frame. $\boldsymbol{r}$ and $\boldsymbol{v}$ is the satellite position and the inertial velocity in the ECEF frame.

The satellite clock batch estimation is implemented based on the fixed ultra-rapid satellite orbit products in observation sessions provided by the three different analysis centers including Wuhan University, Geo Forschungs Zentrum (GFZ) and IGS. These three data sources are abbreviated as WHU, GFU and IGU, respectively. The 1DRMS accuracy of the three ultra-rapid satellite orbit products in the observation session for one-year data are 1.86 cm, 1.31 cm and 0.83 cm, respectively. We use the three different ultra-rapid satellite orbit products to investigate the assimilation ability of the batch-estimated satellite clock to different types of the satellite orbit error [45]. The correlation between the batch-estimated satellite clock error and the fixed satellite orbit error is shown in Figure 3.

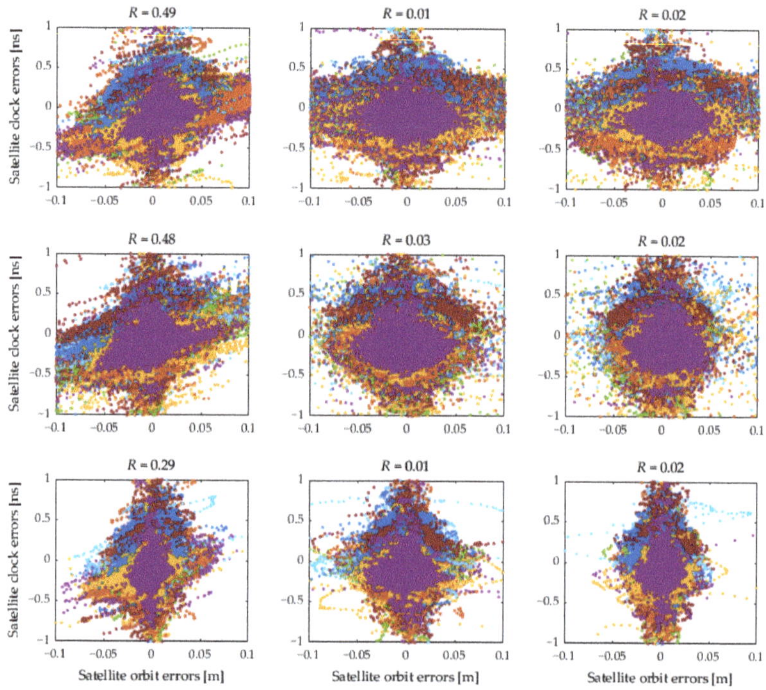


**Figure 3.** Correlation between batch-estimated satellite clock errors and fixed satellite orbit errors for one year. The row panels from top to bottom show the results based on the fixed satellite orbit product provided by WHU, GFU and IGU, respectively. The column panels from left to right show the results of the satellite orbit error in the radial, along-track and cross-track directions. The dots with different colors represent different satellites.

It can be seen from Figure 3 that the batch-estimated satellite clock error and the fixed satellite orbit error in the radial direction are positively correlated, whilst the correlation in the along-track and cross-track directions can be neglected. This demonstrates that the batch-estimated satellite clock error based on the global distributed network can assimilate the satellite orbit radial error more than the other two directions. Moreover, the correlation coefficient of WHU, GFU and IGU is different, especially that of IGU, which is much lower than WHU and GFU. This is because the station selection for the global distributed network is specific, which depends on the station observation quality correlated with the fixed satellite orbit accuracy. Thus, the common satellite orbit radial error absorbed by the batch-estimated satellite clock is different, resulting in the difference in the assimilation ability and the corresponding correlation coefficient.

In addition to the absorbed satellite orbit error, the unabsorbed satellite orbit error is absorbed by the observation residual and also affects the TDB of the batch-estimated satellite clock. In order to investigate the impacts of the satellite orbit error on TDB, we extracted the batch-estimated satellite clock error based on the fixed satellite orbit products provided by WHU, GFU and IGU. The TDB is indicated by the dispersion of the batch-estimated satellite clock error. The bias of the batch-estimated satellite clock error reflects the ICB. Furthermore, the satellite clock batch estimation based on the final satellite orbit product provided by IGS is also performed as the reference for comparison. The batch-estimated satellite clock errors based on the fixed four satellite orbit products, named BEST-WHUO, BEST-GFUO, BEST-IGUO and BEST-IGSO, are shown in Figure 4.

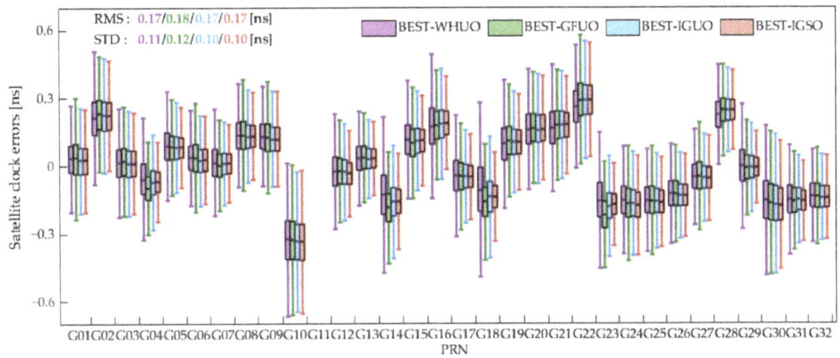


**Figure 4.** Batch-estimated satellite clock errors based on fixed WHUO, GFUO, IGUO and IGSO for one year.

It can be seen from Figure 4 that the batch-estimated satellite clock errors based on these four satellite orbit products have significant biases. This is because the batch-estimated satellite clock error contains the ICB, which is induced by the assimilation of the code hardware delay for satellites. The dispersion of the batch-estimated satellite clock errors based on the three ultra-rapid satellite orbit products is significant relative to the IGS final satellite orbit product. This is because the satellite orbit error, including the absorbed and unabsorbed components, affects the TDB of the batch-estimated satellite clock. Furthermore, the STD difference among the BEST-WHUO, BEST-GFUO and BEST-IGUO indicates that the TDB of the batch-estimated satellite clock is different, in which the STD of BEST-IGUO is slightly lower than BEST-WHUO and BEST-GFUO. This is because the assimilation of the satellite orbit radial error based on the IGUO is less than that of WHUO and GFUO. Meanwhile, the unabsorbed component of IGUO has better accuracy than WHUO and GFUO.

### *3.2. Effect Analysis of PPP Based on Batch-Estimated Satellite Clock*

In order to investigate the impacts of the ICB and TDB for the batch-estimated satellite clock on PPP, the PPP in the kinematic mode will be carried out. The observation model of PPP comes from (9). The elevation-dependent weighting model is used for constructing the stochastic model of PPP. The PPP solution is implemented based on the GAMP software [43]. The PPP positioning errors of the east, north and up directions are calculated epoch-wise by comparing with the reference coordinates provided by IGS. We excluded the positioning results from the last 15 min to avoid the impacts of the satellite orbit extrapolation error on PPP. Furthermore, the convergence time and the positioning accuracy after convergence were used as the indicators to evaluate the PPP positioning performance.

We firstly perform the simulation experiment to clarify the impacts of the ICB on PPP by injecting the artificial constant bias into the satellite clock batch estimation result. The injected bias for each satellite is the time-invariant bias in days, which is consistent with the period of PPP solutions. It is noted that the magnitude of the simulated bias is obtained from the normal distribution with the STD of 3 ns for each satellite, which is consistent with the accuracy of the broadcast clock. One example of the injected bias for each satellite in one day is shown as Figure 5.

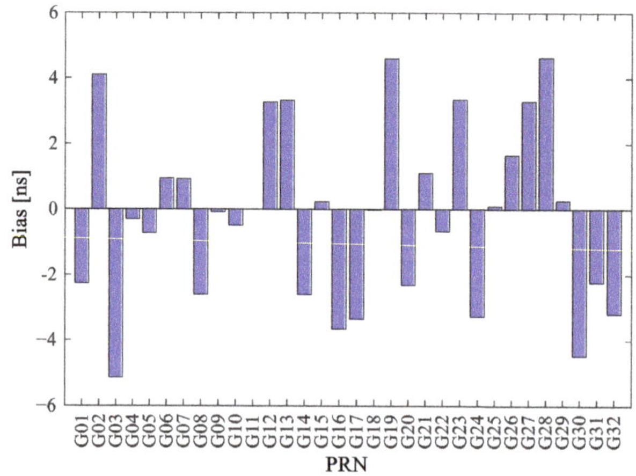


**Figure 5.** One example of injected bias for each satellite in one day.

The IGS final satellite orbit product is used for the satellite clock batch estimation and PPP to avoid the impacts of the satellite orbit error on TDB and PPP. The resulting simulated batch-estimated satellite clock is named BEST + ICB. The PPP experiment based on BEST + ICB can illustrate the impacts of the ICB on PPP. Moreover, we set 20 cm as the convergence threshold to determine the convergence time of PPP. The PPP results for BEST and BEST + ICB are shown in Figure 6.

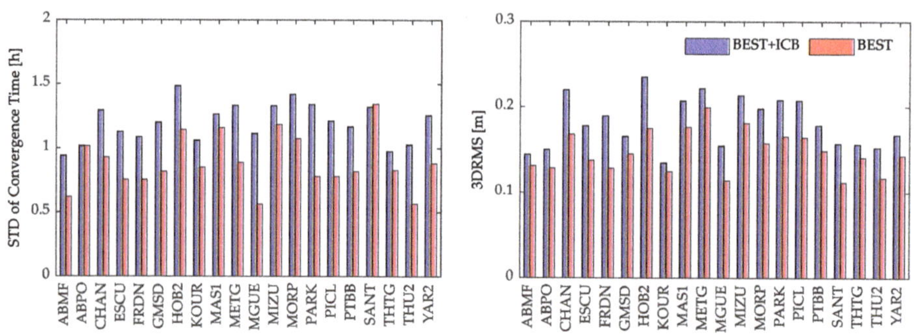


**Figure 6.** PPP results for BEST and BEST + ICB at 20 arbitrary stations for one year. The two subplots on the left show the STD of the convergence time. The two subplots on the right show the 3DRMS of the positioning accuracy after convergence.

The convergence time for each station varies daily. The STD of the convergence time indicates the variation degree for the observation quality from day to day, including the observation accuracy and the satellite distribution. It can be seen from Figure 6 that the STD of the convergence time for BEST + ICB is longer than BEST. This is because the ICB is absorbed by the ambiguity in the PPP observation model and impacts the convergence time of PPP. Furthermore, we can find that the impacts of the ICB on the convergence time across different stations. It can be explained that the convergence time of PPP is affected by the magnitude of the ICB with different satellites. Moreover, the positioning accuracy after convergence for BEST + ICB is slightly worse than BEST, which is attributed to the injected ICB.

In order to explore the impacts of the TDB on PPP, we simulated the random biases epoch-wise and injected the simulated biases into the batch-estimated satellite clock. The injected bias for each satellite is time-variant, which is the main difference from the injected ICBs. It is noted that the STD of the simulated biases is obtained from the normal distribution with the STD of 0.3 ns for all satellites. The value of the simulated biases comes from three times the STD of the satellite clock error. The impacts of the TDBs on PPP can be amplified and verified. Similarly to the injected ICB, the injected random bias for each satellite is also in days. Furthermore, the batch-estimated satellite clock data for simulation is also based on the fixed final satellite orbit product provided by IGS. The resulting simulated batch-estimated satellite clock is named BEST + TDB. The PPP based on BEST + TDB can illustrate the impacts of the TDB on PPP. Moreover, we set the positioning convergence threshold as 50 cm because of the injected simulated random biases. The PPP results for BEST and BEST + TDB are shown in Figure 7.

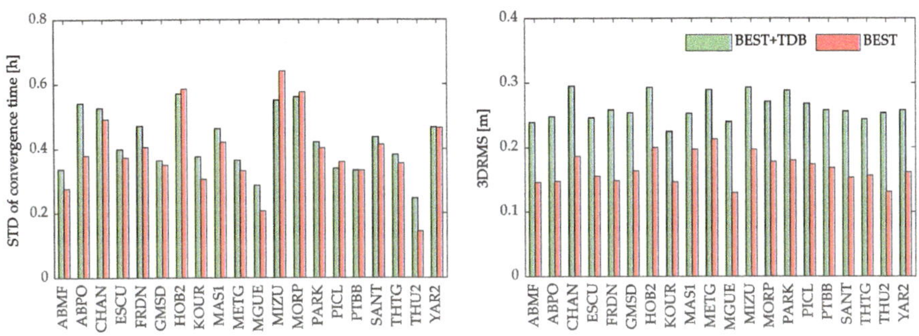


**Figure 7.** PPP results for the BEST and BEST + TDB at 20 arbitrary stations for one year. The two subplots on the left show the STD of the convergence time. The two subplots on the right show the 3DRMS of the positioning accuracy after convergence.

It can be seen from Figure 7 that the STD of the convergence time is certainly shortened. This is because of the expanding convergence threshold, so that the positioning result for BEST + TDB can be regarded as the convergence. From Figure 7, we can see that the positioning accuracy after convergence for BEST + TDB is worse than BEST. This is because the simulated TDB based on the normal distribution is absorbed by the observation residual, which impacts the positioning accuracy after convergence. Furthermore, the STD of the convergence time for BEST + TDB is slightly longer than BEST, which does not exceed 30 min. This is because the absorbed observation residual impacts the PPP observation model strength.

## 4. Discussion

In order to obtain the PPP results at the centimeter level using ultra-rapid satellite products, the satellite clock batch estimation accuracy and its impacts on PPP are analyzed in this contribution.

The traditional observation models for satellite clock estimation generally include undifferenced, epoch-differenced and mixed-differenced models. For satellite clock batch estimation, the undifferenced observation model is typically used. The ICB and TDB are induced by the batch observation model, including the assimilated ranging errors and the model accuracy. The theoretical analysis and experimental results show that the ICB affects the RMS without affecting the STD, whilst the TDB affects the RMS and the STD of the batch-estimated satellite clock. Furthermore, we primarily analyze the impacts of the fixed satellite orbit error on the TDB. The experimental results show that the assimilation of the satellite orbit radial error to the TDB is more than the along-track and the cross-track directions. The assimilation ability depends on the station distribution used for the satellite clock batch estimation.

In addition to the satellite clock biases, the observation redundancy also affects the STD of the batch-estimated satellite clock. Many studies have shown the number of stations suitable for the real-time satellite clock estimation. Such results may not be suitable for the satellite clock batch estimation. We derive the analytic relationship between the observation redundancy and the satellite clock batch estimation accuracy. The experimental results show that the suitable number of stations is suggested to be 40 to achieve the counterbalance between the efficiency and saturable accuracy.

We perform the simulation experiment to clarify the impacts of the ICB and the TDB on PPP. The experiment results show that the convergence time and the positioning accuracy after convergence of PPP are mainly affected by the ICB and TDB of the batch-estimated satellite clock, respectively. Moreover, besides the ICB and the TDB, the impacts of the observation quality on PPP, including the observation accuracy and the satellite distribution, cannot be ignored.

## 5. Conclusions

As one of the typical PPP methods to achieve the centimeter-level positioning by using ultra-rapid satellite clock and orbit products, the satellite clock batch estimation accuracy and its impacts on PPP are analyzed in this contribution.

The source of the satellite clock bias induced by the batch observation model is clarified and divided into the ICB and the TDB. The ICB affects the RMS without affecting the STD, whilst the TDB affects the RMS and the STD of the batch-estimated satellite clock. Furthermore, based on the clarified batch observation model, the observation redundancy suitable for the satellite clock batch estimation is determined by deriving the covariance–variance matrix. The number of stations for GPS satellite clock batch estimation is suggested to be 40 because the accuracy of the batch-estimated satellite clock is saturable. For the application of the batch-estimated satellite clock on PPP, the ICB and TDB are absorbed by the PPP observation model, which impact the convergence time and the positioning accuracy after convergence, respectively.

Based on the experimental results of the satellite clock batch estimation, it can be found that the ICB causes a significant bias for the batch-estimated satellite clock. The TDB is impacted by the assimilation ability of the batch-estimated satellite clock to the satellite orbit error. Furthermore, the simulation experiments of PPP show that the convergence time and the positioning accuracy after convergence are affected by the ICB and TDB of the batch-estimated satellite clock.

**Author Contributions:** Conceptualization, M.L., W.H. and H.L.; methodology, M.L. and H.L.; validation, M.L., R.W. and P.C.; formal analysis, M.L. and H.L.; investigation, M.L. and H.L.; resources, W.H. and H.L.; data curation, M.L. and H.L.; writing—original draft preparation, M.L.; writing—review and editing, M.L.; visualization, M.L. and R.W.; supervision, M.L. and H.L.; project administration, W.H. and H.L.; funding acquisition, W.H. and H.L. All authors have read and agreed to the published version of the manuscript.

**Funding:** This research was jointly funded by the National Key Research and Development Program (No. 2021YFB3901300), the National Natural Science Foundation of China (Nos. 61773132, 61633008, 61803115, 62003108 and 62003109), the 145 High-Tech Ship Innovation Project sponsored by the Chinese Ministry of Industry and Information Technology, the Heilongjiang Province Research Science Fund for Excellent Young Scholars (No. YQ2020F009) and the Fundamental Research Funds for Central Universities (Nos. 3072019CF0401, 3072020CFT0403).

**Data Availability Statement:** The data supporting this research can be found at the hyperlinks as: https://cddis.nasa.gov, accessed on 1 March 2022.

**Acknowledgments:** The authors would like to appreciate the analysis center of WHU, GFZ and IGS for providing the data.

**Conflicts of Interest:** The authors declare no conflict of interest.

## References


1. Nie, Z.; Xu, X.; Wang, Z.; Du, J. Initial Assessment of BDS PPP-B2b Service: Precision of Orbit and Clock Corrections, and PPP Performance. *Remote Sens.* **2021**, *13*, 2050. [CrossRef]
2. Fu, W.; Yang, Y.; Zhang, Q.; Huang, G. Real-time estimation of BDS/GPS high-rate satellite clock offsets using sequential least squares. *Adv. Space Res.* **2018**, *62*, 477–487. [CrossRef]
3. Ogutcu, S.; Farhan, H.T. Assessment of the GNSS PPP performance using ultra-rapid and rapid products from different analysis centres. *Surv. Rev.* **2020**, *54*, 34–47. [CrossRef]
4. Chen, Q.; Song, S.; Zhou, W. Accuracy Analysis of GNSS Hourly Ultra-Rapid Orbit and Clock Products from SHAO AC of iGMAS. *Remote Sens.* **2021**, *13*, 1022. [CrossRef]
5. Jiao, G.; Song, S. High-Rate One-Hourly Updated Ultra-Rapid Multi-GNSS Satellite Clock Offsets Estimation and Its Application in Real-Time Precise Point Positioning. *Remote Sens.* **2022**, *14*, 1257. [CrossRef]
6. Li, H.; Liao, X.; Li, B.; Yang, L. Modeling of the GPS satellite clock error and its performance evaluation in precise point positioning. *Adv. Space Res.* **2018**, *62*, 845–854. [CrossRef]
7. Zhao, L.; Li, N.; Li, H.; Wang, R.; Li, M. BDS Satellite Clock Prediction Considering Periodic Variations. *Remote Sens.* **2021**, *13*, 4058. [CrossRef]
8. Hauschild, A.; Montenbruck, O. Kalman-filter-based GPS clock estimation for near real-time positioning. *GPS Solut.* **2008**, *13*, 173–182. [CrossRef]
9. Huang, G.; Zhang, Q.; Xu, G. Real-time clock offset prediction with an improved model. *GPS Solut.* **2013**, *18*, 95–104. [CrossRef]
10. Ye, S.; Zhao, L.; Song, J.; Chen, D.; Jiang, W. Analysis of estimated satellite clock biases and their effects on precise point positioning. *GPS Solut.* **2017**, *22*, 16. [CrossRef]
11. Heng, L.; Grace, X.G.; Walter, T.; Enge, P. Statistical Characterization of GPS Signal-In-Space Errors. In Proceedings of the 2011 International Technical Meeting of the Institute of Navigation, San Diego, CA, USA, 24–26 January 2011.
12. Montenbruck, O.; Hauschild, A.; Steigenberger, P. Differential Code Bias Estimation using Multi-GNSS Observations and Global Ionosphere Maps. *Navigation* **2014**, *61*, 191–201. [CrossRef]
13. Lou, Y.; Zhang, W.; Wang, C.; Yao, X.; Shi, C.; Liu, J. The impact of orbital errors on the estimation of satellite clock errors and PPP. *Adv. Space Res.* **2014**, *54*, 1571–1580. [CrossRef]
14. Douša, J. The impact of errors in predicted GPS orbits on zenith troposphere delay estimation. *GPS Solut.* **2009**, *14*, 229–239. [CrossRef]
15. Yao, Y.; He, Y.; Yi, W.; Song, W.; Cao, C.; Chen, M. Method for evaluating real-time GNSS satellite clock offset products. *GPS Solut.* **2017**, *21*, 1417–1425. [CrossRef]
16. Gong, X.; Lou, Y.; Zheng, F.; Gu, S.; Shi, C.; Liu, J.; Jing, G. Evaluation and calibration of BeiDou receiver-related pseudorange biases. *GPS Solut.* **2018**, *22*, 98. [CrossRef]
17. Gong, X.; Gu, S.; Zheng, F.; Wu, Q.; Liu, S.; Lou, Y. Improving GPS and Galileo precise data processing based on calibration of signal distortion biases. *Measurement* **2021**, *174*, 108981. [CrossRef]
18. Zhang, J.; Zhao, L.; Yang, F.; Li, L.; Liu, X.; Zhang, R. Integrity monitoring for undifferenced and uncombined PPP under local environmental conditions. *Meas. Sci. Technol.* **2022**, *33*, 065010. [CrossRef]
19. Li, L.; Liu, X.; Jia, C.; Cheng, C.; Li, J.; Zhao, L. Integrity monitoring of carrier phase-based ephemeris fault detection. *GPS Solut.* **2020**, *24*, 43. [CrossRef]
20. Yang, X.; Wang, Q.; Xue, S. Random Optimization Algorithm on GNSS Monitoring Stations Selection for Ultra-Rapid Orbit Determination and Real-Time Satellite Clock Offset Estimation. *Math. Probl. Eng.* **2019**, *2019*, 7579185. [CrossRef]
21. Yang, H.; Xu, C.; Gao, Y. Analysis of GPS satellite clock prediction performance with different update intervals and application to real-time PPP. *Surv. Rev.* **2017**, *51*, 43–52. [CrossRef]
22. Zhou, P.; Yang, H.; Xiao, G.; Du, L.; Gao, Y. Estimation of GPS LNAV based on IGS products for real-time PPP. *GPS Solut.* **2019**, *23*, 27. [CrossRef]
23. Yu, Y.; Huang, M.; Wang, C.; Hu, R.; Duan, T. A New BDS-2 Satellite Clock Bias Prediction Algorithm with an Improved Exponential Smoothing Method. *Appl. Sci.* **2020**, *10*, 7456. [CrossRef]
24. Liu, Q.; Chen, X.; Zhang, Y.; Liu, Z.; Li, C.; Hu, D. A Novel Short-Medium Term Satellite Clock Error Prediction Algorithm Based on Modified Exponential Smoothing Method. *Math. Probl. Eng.* **2018**, *2018*, 7486925. [CrossRef]
25. Lou, Y.; Dai, X.; Song, W. Research on the influence of stations' distance in high-accuracy GPS satellite clock offset estimation. *Geomat. Inf. Sci. Wuhan Univ.* **2011**, *36*, 397–400.
26. Zhang, X.; Li, X.; Guo, F. Satellite clock estimation at 1 Hz for realtime kinematic PPP applications. *GPS Solut.* **2010**, *15*, 315–324. [CrossRef]
27. Chen, L.; Song, W.; Yi, W.; Shi, C.; Lou, Y.; Guo, H. Research on a method of real-time combination of precise GPS clock corrections. *GPS Solut.* **2016**, *21*, 187–195. [CrossRef]
28. Pan, L.; Zhang, X.; Guo, F.; Liu, J. GPS inter-frequency clock bias estimation for both uncombined and ionospheric-free combined triple-frequency precise point positioning. *J. Geod.* **2018**, *93*, 473–487. [CrossRef]
29. Li, X.; Xiong, Y.; Yuan, Y.; Wu, J.; Li, X.; Zhang, K.; Huang, J. Real-time estimation of multi-GNSS integer recovery clock with undifferenced ambiguity resolution. *J. Geod.* **2019**, *93*, 2515–2528. [CrossRef]

30. Yao, J.; Yoon, S.; Stressler, B.; Hilla, S.; Schenewerk, M. GPS satellite clock estimation using global atomic clock network. *GPS Solut.* **2021**, *25*, 106. [CrossRef]
31. Guo, J.; Xu, X.; Zhao, Q.; Liu, J. Precise orbit determination for quad-constellation satellites at Wuhan University: Strategy, result validation, and comparison. *J. Geod.* **2015**, *90*, 143–159. [CrossRef]
32. Xia, F.; Ye, S.; Chen, D.; Tang, L.; Wang, C.; Ge, M.; Neitzel, F. Advancing the Solar Radiation Pressure Model for BeiDou-3 IGSO Satellites. *Remote Sens.* **2022**, *14*, 1460. [CrossRef]
33. Fu, W.; Wang, L.; Chen, R.; Han, Y.; Zhou, H.; Li, T. Combined BDS-2/BDS-3 real-time satellite clock estimation with the overlapping B1I/B3I signals. *Adv. Space Res.* **2021**, *68*, 4470–4483. [CrossRef]
34. Liu, T.; Zhang, B.; Yuan, Y.; Zha, J.; Zhao, C. An efficient undifferenced method for estimating multi-GNSS high-rate clock corrections with data streams in real time. *J. Geod.* **2019**, *93*, 1435–1456. [CrossRef]
35. Xie, W.; Huang, G.; Fu, W.; Li, P.; Cui, B. An efficient clock offset datum switching compensation method for BDS real-time satellite clock offset estimation. *Adv. Space Res.* **2021**, *68*, 1802–1813. [CrossRef]
36. Huang, G.; Xie, W.; Wenju, F.; Li, P.; Wang, H.; Yue, F. BDS Real-time Satellite Clock Offsets Estimation with Three Different Datum Constraints. *J. Glob. Position. Syst.* **2021**, *17*, 34–47. [CrossRef]
37. Zhao, Y.; Cheng, C.; Li, L.; Wang, R.; Liu, Y.; Li, Z.; Zhao, L. BDS signal-in-space anomaly probability analysis over the last 6 years. *GPS Solut.* **2021**, *25*, 49. [CrossRef]
38. Defraigne, P.; Bruyninx, C. On the link between GPS pseudorange noise and day-boundary discontinuities in geodetic time transfer solutions. *GPS Solut.* **2007**, *11*, 239–249. [CrossRef]
39. Bock, H.; Dach, R.; Jäggi, A.; Beutler, G. High-rate GPS clock corrections from CODE: Support of 1 Hz applications. *J. Geod.* **2009**, *83*, 1083–1094. [CrossRef]
40. Gong, X.; Gu, S.; Lou, Y.; Zheng, F.; Ge, M.; Liu, J. An efficient solution of real-time data processing for multi-GNSS network. *J. Geod.* **2017**, *92*, 797–809. [CrossRef]
41. Yang, H.; Gao, Y. GPS Satellite Orbit Prediction at User End for Real-Time PPP System. *Sensors* **2017**, *17*, 1981. [CrossRef]
42. Wang, J.; Huang, G.; Yang, Y.; Zhang, Q.; Gao, Y.; Xiao, G. FCB estimation with three different PPP models: Equivalence analysis and experiment tests. *GPS Solut.* **2019**, *23*, 93. [CrossRef]
43. Zhou, F.; Dong, D.; Li, W.; Jiang, X.; Wickert, J.; Schuh, H. GAMP: An open-source software of multi-GNSS precise point positioning using undifferenced and uncombined observations. *GPS Solut.* **2018**, *22*, 33. [CrossRef]
44. Wang, S.; Zhai, Y.; Zhan, X. Characterizing BDS signal-in-space performance from integrity perspective. *Navigation* **2021**, *68*, 157–183. [CrossRef]
45. Li, X.; Zhu, Y.; Zheng, K.; Yuan, Y.; Liu, G.; Xiong, Y. Precise Orbit and Clock Products of Galileo, BDS and QZSS from MGEX Since 2018: Comparison and PPP Validation. *Remote Sens.* **2020**, *12*, 1415. [CrossRef]

remote sensing

MDPI

*Article*

# GPS, BDS-3, and Galileo Inter-Frequency Clock Bias Deviation Time-Varying Characteristics and Positioning Performance Analysis

**Yibiao Chen [1,2], Jinzhong Mi [2], Shouzhou Gu [2,*], Bo Li [1,2], Hongchao Li [1,3], Lijun Yang [2,4] and Yuqi Pang [2,4]**

1 School of Geomatics, Liaoning Technical University (LNTU), Fuxin 123000, China
2 Chinese Academy of Surveying and Mapping (CASM), Beijing 100830, China
3 Yellow River Conservancy Technical Institute (YRCTI), Kaifeng 475004, China
4 College of Geodesy and Geomatics, Shandong University of Science and Technology (SDUST), Qingdao 266590, China
* Correspondence: gusz@casm.ac.cn

**Abstract:** Multi-frequency observations are now available from GNSSs, thereby bringing new opportunities for precise point positioning (PPP). However, they also introduce new challenges, such as inter-frequency clock bias (IFCB) between the new frequencies and the original dual-frequency observations due to triple-frequency observations, which severely impact the PPP. In this paper, we studied the estimation and correction methods of uncombined inter-frequency clock bias of GPS, BDS-3, and Galileo, analyzed the time-varying characteristics and short-term stability of IFCB, and analyzed the influence of IFCB on the positioning of the GPS, BDS-3, and Galileo, based on a triple-frequency un-differential non-combined PPP model. The obtained results show that the amplitude of Block IIF satellites of the GPS can reach up to 10–20 cm, and the IFCB in BDS-3, Galileo, and GPS Block III satellites can be neglected. After correction by IFCB, the 3D positioning accuracy of the GPS triple-frequency PPP was 1.73 cm and 4.75 cm in the static and kinematic modes, respectively, while the convergence time was 21.64 min and 39.61 min. Compared with the triple-frequency GPS PPP without any correction with IFCB, the static and kinematic 3D positioning accuracy in this work was improved by 27.39% and 17.34%, and the corresponding convergence time was improved by 10.55% and 15.22%, respectively. Furthermore, the delayed IFCB was also used for positioning processing, and it was found that a positioning performance comparable to that of the same day can be obtained. The standard deviation of IFCB for a single satellite was found to be no more than 1 cm, when the IFCB value of a neighboring day was subtracted from the IFCB value of same day, which proves the short-term stability of IFCB.

**Keywords:** inter-frequency clock bias (IFCB); precise point positioning (PPP); time-varying characteristics; short-term stability; positioning performance

**Citation:** Chen, Y.; Mi, J.; Gu, S.; Li, B.; Li, H.; Yang, L.; Pang, Y. GPS, BDS-3, and Galileo Inter-Frequency Clock Bias Deviation Time-Varying Characteristics and Positioning Performance Analysis. *Remote Sens.* **2022**, *14*, 3991. https://doi.org/10.3390/rs14163991

Academic Editors: Chuang Shi, Shengfeng Gu, Yidong Lou and Xiaopeng Gong

Received: 13 July 2022
Accepted: 10 August 2022
Published: 16 August 2022

**Publisher's Note:** MDPI stays neutral with regard to jurisdictional claims in published maps and institutional affiliations.

**Copyright:** © 2022 by the authors. Licensee MDPI, Basel, Switzerland. This article is an open access article distributed under the terms and conditions of the Creative Commons Attribution (CC BY) license (https://creativecommons.org/licenses/by/4.0/).

## 1. Introduction

With the development and application of Global Navigation Satellite Systems (GNSSs), GNSSs are now gradually moving from dual-frequency to multi-frequency, modes and likewise, multi-frequency precise point positioning (PPP) is widely studied by many scholars [1–4]. Among them, the MEO and ISGO satellites of the BeiDou-3 (BDS-3) can provide data in five frequencies (B1C, B1I, B2a, B3I, and B2b), GPS Block IIF and Block III satellites can provide triple-frequency observation data, and the Galileo system currently has 26 satellites providing five-frequency data [5–7]. Information on the available multi-frequency GPS, Galileo, and BDS-3 satellites is provided in Table 1. It is well-known that accurate satellite orbits and clocks are important prerequisites for PPP. Zhou et al. [8]. performed PPP analysis using the orbit and clock products of iGMAS and obtained GNSS kinematic PPPs of 1.4, 1.2, and 2.9 cm in the E, N, and U directions, respectively, along with orbit/clock agreement of 1.5 cm and 60 s, respectively, compared to the orbit/clock of the

IGS. Yang et al. [9] used a triple-frequency ambiguity solution based on undifferentiated observations for satellite clock estimation and compared it with ambiguity floating-point clock and found that the ambiguity-fixed clock was solution improved by 32% and 42.9% in the horizontal and vertical directions. However, with the widespread use of triple-frequency observations, the impact of periodic variation in satellite phase hardware delay on the triple-frequency data is becoming significant [10,11]. Montenbruck et al. [12] found that the carrier phase observations of L1, L2, and L5 of the GPS have an inconsistency of 20 cm, which was labeled as the inter-frequency clock bias (IFCB).

**Table 1.** Information on the available multi-frequency GPS, Galileo, and BDS-3 satellites.

| System | Remark | PRN |
|---|---|---|
| GPS | Block IIF (12) | G01, G03, G06, G08, G09, G10, G24, G25, G26, G27, G30, G32 |
| | Block III (5) | G04, G11, G14, G18, G23 |
| BDS-3 | MEO (24) | C19~C30, C32~C37, C41~C46 |
| | IGSO (3) | C38, C39, C40 |
| Galileo | (26) | E1~E5, E7~E15, E18, E19, E21, E24~E27, E30, E31, E33, E34, E36 |

In order to make better use of the multi-frequency observations, numerous scholars have investigated the IFCB. Montenbruck et al. [13] employed prior correction of the satellite IFCB to weaken the impact of IFCB on PPP. Pan et al. [14] found that the east, north, and up accuracy improved from 3.1 cm, 1.1 cm, and 3.3 cm to 2.1 cm, 0.7 cm, and 2.3 cm, respectively, after taking into account the triple-frequency PPP with IFCB compared to the PPP localization with uncorrected IFCB. In another work, Li et al. [15] observed that correcting the IFCB while performing the triple-frequency uncalibrated phase delay (UPD) estimation can significantly improve the quality of extra-wide-line UPD. Fan et al. [16] analyzed the IFCB of the GPS Block IIF satellite via eight months of observations, and concluded that the inter-peak amplitude could reach up to 10–40 cm and that the IFCB varied more during the eclipse than during the other periods. Furthermore, Zhao and Montenbruck et al. [17,18] analyzed the IFCB of BDS-2 satellites, which are affected by 2–4 cm of IFCB. In contrast, Steigenberger and Zhao [19,20] showed good consistency among the triple frequencies of Galileo and QZSS. In addition to these works, a better understanding of the characteristics of the BDS-3 IFCB is needed. Furthermore, the compatibility of ionosphere-free combination with non-combination estimated IFCB was also focused on in one of the reported studies [21].

Additionally, some scholars further analyzed the cycle variation in IFCB. In this regard, Gong [22] pointed out that due to solar illumination variations, such as the relative Sun–satellite–Earth geometry changes, the internal temperature of the satellite also changes, leading to periodic changes in the satellite phase delay, thereby resulting in the periodic changes in IFCB. Therefore, Li et al. [23] modeled the estimated IFCB using linear and fourth-order harmonic functions and reported more than 89% correction of the IFCB, with an average fitted RMS of 1.35 cm for the GPS IFCB. Moreover, Zhang [24] found that the periodic variation in IFCB is to some extent related to the orbital plane in which the satellite is located, and for the two satellites distributed in same orbital plane, IFCB shows similar amplitudes and waveforms.

Considering the current status of the existing research on IFCB, this paper utilized 117 MGXE (muti-GNSS experiment) stations worldwide for IFCB estimation, and analyzed the intra-day and inter-day time-varying characteristics of IFCB. Furthermore, this work analyzed the impact of IFCB on the GPS, BDS-3, and Galileo multi-frequency precise point positioning in terms of IFCB amplitude, PPP positioning accuracy, and post-test residuals, and investigated the short-term stability of IFCB.

## 2. Methods

### *2.1. Triple-Frequency Uncombined PPP Model with IFCB*

The pseudorange $P^s_{r,i}$ and carrier phase $L^s_{r,i}$ observation equations of GNSSs are given as [15]:

$$\begin{aligned} P^s_{r,i} &= \rho^s_r + dt_r - dt^s + T^s_r + \mu_i I^s_{r,1} + d_{r,i} + d^s_i + \varepsilon^s_{r,i} \\ L^s_{r,i} &= \rho^s_r + dt_r - dt^s + T^s_r - \mu_i I^s_{r,1} + \lambda_i N^s_{r,i} + b_{r,i} + b^s_i + \zeta^s_{r,i} \end{aligned} \tag{1}$$

where $s$ is the satellite and $r$ is the receiver; $i$ ($i$ = 1, 2, 3) is the carrier frequency; $\rho^s_r$ is the geometric distance between satellite and receiver; $dt_r$ and $dt^s$ are the receiver and satellite clock errors, respectively; $T^s_r$ is the tropospheric delay; $I^s_{r,1}$ is the slant ionospheric delay at frequency $f_1$; $\mu_i = f_1^2 / f_i^2$ is the frequency-dependent ionospheric delay amplification factor; $f$ is the frequency; $\lambda_i$ is the carrier wavelength; $N^s_{r,i}$ is the carrier phase ambiguity; $d_{r,i}$ and $d^s_i$ are the code hardware delays at the receiver and satellite, respectively; $b_{r,i}$ and $b^s_i$ are the phase hardware delays from the receiver and satellite, respectively; and $\varepsilon^s_{r,i}$ and $\zeta^s_{r,i}$ are the unmodeled error and the observation noise of the code and carrier phase observations for each frequency.

The phase hardware deviation has obvious time-varying characteristics, and accordingly, it can be decomposed into a constant part and a time-varying part [7,13], as elaborated in Equation (2).

$$\begin{cases} b_{r,i} = \overline{b}_{r,i} + \delta b_{r,i} \\ b^s_i = \overline{b}^s_i + \delta b^s_i \end{cases} \tag{2}$$

where $\overline{b}_{r,i}$ and $\overline{b}^s_i$ are the constant parts of the receiver and satellite phase hardware delays, respectively, while $\delta b_{r,i}$ and $\delta b^s_i$ are the time-varying parts of the receiver and satellite phase hardware delays. Moreover, the following variables are defined herein for the ease of expression:

$$\begin{cases} \alpha_{12} = \frac{f_1^2}{f_1^2 - f_2^2} \\ \beta_{12} = -\frac{f_2^2}{f_1^2 - f_2^2} \\ DCB^s_{12} = d^s_1 - d^s_2 \\ DCB_{r,12} = d_{r,1} - d_{r,2} \\ \delta DPB^s_{12} = \delta b^s_1 - \delta b^s_2 \\ \delta DPB_{r,12} = \delta b_{r,1} - \delta b_{r,2} \\ \delta b^s_{IF12} = \alpha_{12}\delta b^s_1 + \beta_{12}\delta b^s_2 \\ \delta b_{r,IF12} = \alpha_{12}\delta b_{r,1} + \beta_{12}\delta b_{r,2} \\ d^s_{IF12} = \alpha_{12} d^s_1 + \beta_{12} d^s_2 \\ d_{r,IF12} = \alpha_{12} d_{r,1} + \beta_{12} d_{r,2} \end{cases} \tag{3}$$

where $\alpha_{12}$ and $\beta_{12}$ are the frequency factors' ionosphere-free combinations; $DCB^s_{12}$ and $DCB_{r,12}$ are the satellite and receiver differential code bias values, respectively; $\delta DPB^s_{12}$ and $\delta DPB_{r,12}$ are the satellite and receiver time-variant parts of differential phase bias, respectively; $\delta b^s_{IF12}$ and $\delta b_{r,IF12}$ are the ionosphere-free combination time-variant parts of receiver and satellite phase hardware delays, respectively; and $d^s_{IF12}$ and $d_{r,IF12}$ are the IF pseudorange hardware delays at the receiver and satellite, respectively.

After applying the precise satellite clock, track, and DCB product corrections, a triple-frequency uncombined PPP model with IFCB is expressed as:

$$\begin{cases} P^s_{r,1} = u^s_r x + d\bar{t}_r + m^s_r Z_r + \overline{I}^s_{r,1} + \delta b^s_{r,1} + \varepsilon^s_{r,1} \\ P^s_{r,2} = u^s_r x + d\bar{t}_r + m^s_r Z_r + \mu_2 \overline{I}^s_{r,1} + \delta b^s_{r,2} + \varepsilon^s_{r,2} \\ P^s_{r,3} = u^s_r x + d\bar{t}_r + m^s_r Z_r + \mu_3 \overline{I}^s_{r,1} + IFB_r + \delta b^s_{r,3} + \varepsilon^s_{r,3} \\ L^s_{r,1} = u^s_r x + d\bar{t}_r + m^s_r Z_r - \overline{I}^s_{r,1} + \lambda_1 \overline{N}^s_{r,1} + \zeta^s_{r,1} \\ L^s_{r,2} = u^s_r x + d\bar{t}_r + m^s_r Z_r - \mu_2 \overline{I}^s_{r,1} + \lambda_2 \overline{N}^s_{r,2} + \zeta^s_{r,2} \\ L^s_{r,3} = u^s_r x + d\bar{t}_r + m^s_r Z_r - \mu_3 \overline{I}^s_{r,1} + \lambda_3 \overline{N}^s_{r,3} + IFCB + \zeta^s_{r,3} \end{cases} \tag{4}$$

with

$$\begin{cases} d\bar{t}_r = dt_r + d_{r,IF12} + \delta b_{r,IF12} \\ \overline{I}^s_{r,1} = I^s_{r,1} + \beta_{12}DCB_{r,12} - \beta_{12}(\delta DPB^s_{12} + \delta DPB_{r,12}) \\ IFB_r = d_{r,3} - d_{r,IF12} - \mu_3\beta_{12}DCB_{r,12} \\ \delta b^s_{r,j} = \mu_j\beta_{12}(\delta DPB^s_{12} + \delta DPB_{r,12}) - (\delta b^s_{IF12} + \delta b_{r,IF12}) \\ IFCB = (\delta \mathbf{b}^s_3 - \delta \mathbf{b}^s_{IF12} - \mu_3\beta_{12}\delta DPB^s_{12}) + (\delta \mathbf{b}_{r,3} - \delta \mathbf{b}_{r,IF12} - \mu_3\beta_{12}\delta DPB_{r,12}) \\ \lambda_i \overline{N}^s_{r,i} = \lambda_i N^s_{r,i} + \bar{b}_{r,i} + \overline{b}^s_i - d^s_{IF12} - d_{r,IF12} + \mu_i\beta_{12}DCB_{r,12} \end{cases} \tag{5}$$

where $u^s_r$ is the directional cosine of the receiver–satellite linkage and $x$ is the 3D coordinate correction value; $Z_r$ is the wet troposphere delay at the zenith path with a mapping function $m^s_r$; $IFB_r$ is the inter-frequency bias; and $\delta b^s_{r,j}$ is the combined time-varying part of the unparameterized satellite and receiver-side phase hardware deviations, the effect of which can be ignored owing to its small magnitude [7]. Moreover, $d\bar{t}_r$ is the estimated receiver clock error; $\overline{I}^s_{r,1}$ is the estimated slant ionospheric delay at frequency $f_1$; and $\overline{N}^s_{r,i}$ is the estimated carrier phase ambiguity. ALL estimated parameters in our PPP models with IFCB are listed as:

$$X = \left[x, d\bar{t}_r, Z_r, \overline{I}^s_{r,1}, IFB_r, \overline{N}^s_{r,1}, \overline{N}^s_{r,2}, \overline{N}^s_{r,3}\right] \tag{6}$$

## 2.2. IFCB Estimation Method

In this work, IFCB was estimated by using a difference for two ionosphere-free combinations, where the ionosphere-free combined carrier observations of L1 and L2, and L1 and L5 for $L^s_{r,IF12}$ and $L^s_{r,IF13}$, respectively, are

$$\begin{array}{l} L^s_{r,IF12} = \rho^s_r + dt_r - dt^s + T^s_r + \lambda_{IF12}N^s_{r,IF12} + b_{r,IF12} + b^s_{IF12} + \zeta^s_{r,IF12} \\ L^s_{r,IF13} = \rho^s_r + dt_r - dt^s + T^s_r + \lambda_{IF13}N^s_{r,IF13} + b_{r,IF13} + b^s_{IF13} + \zeta^s_{r,IF13} \end{array} \tag{7}$$

Note that the errors in the above equation between the receiver and the satellite antenna, phase winding, etc., have been corrected by the proposed model. Essentially, the difference between $L^s_{r,IF12}$ and $L^s_{r,IF13}$ yields a combination of triple-frequency geometry-free and ionosphere-free phase observations (GFIF).

$$\begin{aligned} GFIF &\equiv L^s_{r,IF12} - L^s_{r,IF13} \\ &= N_{GFIF} + B^s_{r,GFIF} + \delta B \end{aligned} \tag{8}$$

where $N_{GFIF}$ is the carrier phase ambiguity of the GFIF combination, and $B^s_{r,GFIF}$ is the combination of the constant parts of pseudorange and phase hardware delays for GFIF combination. Meanwhile, $\delta B$ is the IFCB value of the ionosphere-free combination, and its relationship with the non-combined $IFCB$ is given as:

$$\delta B = \frac{f_3^2}{f_1^2 - f_3^2} IFCB \tag{9}$$

In order to eliminate the ambiguity $N_{GFIF}$ and the constant term $B^s_{r,GFIF}$ of phase hardware delay, the difference between epoch elements is calculated in the continuous observation arc without cycle slip:

$$\Delta\delta B_{(t,t-1)} = GFIF_t - GFIF_{t-1} \tag{10}$$

where $\Delta\delta B_{(t,t-1)}$ is the variation between the epochs at time $t$ and $t-1$. A weighted average of $\Delta\delta B_{(t,t-1)}$ is usually calculated for the multiple stations of same epoch to improve the data stability and avoid the occasionality of the solution. For $n$ stations, the following expression can be written:

$$\Delta\delta\overline{B}_{(t,t-1)} = \frac{\sum_{r=1}^{n} \Delta\delta B_{(t,t-1)} \cdot \omega_{r,(t,t-1)}}{\sum_{r=1}^{n} \omega_{r,(t,t-1)}} \tag{11}$$

where $\Delta\delta\overline{B}_{(t,t-1)}$ is the weighted average of the variation, and $\omega_{r,(t,t-1)}$ is the corresponding weight of each station, as expressed in Equation (12).

$$\omega_{r,(t,t-1)} = \begin{cases} 0 & E^s_{r(t,t-1)} < 10^\circ \\ \sin E^s_{r(t,t-1)} & 10^\circ \leq E^s_{r(t,t-1)} < 30^\circ \\ 1 & 30^\circ \leq E^s_{r(t,t-1)} \end{cases} \tag{12}$$

where $E^s_{r(t,t-1)} = \frac{E^s_r(t)+E^s_r(t-1)}{2}$ and $E^s_r(t)$ are the elevation angles formed by the station $r$ and the satellite $s$ at moment $t$. It is worthwhile to note that the use of a segmented elevation angle weighting method also reduces the impact of errors caused by the low elevation angles, while avoiding the chance of high elevation angles caused by the fewer moments of measuring stations. Therefore, the value of $\delta B_t$ at moment $t$ is:

$$\delta B_t = \delta\overline{B}_0 + \sum_{k=1}^{k=t} \Delta\delta\overline{B}_{(k,k-1)} \tag{13}$$

In this study, $\delta\overline{B}_0$ was set to 0 to introduce a common deviation for all epoch elements, which has no effect on the PPP floating solution, and the common deviation will be absorbed into the fuzziness parameter. Nevertheless, while conducting PPP-AR, the same benchmark needs to be added to the deviation products of the third frequency to avoid $\delta\overline{B}_0$ effect [14].

## 3. Experiment and Analysis

### 3.1. Data Introduction and Processing Strategy

To ensure the continuity of IFCB series and to avoid the influence of IFCB calculation occasionality, 117 globally distributed MGXE stations for 43–71 days in 2022 were selected for the IFCB estimation, whereas 21 stations for 65–71 days in 2022 were selected for the experimental validation, and the distribution of stations is shown in Figure 1. To analyze the impact of IFCB on the positioning accuracy, two schemes were selected to evaluate the positioning performance of PPP following the IFCB correction, where scheme-1 "PPP" represents triple-frequency PPP positioning without IFCB correction and scheme-2 "PPP + IFCB" represents triple-frequency PPP positioning with IFCB correction. Meanwhile, in order to analyze the short-term stability of IFCB, the PPP accuracy with IFCB correction was analyzed using one-day-delayed and two-day-delayed IFCB products, where the difference in the IFCB of adjacent days was analyzed. In such scenarios, "PPP + IFCB1" means IFCB was delayed by one day and "PPP + IFCB2" means IFCB was delayed by two days. For other data processing strategies, see Table 2. In this work, the PCO and PCV of the GPS Block IIF satellite were corrected with the PCO and PCV of only L2, because the PCO and PCV of the L5 frequency were not available. GPS L1 and L2 were corrected using PCO and PCV information on their respective frequency, and GPS Block III satellite L5 was corrected using PCO and PCV information on L5 frequency. BDS and Galileo satellites were corrected by PCO and PCV at their respective frequencies. In addition, since the frequencies between L2 and L5 are closer, the receiver PCO and PCV of L2 were used to correct the L5. The coordinates in the SINEX file of the IGS were used as the reference coordinates of each station, and the filtering was considered to be converged when the positioning deviations in the three directions of east (E), north (N), and up (U) of the coordinates were less than

10 cm in 30 consecutive epochs. Next, the positioning deviations after the solution filtering were selected for the statistical positioning accuracy.

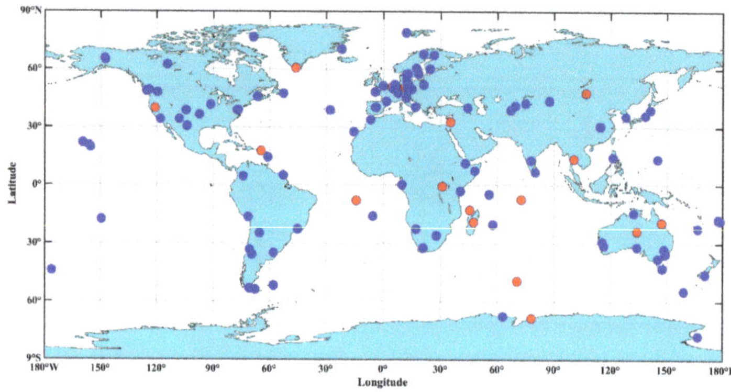


**Figure 1.** MGEX stations used in the experiment. The blue dots represent the stations used to estimate IFCB, and the red dots represent the stations used for PPP validation.

**Table 2.** Triple-frequency PPP positioning processing strategy.

| Type | Processing Strategies |
|---|---|
| Observation data | GPS: L1, L2, L5<br>BDS-3: B1I, B3I, B2a<br>Galileo: E1, E5a, E5b |
| Sampling interval | 30s |
| Cutoff elevation | 10° |
| Clock and orbital products | CODE |
| Satellite antenna correction | igs14.atx |
| Receiver antenna correction | igs14.atx |
| Weight for observations | Elevation-dependent weight |
| Receiver coordinates | Static mode: estimated as constant<br>Kinematic mode: estimated as white noise |
| Receiver clock | Estimated as white noise |
| Inter-frequency bias | Estimated as white noise |
| Ionospheric delay | Estimated as white noise |
| Tropospheric delay | Dry component corrected by Saastamoinen mode; wet component estimated as a random walk |
| Phase ambiguity | Float |

### 3.2. *Time-Varying Feature Analysis of IFCB*

#### 3.2.1. Intraday Time-Varying Characteristics Analysis of IFCB

Figures 2 and 3 show the IFCB time series and the IFCB amplitude for each satellite, respectively. It can be seen that the single-day amplitude of GPS Block IIF satellites was large among all, and the amplitude size was between 10 and 20 cm, which is evidently a non-negligible error for PPP. Alternatively, the single-day amplitudes of GPS Block III and BDS-3 satellites were in the range of 1 to 3 cm, and those of the Galileo satellites were below 2 cm. Meanwhile, the standard deviation of IFCB for Block III satellites of the BDS-3, Galileo, and GPS was about 1.5 mm for a single epoch, which was almost unaffected by the IFCB. Therefore, it was necessary to focus on the variation in IFCB of only GPS Block IIF satellites, and analyze the corresponding impact of IFCB on multi-frequency positioning in terms of both positioning performance and residuals; see Sections 3.3 and 3.4. Since IFCB is considered as a temperature-dependent inter-frequency hardware bias, the different IFCB characteristics of the GPS, BDS-3, and Galileo may be caused by the different designs and payloads of the satellites. However, the IFCBs for GPS Block III and Block IIF satellites express different characteristics and require more information from inside and outside the GPS satellites, for their comprehensive analysis and determination.

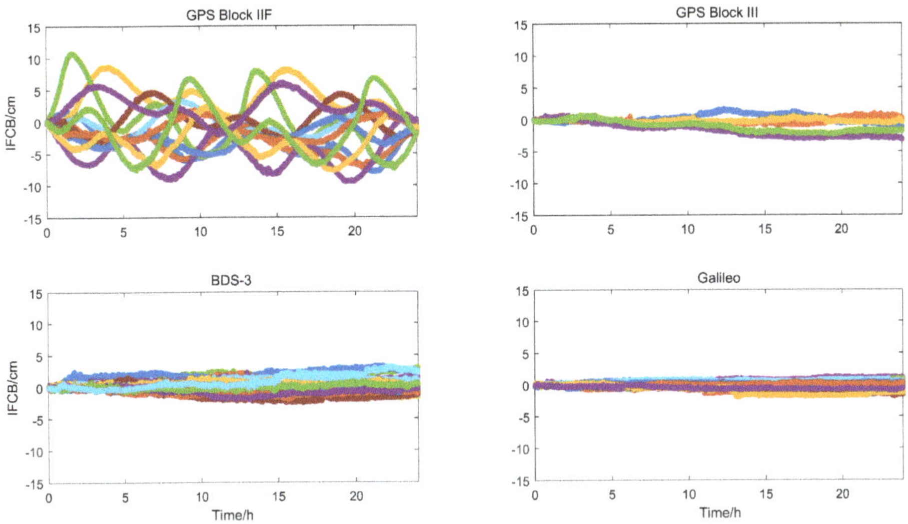


**Figure 2.** IFCB time series plot for GPS, BDS-3, and Galileo, where each color represents a satellite (DOY 65, 2022).

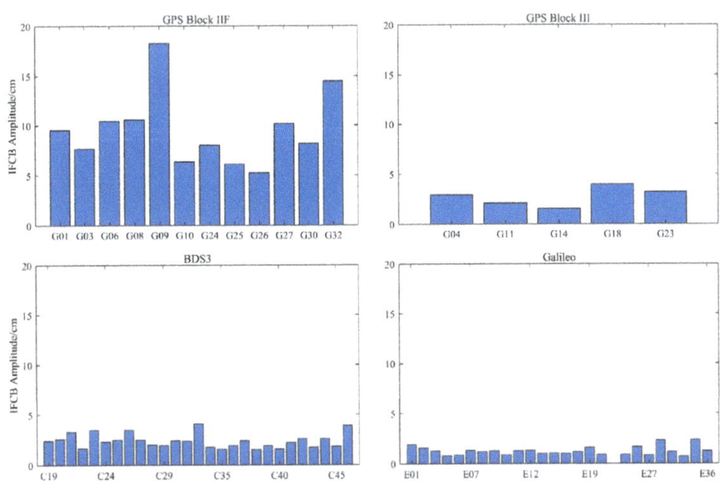


**Figure 3.** IFCB amplitudes of GPS, BDS-3, and Galileo satellites (DOY 65, 2022).

#### 3.2.2. Inter-Day Variation Characteristics of IFCB

Figure 4 shows the IFCB time series plot for DOY 43 to 71 in 2022. The IFCB of the GPS Block IIF satellite varied between −15 cm and 15 cm, and exhibited a clear repetitive feature. Meanwhile, the IFCB of Galileo satellites still exhibited relatively small magnitudes, and the large errors in the IFCB of the BDS-3 in some periods were caused by the small number of observable BDS-3 B2a frequency stations present, which indicates that although the IFCB can maintain a good stability and periodicity in most cases, there still exist serious errors in some periods that need further improvement. Montenbruck and Li et al. [13,25,26] found that the IFCB of the GPS Block IIF satellite had 12 and 6 h periods, where the 12 h period is due to the satellite receiving the same amount of sunlight and the 6 h period is due to the satellite having the same amount of heat at two orbital positions around 6 h. Since IFCB has a 6 h and 12 h periodicity expression, it can be further expressed that the IFCB exhibits a 24 h periodicity, and Figure 4 also shows a characteristic single-day periodicity of IFCB. The single-day periodicity of IFCB further assisted in the analysis of the short-term stability of IFCB provided in Section 3.5.

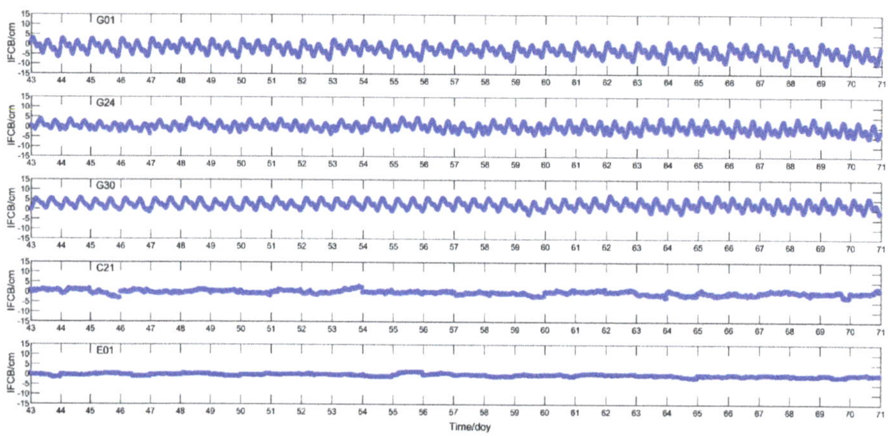


**Figure 4.** IFCB time series plot consisting of G01, G24, G30, C21, and E01 (DOY 43–71, 2022).

### 3.3. *Triple-Frequency PPP Positioning Performance Analysis*

#### 3.3.1. Static Mode

First, the static results of the ASCG station for 2022 DOY 65, from 00:00 UTC to 4:00 UTC, were compared for 1 of the 21 stations. The static PPP positioning error curves under the two solutions of the GPS, BDS-3, and Galileo are shown in Figure 5. During this period, the number of triple-frequency satellites for the ASCG stations of GPS, BDS-3, and Galileo systems was 4.8, 7.6, and 6.6, respectively, indicating that the triple-frequency satellites were involved in the triple-frequency PPP solution.

From Figure 5, it can be observed that the GPS positioning accuracy for the ASCG station was more stable after correcting the IFCB. Meanwhile, for the BDS-3 and Galileo, the change in single-day positioning accuracy was less than 0.1 mm after the IFCB correction, i.e., the positioning accuracy was basically unchanged, which further verifies that the influence of IFCB on the positioning of the BDS-3 and Galileo can simply be ignored. Furthermore, to further analyze the impact of IFCB on GPS positioning, the static PPP accuracy and convergence time under the two scenarios of the GPS at 21 stations for 7 days were recorded, as shown in Table 3. Without correcting the IFCB, the E, N, U, and 3D positioning accuracy of the GPS system was 1.56 cm, 0.6 cm, 1.69 cm, and 2.38 cm, respectively. On the other hand, following the IFCB correction, the positioning accuracy of the GPS improved to 0.99 cm, 0.48 cm, 1.34 cm, and 1.73 cm, respectively, among which the 3D positioning accuracy was improved by 27.39%. The convergence times for the GPS with corrected and uncorrected IFCB were 21.64 min and 24.19 min, respectively, illustrating a 10.55% improvement in the convergence time. It can be clearly seen that IFCB had a serious impact on the GPS static positioning, and the multi-frequency PPP performance of the GPS was improved by adding the IFCB.

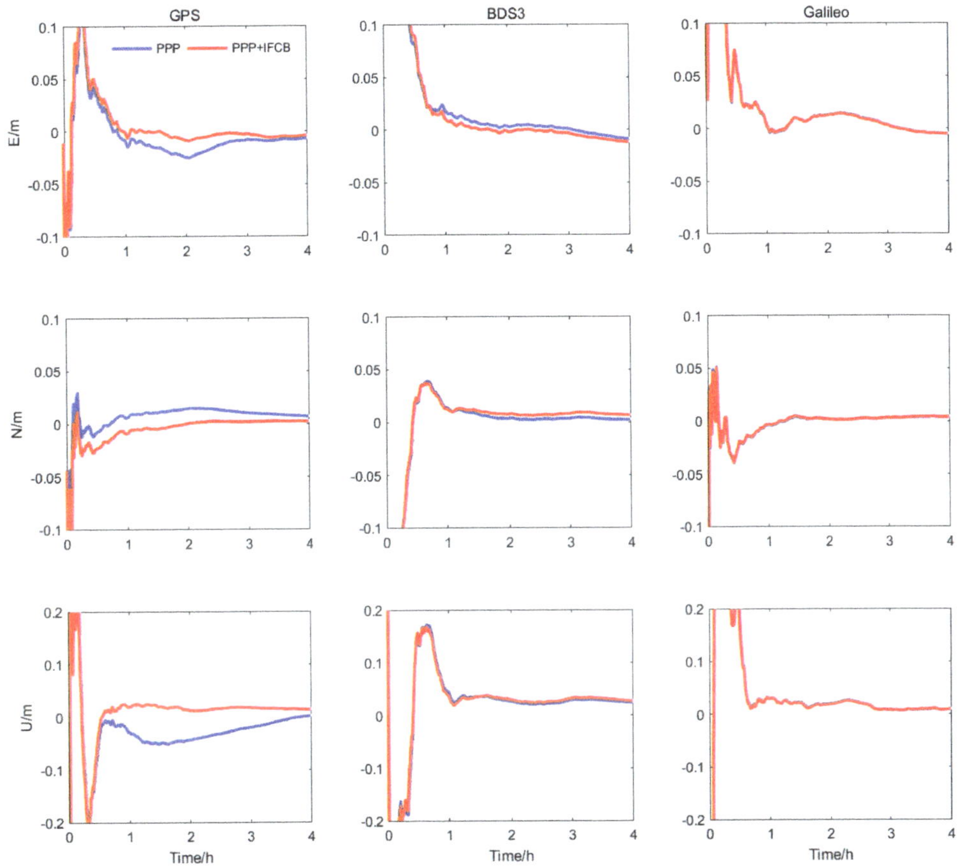


**Figure 5.** ASCG station static mode PPP positioning error curve (DOY 65, 2022; UTC: 00:00 to 4:00).

**Table 3.** Statistics of PPP positioning accuracy and convergence time under 7-day GPS static mode for 21 stations (RMS, unit: cm; convergence time, unit: min).

| Static | E | N | U | 3D | Convergence Time |
|---|---|---|---|---|---|
| PPP | 1.56 | 0.60 | 1.69 | 2.38 | 24.19 |
| PPP + IFCB | 0.99 | 0.48 | 1.34 | 1.73 | 21.64 |
| Improvement | 36.89% | 19.18% | 21.16% | 27.39% | 10.55% |

3.3.2. Imitation Kinetic Mode

Regarding the kinematic mode, the results of the ASCG measurement station for 2022 DOY 65 are compared as an example. The positioning error curves under the two schemes of GPS, BDS-3, and Galileo satellites are plotted in Figure 6. During this period, the number of triple-frequency satellites in the GPS, BDS-3, and Galileo systems was 5.2, 7.6, and 6.5, respectively.

As evident from Figure 6, the kinematic and static modes followed a similar pattern, and likewise, the positioning accuracy of the ASCG station GPS was more stable after the IFCB correction, while for the BDS-3 and Galileo, the single-day 3D positioning accuracy was improved from 6.21 cm and 6.30 cm to 6.20 cm and 6.29 cm, respectively, and the change in positioning accuracy was less than 0.1 mm. The statistics related to PPP positioning accuracy and convergence time in the kinematic mode under the two solutions of the GPS at 21 stations for 7 days are provided in Table 4. Without correcting the IFCB, the E, N, U, and 3D positioning accuracy of the GPS was 2.59 cm, 1.77 cm, 4.81 cm, and 5.74 cm, respectively, whereas after the IFCB correction, the GPS positioning accuracy was enhanced to 2 cm, 1.43 cm, 4.06 cm, and 4.75 cm, respectively, where the 3D positioning accuracy was improved by 17.34%. Furthermore, the convergence times for the GPS with corrected and uncorrected IFCB were 21.64 min and 24.19 min, respectively, indicating a 15.22% improvement. Similar to the static mode, the impact of IFCB on the multi-frequency precise point positioning in the GPS kinematic mode was also significant, and the multi-frequency precise point positioning performance of the GPS was further improved by the addition of IFCB.

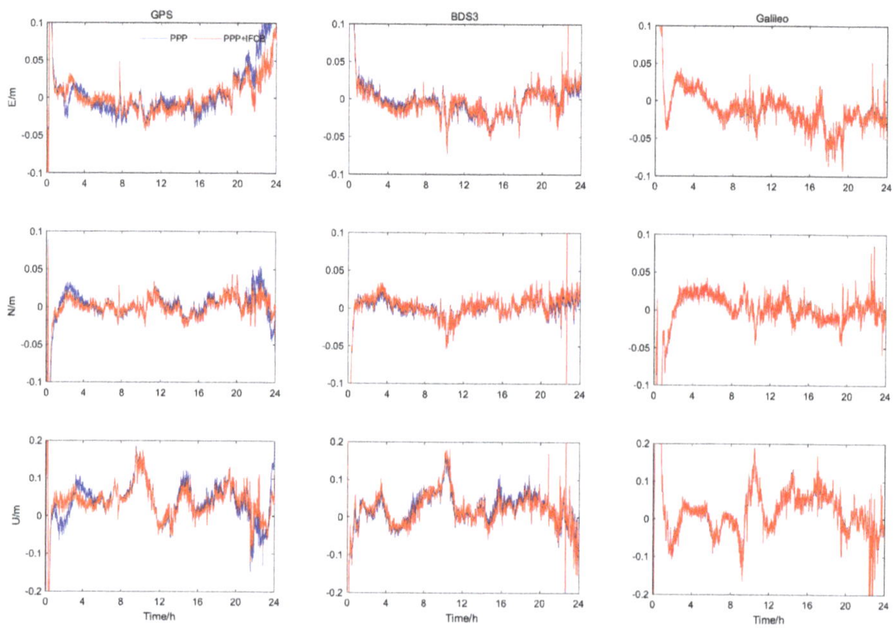


**Figure 6.** ASCG station kinematic mode PPP positioning error curve (DOY 65, 2022; UTC: 00:00 to 24:00).

**Table 4.** Statistics of PPP positioning accuracy and convergence time under 7-day GPS kinematic mode for 21 stations (RMS, unit: cm; convergence time, unit: min).

| Kinematic | E | N | U | 3D | Convergence Time |
|---|---|---|---|---|---|
| PPP | 2.59 | 1.77 | 4.81 | 5.74 | 46.72 |
| PPP + IFCB | 2.00 | 1.43 | 4.06 | 4.75 | 39.61 |
| Promote | 22.86% | 19.45% | 15.53% | 17.34% | 15.22% |

### *3.4. Model Deviation and Residual Analysis*

In addition to the observation noise, some non-modeled errors (e.g., IFCB) were reflected in the post-test residuals of the observation equations, and the time series of the post-test residuals of corrected IFCB and uncorrected IFCB third-frequency phases for the ASCG stations of the GPS, BDS-3, and Galileo are presented in Figure 7. As expected, the GPS without the corrected IFCB exhibited a significant systematic bias effect in the L5 phase residuals, while the IFCB-corrected L5 eliminated this bias effect. Contrary to the GPS, for the BDS-3 and Galileo, the residuals did not show any influence of IFCB, further demonstrating that the IFCB can be neglected for the BDS-3 and Galileo. Meanwhile, to further analyze the effect of IFCB on the GPS L5 phase residuals, the root mean square error of the L5 phase residuals of the GPS at 21 stations for 7 days is given in Table 5. The standard deviation of the L5 phase residuals of the GPS before and after the IFCB correction was 1.18 cm and 0.41 cm, respectively, with a 65.12% reduction. Accordingly, it was concluded that after correcting the IFCB, the effect of apparent systematic bias in the L5 residuals of the GPS can be eliminated, and thus, the rejection in the positioning solution process due to excessive residuals can be avoided.

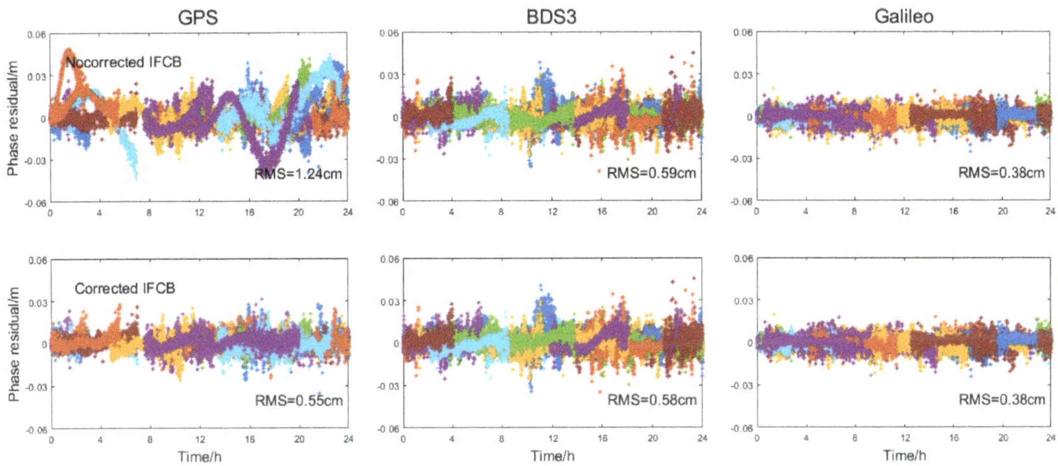


**Figure 7.** Post-check residual time series of GPS, BDS-3, and Galileo (corrected and uncorrected IFCB) third-frequency phase for ASCG stations (DOY 65, 2022).

**Table 5.** Root mean square error of GPS L5 phase residuals for 21 stations in 7 days (RMS, unit: cm).

| Stations | PPP | PPP + IFCB | Promote | Stations | PPP | PPP + IFCB | Promote |
|---|---|---|---|---|---|---|---|
| ABPO | 1.09 | 0.43 | 60.05% | BRUX | 1.33 | 0.23 | 82.74% |
| ALIC | 1.20 | 0.35 | 70.97% | BSHM | 1.37 | 0.30 | 78.46% |
| ASCG | 1.14 | 0.55 | 52.07% | DAV1 | 0.82 | 0.36 | 56.46% |
| CRO1 | 1.29 | 0.46 | 64.79% | DGAR | 1.12 | 0.48 | 57.16% |
| CUSV | 1.05 | 0.35 | 66.12% | MBAR | 1.34 | 0.44 | 66.81% |
| FAA1 | 1.35 | 0.55 | 59.49% | MDO1 | 1.18 | 0.33 | 71.57% |
| FFMJ | 1.34 | 0.22 | 83.88% | MET3 | 1.26 | 0.30 | 76.27% |
| KRGG | 1.03 | 0.43 | 58.59% | QAQ1 | 1.13 | 0.38 | 66.67% |
| MAYG | 1.19 | 0.55 | 53.61% | QUIN | 1.25 | 0.35 | 71.56% |
| TOW2 | 1.17 | 0.42 | 63.82% | SUTM | 1.19 | 0.51 | 57.28% |
| ULAB | 0.98 | 0.53 | 45.84% | | | | |

### 3.5. Short-Term Stability of IFCB

To investigate the short-term stability of IFCB, Figure 8 provides the average STD of IFCB with a one-day delay versus a two-day delay in 2022 for 65 to 71 days. From Figure 8, the average STD of the one-day-delayed and two-day-delayed IFCB was 0.6 and 0.7 cm, respectively. Essentially, the average STD of IFCB of a single satellite did not exceed 1 cm. The statistics for 7-day positioning of 21 stations were recorded using both one-day-delayed and two-day-delayed IFCB, as shown in Table 6. It can be seen that IFCB products with a one-day delay and a two-day delay could both obtain the same positioning performance as that of the same-day IFCB products to a certain extent, thereby validating the short-term stability of IFCB.

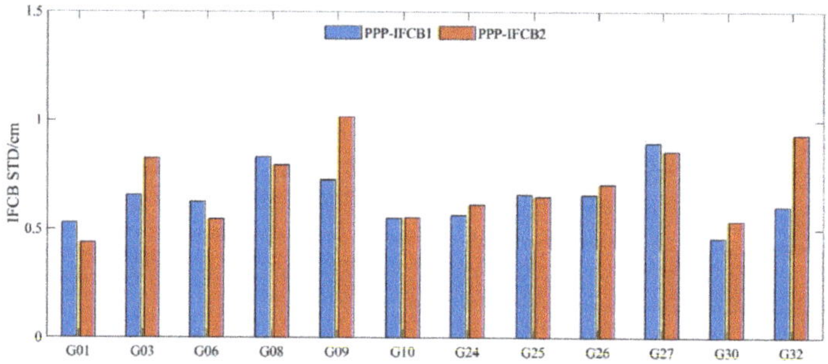


**Figure 8.** IFCB-STD statistics; blue is the STD of the difference between one-day delay and same day, and red is the STD of two-day delay and same day (DOY 65–71, 2022).

**Table 6.** Statistics of 7-day positioning accuracy and convergence time for 21 stations (RMS, unit: cm; convergence time, unit: min).

| Mode | | E | N | U | 3D | Convergence Time |
|---|---|---|---|---|---|---|
| Static | PPP + IFCB | 0.99 | 0.48 | 1.34 | 1.73 | 21.64 |
| | PPP + IFCB1 | 0.98 | 0.48 | 1.34 | 1.73 | 22.16 |
| | PPP + IFCB2 | 0.99 | 0.49 | 1.36 | 1.76 | 21.96 |
| Kinematic | PPP + IFCB | 2.00 | 1.43 | 4.06 | 4.75 | 39.61 |
| | PPP + IFCB1 | 2.01 | 1.44 | 4.06 | 4.76 | 39.66 |
| | PPP + IFCB2 | 2.04 | 1.46 | 4.09 | 4.79 | 39.70 |

## 4. Conclusions

IFCB is crucial for high-precision triple-frequency PPP. In this paper, the time-varying characteristics of IFCB for the GPS, BDS-3, and Galileo were analyzed using 117 MGEX station observations, and it was found that the amplitude of GPS Block IIF satellites could reach 10–20 cm, the amplitude of Block III and BDS-3 satellites of the GPS was around 1–3 cm, and the amplitude of Galileo satellites was below 2 cm.

Then, the positioning performance of triple-frequency PPP before and after the IFCB correction was analyzed using the 7-day data from 21 MGEX stations. After the IFCB correction, the positioning performance of BDS-3 and Galileo systems changed negligibly, whereas for the GPS, the 3D positioning accuracies of triple-frequency PPP in static and kinematic modes were improved to 1.73 cm and 4.75 cm, respectively. Compared with the GPS triple-frequency PPP without any IFCB correction, the 3D accuracy post-IFCB-correction improved by 27.39% and 17.34% (static mode and dynamic mode), and the convergence time improved by 10.55% and 15.22% (static mode and dynamic mode), respectively. In addition, the L5 phase post-check residuals of the GPS showed obvious systematic errors. However, the influence of bias could be eliminated by L5 after the IFCB correction. That is to say, the implementation of IFCB estimation can effectively solve the systematic bias problem arising from the multi-frequency positioning results, and realize the unification of traditional clock-difference products and multi-frequency precision positioning.

Since IFCB exhibits obvious periodic characteristics, the short-term stability of IFCB was also investigated in this paper, and the same positioning performance as that of the same day was obtained by using the IFCB products with a one-day delay and a two-day delay.

**Author Contributions:** Y.C., J.M. and S.G. conceived the idea and designed the experiments; Y.C., B.L. and Y.P. performed the experiments and analyzed the data; Y.C. wrote the main manuscript; J.M., S.G., L.Y. and H.L. reviewed the paper. All authors have read and agreed to the published version of the manuscript.

**Funding:** This research was funded by the National Key Research and Development Program of China (2021YFC3000504), the National Key Research and Development Program of China (2021YFB3900802), and the National Natural Science Foundation of China (41930535).

**Data Availability Statement:** The datasets analyzed in this study are managed by the IGS.

**Acknowledgments:** All authors gratefully acknowledge CODE and the IGS for providing the data, orbit, and clock products. The authors gratefully acknowledge the Chinese Academy of Surveying and Mapping, which provided the experimental environment.

**Conflicts of Interest:** The authors declare no conflict of interest.

## References

1. Zumberge, J.F.; Heflin, M.B.; Jefferson, D.C.; Watkins, M.M.; Webb, F.H. Precise point positioning for the efficient and robust analysis of GPS data from large networks. *J. Geophys. Res. Solid Earth* **1997**, *102*, 5005–5017. [CrossRef]
2. Guo, J.; Geng, J. GPS satellite clock determination in case of inter-frequency clock biases for triple-frequency precise point posi-tioning. *J. Geod.* **2017**, *92*, 1133–1142. [CrossRef]
3. Liu, G.; Zhang, X.; Li, P. Improving the Performance of Galileo Uncombined Precise Point Positioning Ambiguity Resolution Using Triple-Frequency Observations. *Remote Sens.* **2019**, *11*, 341. [CrossRef]
4. Zhang, X.; Hu, J.; Ren, X. New progress of PPP/PPP-RTK and positioning performance comparison of BDS/GNSS PPP. *Acta Geod. Cartogr. Sin.* **2020**, *49*, 1084–1100.
5. Su, K.; Jin, S. Analytical performance and validations of the Galileo five-frequency precise point positioning models. *Measurement* **2021**, *172*, 108890. [CrossRef]
6. Cai, H.; Meng, Y.; Geng, C.; Weiguang, G.; Tianqiao, Z.; Gang, L.; Bo, S.; Jie, X.; Hongyang, L.; Yue, M.; et al. BDS-3 performan ceassessment: PNT, SBAS, PPP, SMC and SAR. *Acta Geod. Cartogr. Sin.* **2021**, *50*, 427–435.
7. Zhou, F.; Xu, T. Modeling and assessment of GPS/BDS/Galileo treple-frequency precise point positioning. *Acta Geod. Cartogr. Sin.* **2021**, *50*, 61–70.
8. Zhou, W.; Cai, H.; Chen, G.; Jiao, W.; He, Q.; Yang, Y. Multi-GNSS Combined Orbit and Clock Solutions at iGMAS. *Sensors* **2022**, *22*, 457. [CrossRef]
9. Yang, X.; Gu, S.; Gong, X.; Song, W.; Lou, Y.; Liu, J. Regional BDS satellite clock estimation with triple-frequency ambiguity resolution based on un-differenced observation. *GPS Solut.* **2019**, *23*, 33. [CrossRef]
10. Montenbruck, O.; Steigenberger, P.; Prange, L.; Deng, Z.; Zhao, Q.; Perosanz, F.; Romerof, I.; Nollg, C.; Stürzeh, A.; Weber, G.; et al. The Multi-GNSS Experiment (MGEX) of the International GNSS Service (IGS)—Achievements, prospects and challenges. *Adv. Space Res.* **2017**, *59*, 1671–1697. [CrossRef]
11. Li, B.; Shen, Y.; Zhang, X. Three frequency GNSS navigation prospect demonstrated with semi-simulated data. *Adv. Space Res.* **2013**, *51*, 1175–1185. [CrossRef]
12. Montenbruck, O.; Hauschild, A.; Steigenberger, P.; Langley, R. Three's the Challenge: A Close Look at GPS SVN62 Triple-frequency Signal Combinations Finds Carrier-phase Variations on the New L5. *GPS World* **2010**, *21*, 8–19.
13. Montenbruck, O.; Hugentobler, U.; Dach, R.; Steigenberger, P.; Hauschild, A. Apparent clock variations of the Block IIF-1 (SVN62) GPS satellite. *GPS Solut.* **2012**, *16*, 303–313. [CrossRef]
14. Pan, L.; Zhang, X.; Li, X.; Liu, J.; Li, X. Characteristics of inter-frequency clock bias for Block IIF satellites and its effect on triple-frequency GPS precise point positioning. *GPS Solut.* **2017**, *21*, 811–822. [CrossRef]
15. Li, P.; Jiang, X.; Zhang, X.; Ge, M.; Schuh, H. GPS + Galileo + BeiDou precise point positioning with triple-frequency ambiguity resolution. *GPS Solut.* **2020**, *24*, 78. [CrossRef]
16. Fan, L.; Shi, C.; Li, M.; Wang, C.; Zheng, F.; Jing, G.; Zhang, J. GPS satellite inter-frequency clock bias estimation using triple-frequency raw observations. *J. Geod.* **2019**, *93*, 2465–2479. [CrossRef]
17. Zhao, Q.; Wang, G.; Liu, Z.; Hu, Z.; Dai, Z.; Liu, J. Analysis of BeiDou Satellite Measurements with Code Multipath and Geometry-Free Iono-sphere-Free Combinations. *Sensors* **2016**, *16*, 123. [CrossRef] [PubMed]
18. Montenbruck, O.; Hauschild, A.; Steigenberger, P.; Hugentobler, U.; Teunissen, P.; Nakamura, S. Initial assessment of the COMPASS/BeiDou-2 regional navigation satellite system. *GPS Solut.* **2013**, *17*, 211–222. [CrossRef]
19. Steigenberger, P.; Hauschild, A.; Montenbruck, O.; Rodriguez-Solano, C.; Hugentobler, U. Orbit and Clock Determination of QZS-1 Based on the CONGO Network. *Navig. J. Inst. Navig.* **2013**, *60*, 31–40. [CrossRef]
20. Zhao, L.; Ye, S.; Song, J. Handling the satellite inter-frequency biases in triple-frequency observations. *Adv. Space Res.* **2017**, *59*, 2048–2057. [CrossRef]
21. Su, K.; Jin, S.; Jiao, G. GNSS carrier phase time-variant observable-specific signal bias (OSB) handling: An absolute bias perspective in multi-frequency PPP. *GPS Solut.* **2022**, *26*, 71. [CrossRef]
22. Gong, X.; Gu, S.; Lou, Y.; Zheng, F.; Yang, X.; Wang, Z.; Liu, J. Research on empirical correction models of GPS Block IIF and BDS satellite inter-frequency clock bias. *J. Geod.* **2020**, *94*, 36. [CrossRef]
23. Li, H.; Li, B.; Xiao, G.; Wang, J.; Xu, T. Improved method for estimating the inter-frequency satellite clock bias of triple-frequency GPS. *GPS Solut.* **2016**, *20*, 751–760. [CrossRef]

24. Zhang, F.; Chai, H.; Li, L.; Xiao, G.; Du, Z. Estimation and analysis of GPS inter-fequency clock biases from long-term triple-frequency ob-servations. *GPS Solut.* **2021**, *25*, 126. [CrossRef]
25. Li, P.; Zhang, X.; Ge, M.; Schuh, H. Three-frequency BDS precise point positioning ambiguity resolution based on raw observables. *J. Geod.* **2018**, *92*, 1357–1369. [CrossRef]
26. Li, H.; Zhou, X.; Wu, B. Fast estimation and analysis of the inter-frequency clock bias for Block IIF satellites. *GPS Solut.* **2013**, *17*, 347–355. [CrossRef]

*remote sensing*

MDPI

*Article*

# Integration of Multi-GNSS PPP-RTK/INS/Vision with a Cascading Kalman Filter for Vehicle Navigation in Urban Areas

**Shengfeng Gu, Chunqi Dai *, Feiyu Mao and Wentao Fang**

GNSS Research Center, Wuhan University, 129 Luoyu Road, Wuhan 430079, China
* Correspondence: cqdai@whu.edu.cn

**Abstract:** Precise point positioning (PPP) has received much attention in recent years for its low cost, high accuracy, and global coverage. Nowadays, PPP with ambiguity resolution and atmospheric augmentation is widely regarded as PPP-RTK (real-time kinematic), which weakens the influence of the long convergence time in PPP and regional service coverage in RTK. However, PPP-RTK cannot work well in urban areas due to limitations of non-line-of-sight (NLOS) conditions. Inertial navigation systems (INS) and vision can realize continuous navigation but suffer from error accumulation. Accordingly, the integration model of multi-GNSS (global navigation satellite system) and PPP-RTK/INS/vision with a cascading Kalman filter and dynamic object removal model was proposed to improve the performance of vehicle navigation in urban areas. Two vehicular tests denoted T01 and T02 were conducted in urban areas to evaluate the navigation performance of the proposed model. T01 was conducted in a relatively open-sky environment and T02 was collected in a GNSS-challenged environment with many obstacles blocking the GNSS signals. The positioning results show that the dynamic object removal model can work well in T02. The results indicate that multi-GNSS PPP-RTK/INS/vision with a cascading Kalman filter can achieve a positioning accuracy of 0.08 m and 0.09 m for T01 in the horizontal and vertical directions and 0.83 m and 0.91 m for T02 in the horizontal and vertical directions, respectively. The accuracy of the velocity and attitude estimations is greatly improved by the introduction of vision.

**Keywords:** PPP; multi-GNSS; PPP-RTK/INS/vision integration; cascading Kalman filter; urban vehicle navigation; dynamic object removal

**Citation:** Gu, S.; Dai, C.; Mao, F.; Fang, W. Integration of Multi-GNSS PPP-RTK/INS/Vision with a Cascading Kalman Filter for Vehicle Navigation in Urban Areas. *Remote Sens.* **2022**, *14*, 4337. https://doi.org/10.3390/rs14174337

Academic Editor: Andrzej Stateczny

Received: 13 August 2022
Accepted: 26 August 2022
Published: 1 September 2022

**Publisher's Note:** MDPI stays neutral with regard to jurisdictional claims in published maps and institutional affiliations.

**Copyright:** © 2022 by the authors. Licensee MDPI, Basel, Switzerland. This article is an open access article distributed under the terms and conditions of the Creative Commons Attribution (CC BY) license (https://creativecommons.org/licenses/by/4.0/).

## 1. Introduction

Autonomous driving, unmanned aerial vehicles (UAVs), and the Internet of Things (IoT) are technologies that have developed rapidly in recent years. Precise navigation and positioning in complex environments are receiving increasing attention. However, any one sensor alone is not able to provide position solutions with high accuracy, availability, reliability, and continuity at any time and in all environments [1]. The integration of different sensors, for example, the integration of GNSS, INS, and vision, has become a trend [2–4].

GNSS (global navigation satellite system) is an efficient tool for providing precise positioning regardless of time and location and it is widely used in transportation. Zumberge et al. [5] proposed the precise point positioning (PPP) technique, which has received much attention in recent years for its low costs, global coverage, and high accuracy [6,7]. Though PPP can provide centimeter-level positioning for real-time kinematic applications, a nearly 30 min convergence time has limited its applications in UAVs and other technologies. Thus, great efforts have been focused on improving the PPP performance, especially to accelerate its convergence, and promoting various methods, e.g., multi-GNSS combination, ambiguity resolution, and atmospheric augmentation. Lou et al. [8] presented a comprehensive analysis of quad-constellations with PPP. The results showed that in comparison with the GPS-only solution, the four-system combined PPP can reduce the convergence time by more than 60% on average in kinematic mode. For ambiguity resolution, Ge et al. [9] proposed the uncalibrated phase delay (UPD) method. Then, Laurichesse et al. [10] proposed the integer phase clock method, and Collins et al. [11] proposed the decoupled clock model to facilitate PPP ambiguity resolution (PPP-AR). It was proved that these three PPP-AR methods can dramatically accelerate the convergence and improve the positioning accuracy of PPP [9–12]. The undifferenced and uncombined data processing strategy has received increasing interest [13–18]. First proposed by Gerhard Wübbena et al. [19], PPP with ambiguity resolution and atmospheric augmentation is nowadays widely regarded as PPP-RTK (real-time kinematic). As PPP-RTK weakens the influence of the long convergence time in PPP and regional service coverage in RTK, it is regarded as a promising technique for high-precision navigation in mass market applications, including vehicle platforms. As a result, some regional authorities have developed their own PPP-RTK augmentation services, e.g., QZSS centimeter-level accuracy service (QZSS CLAS) began offering PPP-RTK services in 2018 [20]. Chinese BDS also intends to provide its satellite-based PPP-RTK service in the future.

However, the performance of GNSS is limited by non-line-of-sight (NLOS) conditions, which means PPP-RTK cannot work very well in challenging environments such as urban areas [21]. When the satellite signals are blocked by buildings or other structures, PPP-RTK fails to provide positioning results if there are less than four satellites available and the performance is terrible due to frequent re-convergence and gross errors. Inertial navigation systems (INS) are immune to interference and can output navigation states continuously without external information. However, the accuracy degrades fast over time due to the accumulated errors. Integrating GNSS with INS can minimize their respective drawbacks and improve the performance of GNSS or INS alone. There are two common integration strategies for PPP with INS, tightly coupled integration and loosely coupled integration [22]. Moreover, it has been proved that the tightly coupled integration of PPP/INS performs better than loosely coupled integration, especially under GNSS-challenged environments [23]. Furthermore, Rabbou M A [24] studied the integration of GPS PPP and MEMS (micro-electro-mechanical System)-based inertial system, the results of which suggested that decimeter-level positioning accuracy was achievable for GPS outages within 30 s. Gao et al. [23] analyzed the integration of multi-GNSS PPP with MEMS IMU. The results showed that the position RMS improved from 23.3 cm, 19.8 cm, and 14.9 cm for the GPS PPP/INS tightly coupled integration to 7.9 cm, 3.3 cm, and 5.1 cm for the multi-GNSS PPP/INS in the north, east, and up components, respectively. PPP-AR/INS tightly coupled integration is able to realize stable centimeter-level positioning after the first AR and achieve fast re-convergence and re-resolving after a short period of GNSS outage [25]. Han et al. [26] analyzed the performance of the tightly coupled integration of RTK/INS constrained with the ionospheric and tropospheric models. Gu et al. [27] realized the tightly coupled integration of multi-GNSS PPP/INS with atmospheric augmentation. Taking the advantages of PPP-RTK over PPP and RTK into consideration, the integration of multi-GNSS PPP-RTK and MEMS IMU still needs further research.

The performance of GNSS/INS tightly coupled integration could deteriorate even if there were short periods of GNSS signal outages and the positioning accuracy was terrible during long periods of GNSS signal outages, as the drift of INS accumulates rapidly. Therefore, other aiding sensors are required to limit the drift errors of INS when the GNSS signals are blocked. On the one hand, a camera is suitable for solving this problem since visual odometry (VO) can estimate the motion of a vehicle with a slow drift. On the other hand, the model of the monocular camera is relatively simple, but it lacks the metric scale, which can be recovered by IMU. Consequently, a monocular camera is usually integrated with IMU to achieve accurate pose estimations. The fusion algorithms of IMU and vision are usually based on an extended Kalman filter (EKF) [28–30] or nonlinear optimization [31,32]. The former method usually carries out linearization only once, so there may be obvious linearization errors for vision data processing. The latter method utilizes iterative linearization, which can achieve higher estimation accuracy but is subject to an increased computational burden. The multi-state constraint Kalman filter (MSCKF) is a popular EKF-based visual–inertial odometry (VIO) approach, which is capable of high-precision pose estimations in large-scale real-world environments [28]. MSCKF maintains several previous camera poses in the state vector by a sliding window and forms the constraints among multiple camera poses by using visual measurements of the same feature point across multiple camera views. Accordingly, the computational complexity is linear with the number of feature points.

On the one hand, VIO can provide accurate pose estimations when GNSS is unavailable. On the other hand, VIO or VI-SLAM (simultaneous localization and mapping) can only achieve an estimation of motion and provide the relative position and attitude and there are unavoidable accumulated drifts over time. Consequently, the integration of GNSS, INS, and vision is receiving increasing interest [33–36]. Kim et al., 2005 used a six-degrees-of-freedom (DOF) SLAM to aid GNSS/INS navigation by providing reliable navigation solutions in denied and unknown environments GNSS. Then, Won et al. integrated GNSS with vision for low GNSS visibility [34], and proposed the selective integration of GNSS, INS, and vision under GNSS-challenged environments [2], which was able to improve the positioning accuracy compared with nonselective integration. However, in most of these studies, only the position provided by the GNSS or pseudo-range measurements were utilized in the fusion of GNSS, INS, and vision. The application of the carrier phase in multi-sensor fusion is less studied. More recently, Liu [35] proposed the tightly coupled integration of a GNSS/INS/stereo vision/map matching system for land vehicle navigation, but only the positioning results of PPP were integrated with the INS, stereo vision, and map matching system. Li et al. [36] further conducted the tightly coupled integration of RTK, MEMS-IMU, and monocular cameras. Obviously, more efforts should be focused on PPP-RTK/INS/vision integration to fully explore the potential of GNSS for further research.

There are many dynamic objects in urban areas which interfere with VIO. Dynamic objects can provide dynamic feature points, but mainstream SLAM uses static feature points to recover the motion. There are a lot of researches about dynamic object removal in VIO or SLAM, but most of them mainly focus on vision [37,38]. Thus, a simple dynamic feature points removal model based on position is proposed in this paper with the help of GNSS. As VIO has accumulation errors, a model based on position does not work well without GNSS.

This paper aims to evaluate the navigation performance of the integration of multi-GNSS PPP-RTK, MEMS-IMU, and monocular cameras with a cascading filter and the dynamic object removal algorithm in urban areas. The remainder of this paper is organized as follows: first, Section 2 presents the mathematical models of PPP-RTK, the MEMS-IMU, and monocular camera integration based on the MSCKF, integration of multi-GNSS PPP-RTK, INS, and vision as well as the dynamic object removal model, and introduces the structure of the proposed model. Then, the details of the test are demonstrated in Section 3 and the efficiency of different techniques in urban vehicle navigation is analyzed in Section 4. Finally, Section 5 presents the conclusions.

## 2. Methods

The undifferenced and uncombined PPP-RTK, INS model, PPP-RTK/INS tightly coupled integration model, as well as the vision model, are presented in this section in order to derive the integration model of multi-GNSS PPP-RTK/INS/vision with a cascading filter. According to the suggestion in RINEX 3.02 (https://kb.igs.org/hc/en-us/articles/115003980628-RINEX-3-02 (accessed on 12 August 2022)), the GPS and BDS systems are denoted as G and C, respectively.

### 2.1. PPP-RTK Model

The raw observations of the GNSS pseudo-range and carrier phase can be expressed as follows [39]:

$$\left.\begin{array}{l} P^s_{r,f} = \rho^s_r + t_{r,sys} + \alpha^s_r T_Z + \frac{40.3}{f^2}\gamma^s_r I^s_r - b^{s,f} + b_{r,f} + \varepsilon_p \\ \Phi^s_{r,f} = \rho^s_r + t_{r,sys} + \alpha^s_r T_Z - \frac{40.3}{f^2}\gamma^s_r I^s_r + \lambda N^s_{r,f} + \varepsilon_\Phi \end{array}\right\} \tag{1}$$

in which $P^s_{r,f}$ and $\Phi^s_{r,f}$ are the pseudo-range and carrier phase at frequency $f$ corresponding to receiver $r$ and satellite $s$ in length units, respectively; $\rho^s_r$ is the geometric distance between receiver $r$ and satellite $s$; $t_{r,sys}$ is the receiver clock error corresponding to the system $sys \in (G\ C)$ in the length units, respectively; $\alpha^s_r$ and $\gamma^s_r$ are the mapping functions of the tropospheric and ionospheric delays, respectively; $T_Z$ and $I^s_r$ stand for the zenith tropospheric delay and the zenith total electron content (TEC); $b^{s,f}$ and $b_{r,f}$ denote the hardware delay for satellite $s$ and receiver $r$, respectively; $\lambda$ and $N^s_{r,f}$ are the carrier phase wavelength and float ambiguity; $\varepsilon_p$ and $\varepsilon_\Phi$ represent the measurement noise of pseudo-range and carrier phase including the unmodeled multipath error, respectively. Additionally, it is assumed that other errors, such as satellite orbit and clock errors and relativistic effects, are corrected in advance.

After correcting the hardware delay for the satellite and linearization, Equation (1) can be written as

$$\left.\begin{array}{l} \Delta P^s_{r,f} = \boldsymbol{h}^r_s \delta \boldsymbol{x}^e_{GNSS} + t_{r,sys} + \alpha^s_r \delta T_w + \frac{40.3}{f^2}\gamma^s_r I^s_r + b_{r,f} + \varepsilon_p \\ \Delta\Phi^s_{r,f} = \boldsymbol{h}^r_s \delta \boldsymbol{x}^e_{GNSS} + t_{r,sys} + \alpha^s_r \delta T_w - \frac{40.3}{f^2}\gamma^s_r I^s_r + \lambda N^s_{r,f} + \varepsilon_\Phi \end{array}\right\} \tag{2}$$

where $\Delta P^s_{r,f}$ and $\Delta\Phi^s_{r,f}$ are the OMC (observed-minus-computed) of the pseudo-range and carrier phase, respectively; superscript $\cdot^e$ represents the $e$-frame (earth-centered earth-fixed frame); $\delta \boldsymbol{x}^e_{GNSS}$ and $\boldsymbol{h}^s_r$ are the correction vectors of the receiver position and the corresponding direction cosine vector; $\delta T_w$ denotes the residual of the zenith tropospheric wet delay. Additionally, the DESIGN (deterministic plus stochastic ionosphere model for GNSS) model is adopted in this study as [39,40]

$$\left.\begin{array}{l} I^s_r = a_0 + a_1 dL + a_2 dB + a_3 dL^2 + a_4 dB^2 + r^s_r \\ \widetilde{I}^s_r = a_0 + a_1 dL + a_2 dB + a_3 dL^2 + a_4 dB^2 + r^s_r + \varepsilon_{\widetilde{I}^s_r} \end{array}\right\} \tag{3}$$

in which $\widetilde{I}^s_r$ is the virtual observation of ionospheric delay and can be obtained from the ionospheric delay prior models of high-precision ionospheric products; $a_i$ ($i$ = 0, 1, 2, 3, 4) describes the spatial distribution of the ionospheric delay; and $dL$ and $dB$ represent the difference in longitude and latitude between the approximate location of the station and the ionospheric pierce point (IPP), respectively. $r^s_r$ describes the stochastic behavior of the ionospheric delay in the time domain and $\varepsilon_{\widetilde{I}^s_r}$ is the corresponding noise of the virtual observation.

Then, the state vector $x_{PPP-RTK}$ can be written as

$$x_{PPP-RTK} = \begin{pmatrix} \delta \boldsymbol{x}^e_{GNSS} & t_r & \delta T_w & b_r & \boldsymbol{N}_r & \boldsymbol{a}_r \boldsymbol{r}_r \end{pmatrix}^T \tag{4}$$

where $\boldsymbol{N}_r = \begin{pmatrix} \boldsymbol{N}_{r,1} & \boldsymbol{N}_{r,2} \end{pmatrix}^T$ denotes the float ambiguity on frequency $f_1$ and $f_2$; $\boldsymbol{a}_r = (a_0\ a_1\ a_2\ a_3\ a_4)^T$ and $\boldsymbol{r}_r = \left(r^1_r \ldots r^j_r\right)^T$ means the deterministic and stochastic parameters of the DESIGN, respectively.

The float ambiguity $N^s_{r,f}$ should be further formulated for the PPP ambiguity resolution. It can be expressed as

$$N^s_{r,f} = n - d_r + d^s \tag{5}$$

in which $n$ means the integer ambiguity and $d_r$ and $d^s$ are the UPD for the receiver and satellite. After the float ambiguity is obtained by Equation (2), the UPD can be removed and then the integer property of the ambiguity can be recovered. Moreover, the LAMBDA (least-squares ambiguity decorrelation adjustment) method is applied to search for the optimal fixed value of the ambiguity [41]. Finally, the integer ambiguity is used as constraints to obtain the PPP solution with fixed ambiguity.

## 2.2. INS Model

In this paper, the mechanization is conducted in the *e*-frame (earth-centered earth-fixed frame) for easily integrating the state of INS with the GNSS observables. Then, the dynamic equation of INS can be described as

$$\begin{pmatrix} \dot{x}_{INS}^e \\ \dot{v}_{INS}^e \\ \dot{C}_b^e \end{pmatrix} = \begin{pmatrix} v_{INS}^e \\ C_b^e f^b - 2\omega_{ie}^e \times v_{INS}^e + g^e \\ C_b^e [\omega_{eb}^b \times] \end{pmatrix} \tag{6}$$

where $x_{INS}^e$ is the position vector in the *e*-frame, respectively; $v_{INS}^e$ is the velocity vector in the *e*-frame; $C_b^e$ represents the rotation matrix from the *b*-frame (body frame) to the *e*-frame; $f^b$ is the specific force vector generated by the accelerometers in the *b*-frame; $\omega_{ie}^e$ is the earth rotation vector of the *e*-frame against the *i*-frame (inertial frame) in the *e*-frame; $g^e$ denotes the local gravity vector in the *e*-frame; $\omega_{eb}^b$ denotes the rotation rate vector of the *b*-frame against the *e*-frame projected to the *b*-frame; and $[\cdot\times]$ denotes the skew-symmetric matrix.

By using the Phi-angle error model, the INS error model can be written as [42]

$$\begin{pmatrix} \delta\dot{x}_{INS}^e \\ \delta\dot{v}_{INS}^e \\ \dot{\phi} \end{pmatrix} = \begin{pmatrix} \delta v_{INS}^e \\ -2\omega_{ie}^e \times \delta v_{INS}^e + C_b^e f^b \times \phi + C_b^e \delta f^b + \delta g^e \\ -\omega_{ie}^e \times \phi - C_b^e \delta\omega_{ib}^b \end{pmatrix} \tag{7}$$

in which $\phi$ indicates the correction vector of attitude; $\delta g^e$ represents the gravity error in the e-frame; and $\delta f^b$ and $\delta\omega_{ib}^b$ are the sensor errors of the accelerometer and gyroscope, respectively. Bias and scale factor errors along with white noise can be used to model the sensor error [42], which can be expressed as

$$\begin{cases} \delta f^b = B_a + diag\left(f^b\right) S_a + w_v \\ \delta\omega_{ib}^b = B_g + diag\left(\omega_{ib}^b\right) S_g + w_\phi \end{cases} \tag{8}$$

in which, $B_a$ and $S_a$ indicate the bias and scale factor errors of the accelerometer, respectively; *diag* denotes the diagonal matrix; $B_g$ and $S_g$ indicate the bias and scale factor errors of the gyroscope, respectively; and $w_v$ and $w_\phi$ indicate the corresponding random white noise. Bias and scale factor errors can be modeled as first-order Gauss–Markov processes and expressed as [42]

$$\begin{cases} \begin{pmatrix} \dot{B}_a \\ \dot{B}_g \end{pmatrix} = \begin{pmatrix} \frac{-1}{\tau_{ba}} B_a \\ \frac{-1}{\tau_{bg}} B_g \end{pmatrix} + \begin{pmatrix} w_{ba} \\ w_{bg} \end{pmatrix} \\ \begin{pmatrix} \dot{S}_a \\ \dot{S}_g \end{pmatrix} = \begin{pmatrix} \frac{-1}{\tau_{sa}} S_a \\ \frac{-1}{\tau_{sg}} S_g \end{pmatrix} + \begin{pmatrix} w_{sa} \\ w_{sg} \end{pmatrix} \end{cases} \tag{9}$$

where $\tau_{(\bullet)}$ and $w_{(\bullet)}$ (denotes the subscript *ba*, *bg*, *sa*, or *sg*) denote the corresponding correlation time and driving white noise, respectively.

Finally, the INS error state can be modeled as

$$x_{INS} = \begin{pmatrix} \delta x_{INS}^e & \delta v_{INS}^e & \phi & B_a & B_g & S_a S_g \end{pmatrix}^T \tag{10}$$

## 2.3. PPP-RTK/INS Tightly Coupled Integration Model

In the error state of PPP-RTK and INS, $\boldsymbol{x}^e_{GNSS}$ and $\boldsymbol{x}^e_{INS}$ denote the position of the GNSS receiver antenna reference point (ARP) and IMU center, respectively. They do not represent the same position and their spatial relationship in the *e*-frame can be expressed as [42]

$$\boldsymbol{x}^e_{GNSS} = \boldsymbol{x}^e_{INS} + \boldsymbol{C}^e_b \boldsymbol{l}^b \tag{11}$$

in which $\boldsymbol{l}^b$ means the lever-arm correction vector in the *b*-frame. As for the approximate coordinates $\widetilde{\boldsymbol{x}}^e_{GNSS}$ and $\widetilde{\boldsymbol{x}}^e_{INS}$, their relationship can be described as

$$\widetilde{\boldsymbol{x}}^e_{GNSS} = \widetilde{\boldsymbol{x}}^e_{INS} + \widetilde{\boldsymbol{C}}^e_b \boldsymbol{l}^b \tag{12}$$

where $\widetilde{\boldsymbol{C}}^e_b$ is the approximation of $\boldsymbol{C}^e_b$, and satisfies

$$\widetilde{\boldsymbol{C}}^e_b = (\boldsymbol{I} - \boldsymbol{\phi}\times)\boldsymbol{C}^e_b \tag{13}$$

Then, the following equation of $\delta\boldsymbol{x}^e_{GNSS}$ and $\delta\boldsymbol{x}^e_{INS}$ can be derived from Equations (11) to (13):

$$\delta\boldsymbol{x}^e_{GNSS} = \delta\boldsymbol{x}^e_{INS} + \boldsymbol{C}^e_b \boldsymbol{l}^b \times \boldsymbol{\phi} \tag{14}$$

The state error in the integrated navigation is defined as the observation minus the true value, whereas the state error in the GNSS is defined as the true value minus the observation. Thus, the signs of $\delta\boldsymbol{x}^e_{GNSS}$ and $\delta\boldsymbol{x}^e_{INS}$ are opposite. After adding a minus sign to Equation (14) and substituting it into Equation (2), the observation equation of the PPP-RTK/INS can be further expressed as

$$\left.\begin{aligned} \Delta P^s_{r,f} &= -\boldsymbol{h}^s_r \delta\boldsymbol{x}^e_{INS} - \boldsymbol{h}^s_r \boldsymbol{C}^e_b \boldsymbol{l}^b \times \boldsymbol{\phi} + t_{r,sys} + \alpha^s_r \delta T_w + b_{r,f} + \varepsilon_P \\ &\quad + \tfrac{40.3}{f^2}\gamma^s_r(a_0 + a_1 dL + a_2 dB + a_3 dL^2 + a_4 dB^2 + r^s_r + \varepsilon_{\widetilde{I}^s_r}) \\ \Delta\Phi^s_{r,f} &= -\boldsymbol{h}^s_r \delta\boldsymbol{x}^e_{INS} - \boldsymbol{h}^s_r \boldsymbol{C}^e_b \boldsymbol{l}^b \times \boldsymbol{\phi} + t_{r,sys} + \alpha^s_r \delta T_w + \lambda N^s_{r,f} + \varepsilon_\Phi \\ &\quad - \tfrac{40.3}{f^2}\gamma^s_r(a_0 + a_1 dL + a_2 dB + a_3 dL^2 + a_4 dB^2 + r^s_r + \varepsilon_{\widetilde{I}^s_r}) \\ \widetilde{I}^s_r &= a_0 + a_1 dL + a_2 dB + a_3 dL^2 + a_4 dB^2 + r^s_r + \varepsilon_{\widetilde{I}^s_r} \end{aligned}\right\} \tag{15}$$

Combining the state vector $\boldsymbol{x}_{PPP-RTK}$ in Equation (4) and $\boldsymbol{x}_{INS}$ in Equation (10), the state vector of the PPP-RTK/INS can be described as

$$\boldsymbol{x} = \begin{pmatrix} \delta\boldsymbol{x}^e_{INS} & \delta\boldsymbol{v}^e_{INS} & \boldsymbol{\phi} & \boldsymbol{B} & \boldsymbol{S} & t_r & \delta T_w & b_r & \boldsymbol{N_r} & \boldsymbol{a_r} & \boldsymbol{r_r} \end{pmatrix}^T \tag{16}$$

## 2.4. INS/Vision Tightly Coupled Integration Model

By denoting the error state of the camera as $\left(\delta\boldsymbol{p}^e_{c_i}\ \boldsymbol{\phi}_{c_i}\right)$ for the *i*th image, the error state vector of INS/vision tightly coupled integration with the MSCKF at the time when the *k*th image is captured is expressed as

$$\boldsymbol{x}_{i,c} = \begin{pmatrix} \boldsymbol{x}_{INS} & | & \delta\boldsymbol{p}^e_{c_j} & \boldsymbol{\phi}_{c_j} & \cdots & \delta\boldsymbol{p}^e_{c_k} & \boldsymbol{\phi}_{c_k} \end{pmatrix}^T \tag{17}$$

in which $\delta\boldsymbol{p}^e_{c_i}$ and $\boldsymbol{\phi}_{c_i}$ ($i = j, j + 1, \ldots, k$) indicate the error states of the camera position and attitude for epoch *i*. The above error state vector is augmented when new camera data is introduced.

Visual measurements of the same feature point from multiple camera views are used to construct the geometric constraints. At the time of taking the *i*th ($i < k + 1$) image, the transformation of the static feature point $P_k$ can be expressed as

$$\boldsymbol{p}^{c_i}_{P_k,i} = \boldsymbol{R}^{c_i}_e \left(\boldsymbol{p}^e_{P_k} - \boldsymbol{p}^e_{c_i}\right) \tag{18}$$

in which, $\boldsymbol{p}^{c_i}_{P_k,i} = \begin{pmatrix} x^{c_i}_{P_k,i} & y^{c_i}_{P_k,i} & z^{c_i}_{P_k,i} \end{pmatrix}$ indicates the position of $P_k$ in the camera frame; $\boldsymbol{R}^{c_i}_e$ and $\boldsymbol{p}^e_{c_i}$ are the attitude rotation matrix and position vector against the global frame (*e*-frame), respectively; $\boldsymbol{p}^e_{P_k}$ is the estimated position of $P_k$ in the *e*-frame, which can be calculated by triangulation. By differentiating Equation (18), the equation of the camera state can be obtained as follows:

$$\delta \boldsymbol{p}^{c_i}_{P_k,i} = -\boldsymbol{R}^{c_i}_e \left[ \left( \boldsymbol{p}^e_{P_k} - \boldsymbol{p}^e_{c_i} \right) \times \right] \boldsymbol{\phi}_{c_i} - \boldsymbol{R}^{c_i}_e \delta \boldsymbol{p}^e_{c_i} + \boldsymbol{R}^{c_i}_e \delta \boldsymbol{p}^e_{P_k} \tag{19}$$

in which $\delta \boldsymbol{p}^e_{P_k}$ indicates the error of the approximate position of the feature point.

Concerning the camera measurement residual vector, it can be written as

$$z_{P_k,i} = \begin{pmatrix} u^0_{P_k,i} - \widetilde{u}_{P_k,i} \\ v^0_{P_k,i} - \widetilde{v}_{P_k,i} \end{pmatrix}, \ \varepsilon_{P_k,i} = \begin{pmatrix} \varepsilon_{\widetilde{u}_{P_k,i}} \\ \varepsilon_{\widetilde{v}_{P_k,i}} \end{pmatrix} \tag{20}$$

where $\left(u^0_{P_k,i}, v^0_{P_k,i}\right)^T$ is the estimated pixel coordinate of $P_k$ by back projection and $\left(\widetilde{u}_{P_k,i}, \widetilde{v}_{P_k,i}\right)^T$ is the observation of $P_k$ in the *i*th image and $\boldsymbol{\varepsilon}_{P_k,i}$ is the measurement noise. Then, based on the chain rule, the residual formula can be expressed as [27]

$$\boldsymbol{z}_{P_k,i} = \delta\left(u_{P_k,i}, v_{P_k,i}\right)^T = \frac{\partial\left(u_{P_k,i}, v_{P_k,i}\right)^T}{\partial \boldsymbol{p}^{c_i}_{P_k,i}} \delta \boldsymbol{p}^{c_i}_{P_k,i} = \boldsymbol{H}_{x,i} \boldsymbol{x}_{i,c} + \boldsymbol{H}_{f,i} \delta \boldsymbol{p}^e_{P_k} + \boldsymbol{\varepsilon}_{P_k,i} \tag{21}$$

where

$$\boldsymbol{H}_{x,i} = \begin{pmatrix} 0_{2\times21} & 0_{2\times6} & \cdots & -\boldsymbol{J}\boldsymbol{R}^{c_i}_e & -\boldsymbol{J}\boldsymbol{R}^{c_i}_e \left[ \left( \boldsymbol{p}^e_{P_k} - \boldsymbol{p}^e_{c_i} \right) \times \right] & \cdots \end{pmatrix} \boldsymbol{H}_{f,i} = \boldsymbol{J}\boldsymbol{R}^{c_i}_e \boldsymbol{J} = \frac{\partial\left(u_{P_k,i}, v_{P_k,i}\right)^T}{\partial \boldsymbol{p}^{c_i}_{P_k,i}} = \begin{pmatrix} \frac{f_x}{z^{c_i}_{P_k,i}} & 0 & -\frac{f_x x^{c_i}_{P_k,i}}{{z^{c_i}_{P_k,i}}^2} \\ 0 & \frac{f_y}{z^{c_i}_{P_k,i}} & -\frac{f_y y^{c_i}_{P_k,i}}{{z^{c_i}_{P_k,i}}^2} \end{pmatrix}$$

in which $(f_x, f_y)$ means the focal length and $\begin{pmatrix} x^{c_i}_{P_k,i} & y^{c_i}_{P_k,i} & z^{c_i}_{P_k,i} \end{pmatrix}$ means the position of a feature point in the camera frame.

Because $\delta \boldsymbol{p}^e_{P_k}$ is not the state that needs to be estimated and $\boldsymbol{H}_{f,i}$ is known, we can calculate the left null space A, which satisfies the equation as follows:

$$\boldsymbol{A}^T \boldsymbol{H}_{f,i} = 0 \tag{22}$$

Then, multiplying $\boldsymbol{A}^T$ at both sides of Equation (21), the measurement model can be described as

$$\boldsymbol{z}_{a,i} = \boldsymbol{A}^T \boldsymbol{z}_{P_k,i} = \boldsymbol{A}^T \boldsymbol{H}_{x,i} \boldsymbol{x}_{i,c} + \boldsymbol{A}^T \boldsymbol{\varepsilon}_{P_k,i} = \boldsymbol{H}_{a,x,i} \boldsymbol{x}_{i,c} + \boldsymbol{\varepsilon}_{a,i} \tag{23}$$

where $\boldsymbol{z}_{a,i} = \boldsymbol{A}^T \boldsymbol{z}_{P_k,i}$, $\boldsymbol{H}_{a,x,i} = \boldsymbol{A}^T \boldsymbol{H}_{x,i}$ and $\boldsymbol{\varepsilon}_{a,i} = \boldsymbol{A}^T \boldsymbol{\varepsilon}_{P_k,i}$, respectively.

The pixel coordinate of $P_k$ can be described as follows:

$$\begin{pmatrix} \widetilde{u}_{P_k,i} \\ \widetilde{v}_{P_k,i} \\ 1 \end{pmatrix} = \frac{1}{z^{c_i}_{P_k,i}} K \boldsymbol{p}^{c_i}_{P_k,i} \tag{24}$$

in which K denotes the camera intrinsic parameter.

Substituting Equation (18) into (25), we have:

$$\begin{pmatrix} \widetilde{u}_{P_k,i} \\ \widetilde{v}_{P_k,i} \\ 1 \end{pmatrix} = \frac{1}{z^{c_i}_{P_k,i}} K \boldsymbol{R}^{c_i}_e \left( \boldsymbol{p}^e_{P_k} - \boldsymbol{p}^e_{c_i} \right) \tag{25}$$

Thus,

$$\begin{pmatrix} \Delta\widetilde{u}_{P_k} \\ \Delta\widetilde{v}_{P_k} \\ 0 \end{pmatrix} = K\left[\frac{1}{z^{c_{i+1}}_{P_k,i+1}} \boldsymbol{R}^{c_{i+1}}_{e}\left(\boldsymbol{p}^{\acute{e}}_{P_k} - \boldsymbol{p}^{e}_{c_{i+1}}\right) - \frac{1}{z^{c_i}_{P_k,i}} \boldsymbol{R}^{c_i}_{e}\left(\boldsymbol{p}^{e}_{P_k} - \boldsymbol{p}^{e}_{c_i}\right)\right] \tag{26}$$

where $\Delta\widetilde{u}_{P_k} = \widetilde{u}_{P_k,i+1} - \widetilde{u}_{P_k,i}$, $\Delta\widetilde{v}_{P_k} = \widetilde{v}_{P_k,i+1} - \widetilde{v}_{P_k,i}$. $\boldsymbol{p}^{\acute{e}}_{P_k}$ is the position of $P_k$ for the epoch $i+1$. $\left(\widetilde{u}_{P_k,i+1}, \widetilde{v}_{P_k,i+1}\right)^T$ is the observation of $P_k$ in the i+1th image.

Considering that the time interval is relatively small, we assume that the pose and position of the camera have no obvious changes. Because $z^{c_{i+1}}_{P_k,i+1}$ is relatively large, we can make assumptions: $\boldsymbol{R}^{c_{i+1}}_{e} \approx \boldsymbol{R}^{c_i}_{e}$, $\frac{1}{z^{c_{i+1}}_{P_k,i+1}} \approx \frac{1}{z^{c_i}_{P_k,i}}$. For the static feature point, $\boldsymbol{p}^{e}_{P_k} - \boldsymbol{p}^{\acute{e}}_{P_k} = 0$, thus we can obtain

$$z^{c_{i+1}}_{P_k,i+1}\begin{pmatrix} \Delta\widetilde{u}_{P_k} \\ \Delta\widetilde{v}_{P_k} \\ 0 \end{pmatrix} = -K\boldsymbol{R}^{c_{i+1}}_{e}\left(\boldsymbol{p}^{e}_{c_{i+1}} - \boldsymbol{p}^{e}_{c_i}\right) \tag{27}$$

As we know, $\boldsymbol{p}^{e}_{c_{i+1}}$ and $\boldsymbol{p}^{e}_{c_i}$ can be obtained by PPP-RTK/INS. For dynamic objects on the road, $[\left(\boldsymbol{p}^{e}_{P_k} - \boldsymbol{p}^{e}_{c_{i+1}}\right) - \left(\boldsymbol{p}^{\acute{e}}_{P_k} - \boldsymbol{p}^{e}_{c_i}\right)] < \left(\boldsymbol{p}^{e}_{c_{i+1}} - \boldsymbol{p}^{e}_{c_i}\right)$. By setting a threshold for $z^{c_{i+1}}_{P_k,i+1}$, the dynamic objects can be removed.

### 2.5. PPP-RTK/INS/Vision Integration Model with a Cascading Filter

The PPP-RTK/INS/vision integration model with a cascading filter is realized by integrating the output of the tightly coupled integration of PPP-RTK/INS with the tightly coupled integration of INS/vision. The difference between the position provided by the PPP-RTK/INS and the position predicted by INS/vision constitutes the observation of position. Based on Equations (14) and (23), the measurement model of the PPP-RTK/INS/Vision integration model can be expressed as

$$\left.\begin{aligned} \hat{\boldsymbol{x}}^{e}_{GNSS} - \widetilde{\boldsymbol{x}}^{e}_{GNSS} &= \delta \boldsymbol{x}^{e}_{INS} + \boldsymbol{C}^{e}_{b}\boldsymbol{l}^{b} \times \boldsymbol{\phi} + \boldsymbol{\varepsilon}_{x} \\ \boldsymbol{z}_{a,i} &= \boldsymbol{H}_{a,x,i}\boldsymbol{x}_{i,c} + \boldsymbol{\varepsilon}_{a,i} \end{aligned}\right\} \tag{28}$$

in which $\hat{\boldsymbol{x}}^{e}_{GNSS}$ is the position predicted by INS/vision and $\widetilde{\boldsymbol{x}}^{e}_{GNSS}$ is the positioning results of PPP-RTK/INS. $\boldsymbol{\varepsilon}_{x}$ is the measurement noise.

The integration model of multi-GNSS PPP-RTK/INS/vision with a cascading filter is derived from the description above. An overview of the proposed model is shown in Figure 1. First, the position is obtained by the GNSS to assist the navigation initialization, e.g., IMU alignment with GNSS. Then, the INS begins to provide high-rate navigation. When the INS synchronizes with the GNSS, tightly coupled integration is performed based on either PPP/INS or PPP-RTK/INS. As for the latter, high-precision atmospheric correction is applied and AR is carried out with the UPD products. Furthermore, feature points are extracted and tracked in each image. When the time of the camera synchronizes with the INS, the state vector is augmented and the MSCKF is adopted to calculate the relative position of the vehicle platform. The positioning results of the GNSS/INS are then integrated with the MSCKF to produce the final navigation information. There are two Kalman filters: the filter of PPP-RTK/INS and the filter of INS/Vision. The positioning results of PPP-RTK/INS are added into the filter of the INS/vision so it becomes a cascading Kalman filter. All the IMU sensor errors are fed back in time in the process.

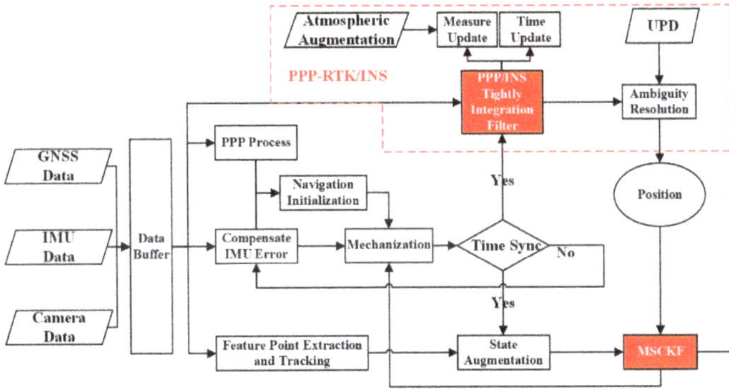


**Figure 1.** The algorithm structure of PPP-RTK/INS/VISION with a cascading filter.

## 3. Experiment

To evaluate the positioning accuracy and performance of the proposed integration model of multi-GNSS PPP-RTK/INS/vision with a cascading filter in urban areas, based on the FUSING (FUSing IN Gnss) software [7,40,41], this algorithm was further developed by us. At present, FUSING has been developed into an integrated software platform that can deal with real-time multi-GNSS precise orbit determination, atmospheric delay modeling and monitoring, satellite clock estimation, as well as multi-sensor fusion navigation.

Two datasets were collected based on a vehicle platform as shown in Figure 2. One is in the suburban area of Wuhan City on 1 January 2020, and the other is on the Second Ring Road of Wuhan city on 2 January 2020. For simplicity, according to the DOY (day of the year), the two-vehicle tests are denoted as T01 and T02, respectively.

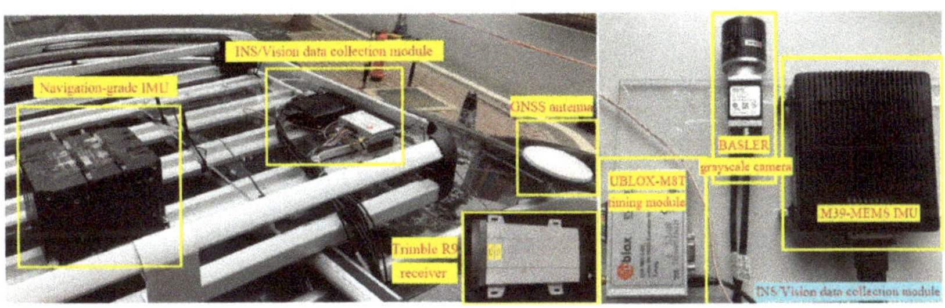


**Figure 2.** GNSS/INS/vision data collection platform.

As shown in Figure 2, the raw data was collected by the IMU of two different grades. A MEMS-grade IMU was used to integrate with PPP-RTK and vision to evaluate the performance of the proposed model. The navigation-grade IMU is of high accuracy and is integrated with RTK to calculate the reference solution, which was regarded as the true value. Both IMUs collected the data at a sampling rate of 200 Hz and their performance parameters are listed in Table 1. The grayscale Basler acA640-90gm camera was equipped to collect the raw images at a sampling rate of 10 Hz with a resolution of 659 × 494. The UBLOX-M8T was used for generating the pulses per second (PPS) to trigger the camera exposure and it also recorded the time of the pulse at the same time. GNSS data were collected by Trimble Net R9 at a sampling rate of 1 Hz. The camera-IMU extrinsic parameters were calibrated offline by utilizing the Kalibr tool (https://github.com/ethz-asl/kalibr/ (accessed on 3 January 2020)). The lever-arm correction vector was measured manually.

**Table 1.** Performance parameters of the IMU sensors.

| IMU Sensors | Random Walk | | Bias | |
|---|---|---|---|---|
| | Velocity $(m/s/\sqrt{h})$ | Angular $(°/\sqrt{h})$ | Gyro. (°/h) | Acce. (mGal) |
| MEMS-grade | 0.03 | 0.17 | 8 | 200 |
| Navigation-grade | 0.03 | 0.003 | 0.027 | 15 |

The test trajectory and the true scenarios for these two tests are shown in the left and right panels of Figure 3, respectively. It can be seen that dataset T01 was collected in a relatively open sky and only a few obstacles were blocking the GNSS signals. T02 was collected in a GNSS-challenged environment and there were many tall buildings on both sides of the narrow road, including some viaducts and tunnels, which could have totally blocked the GNSS signals. The vehicle speeds of T01 and T02 were about 10 m/s and 15 m/s, respectively, which is shown in Figure 4. It can be seen that there were significant changes in velocity and direction. The number of visible satellites and the PDOP (precision dilution of positioning) with a cutoff angle of 10° are shown in Figure 5. Taking, for instance, the GPS, the average number of tracking satellites of T01 and T02 were 9 and 7, respectively, and the average PDOP values were 1.8 and 10.9, respectively, which demonstrates the difference in the observation environment between T01 and T02.

**Figure 3.** The test trajectory and typical scenarios of T01 (left panel) and T02 (right panel).

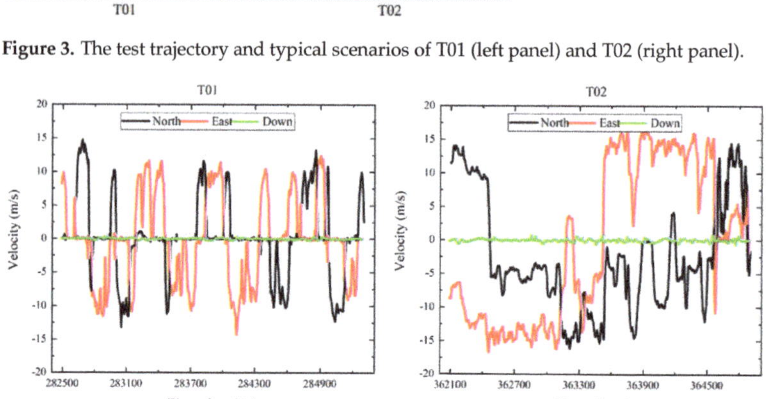


**Figure 4.** Velocity of the vehicle for T01 (left panel) and T02 (right panel), respectively.

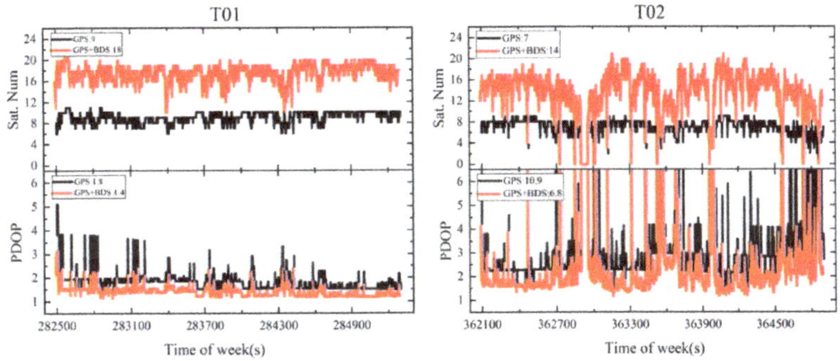


**Figure 5.** Satellite number of GPS/BDS and PDOP for T01 (left panel) and T02 (right panel), respectively.

Considering the high-precision ionospheric and tropospheric delay augmentation and ambiguity resolution to support PPP-RTK, the measurement data of seven reference stations as distributed in Figure 6 were also collected. They were processed to generate high-precision atmospheric delay corrections and UPD products. The average distance between the seven reference stations is 40 km and the green trajectories are the trajectories of the two tests as shown in Figure 3.

**Figure 6.** Distribution of seven reference stations for generating the atmospheric and UPD products. The green lines denote the trajectories of the two experiments.

The details of the GNSS data processing strategy are presented in Table 2. The positioning performance was evaluated by RMS (root mean square). The reference solution was calculated by a loosely coupled RTK/INS solution with a bi-directional smoothing algorithm, in which navigation-grade IMU and GNSS data collected by Trimble R9 were adopted. The ground truth was calculated using commercial software named GINS (http://www.whmpst.com/cn/ (accessed on 15 April 2021)). The nominal positioning accuracy of the RTK/INS loosely coupled solution provided by GINS was at the level of 2 cm for horizontal and 3 cm for vertical.

**Table 2.** GNSS data processing strategy of PPP and PPP-RTK.

| Items | PPP | PPP-RTK |
|---|---|---|
| Ambiguity | float | fixed |
| Troposphere | GPT2w model and VMF1_HT [43] and the residuals are estimated as random walk | regional model |
| Ionosphere | DESIGN-5 model [44] with GIM served as a prior constraint | regional model |
| Observation model | undifferenced and uncombined | |
| Frequency band | GPS: L1/L2; BDS: B1/B2 | |
| Cutoff angle | 10° | |
| PCO/PCV | igs14.atx | |
| Solid earth tides | IERS 2010 | |
| Receiver clock | estimated as white noise | |
| Ephemeris | precise products provided by GFZ | |
| Code bias | receiver: estimated as random walk; satellite: corrected with IGS product | |
| Sigma of code | 0.3 m | |
| Sigma of phase | 0.003 m | |

## 4. Results

All results were output at a frequency of 1 Hz. In the following analysis, PPP-RTK/INS means the tightly coupled integration of PPP-RTK and INS. PPP-RTK/INS/vision means the integration of PPP-RTK, INS, and vision with a cascading filter. Before analyzing the performance of our proposed integration system in urban areas, the effects of the dynamic feature point removal algorithm aided by position are presented in Figure 7. It can be seen that the feature points on the car (in the red box) were removed. Some points that were extracted from unobvious places were also removed and the most static obvious feature points were saved, which can be used for visual localization. Figure 8 shows the positioning results of PPP-RTK/INS/vision for T02 before and after dynamic feature point removal. The positioning accuracy was improved by dynamic feature point removal, which is demonstrated in the green box in Figure 8. When PPP-RTK/INS provided stable, high-accuracy position information, the error caused by the dynamic feature points was constrained. However, the positioning performance was obviously influenced by the dynamic feature points when the GNSS signals were interfered with. Combining Figures 5 and 8, the GNSS observation conditions were poor and PPP-RTK/INS performed poorly in positioning around time 362,900 s, 364,000 s, and 364,600 s, thus PPP-RTK/INS was not able to restrain the interference of the dynamic feature points. Therefore, the dynamic feature points removal algorithm based on position improved the positioning accuracy when the GNSS signals were severely disturbed. The statistics of the positioning results show that the positioning accuracy was improved by 3 cm and 1 cm in the horizontal and vertical directions, respectively.

**Figure 7.** The effect of dynamic feature points removal.

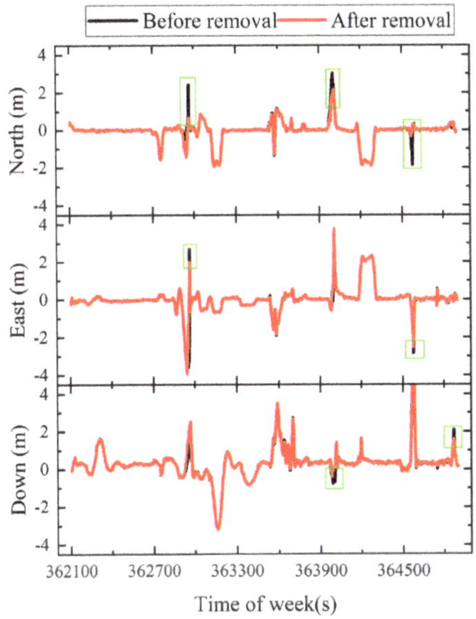


**Figure 8.** The effects of dynamic feature points removal on positioning accuracy.

The comparison of the performance of different positioning solutions is presented in Figures 9 and 10 for T01 and T02, respectively. It can be seen in the left section in Figure 9 that PPP-RTK converged much faster than PPP, though the ambiguity sometimes failed to fix. It took about 30 s to converge for PPP-RTK and more than 10 min for PPP. Moreover, PPP-RTK had higher accuracy than PPP after convergence. The inclusion of BDS made the series more stable, especially for the vertical direction. Although the contribution of the INS was rather limited in the horizontal direction as shown in the GC-PPP-RTK/INS solution, the outliers may have been inhibited, e.g., around the time 28,400 s. As for the vertical direction, the INS significantly contributed to the improvement of the positioning accuracy. The INS helped the GNSS to converge to and maintain a higher level of positioning accuracy. Additionally, the introduction of vision reduced the fluctuation, although overall, there was no big difference. As for T02, there was no obvious convergence in Figure 10 as the observation environment was complicated and the positioning accuracy was relatively low. However, it still can be seen in the left section in Figure 10 that PPP-RTK performed better than PPP. In order to further demonstrate the convergence and reconvergence effects of PPP and PPP-RTK, enlarged images of parts of Figure 10 are shown in Figure 11 and the correspondence can be seen from the time of the week. PPP-RTK converged and reconverged much faster than PPP and achieved higher positioning accuracy, though the observational environment was challenging. Figure 10 shows that the series of G-PPP and G-PPP-RTK were interrupted many times because the GNSS signals were blocked out, which is embodied in the tracking number of the visible satellites in Figure 5. Both the continuity and accuracy of the positioning were improved with the BDS included. However, the GNSS was still unable to provide positioning results from 362,906 s to 362,965 s because the vehicle was in the tunnel at that time and there was no GNSS signal at all. GC-PPP-RTK/INS provided continuous and more stable positioning information with the integration with the INS, but there was also obvious fluctuation and the existence of epochs with large positioning errors. The positioning errors diverged to 21.74 m, 15.82 m, and 3.82 m in the north, east, and down directions, respectively. The three-dimensional positioning error at 362,965 s was 27.16 m, which is not suitable for vehicle navigation. Furthermore, the positioning error was obviously reduced when vision was included, especially around the time of the week at 362,900 s, 363,600 s, and 364,000 s. The cumulative errors of the INS were effectively constrained by vision. Thus, in the period from 362,906 s to 362,965 s, the maximum errors of the INS/vision were −1.43 m, −3.94 m, and 2.50 m in the north, east, and down directions, respectively. The corresponding three-dimensional error was 4.88 m, which was 0.49% of the traveled distance. It can be concluded that the integration of multi-GNSS PPP-RTK/INS/vision with a cascading filter performed best in comparison with the other four solutions.

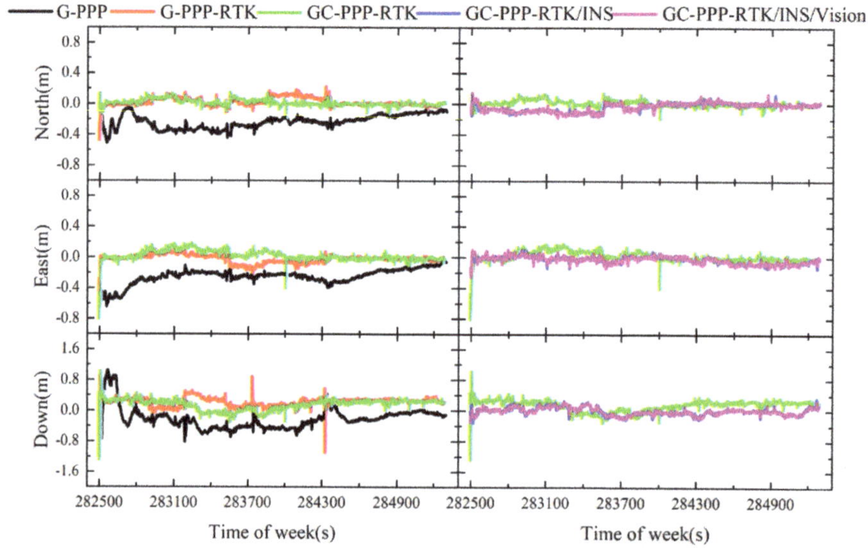


**Figure 9.** Position difference series of T01 in north, east, and down directions, respectively.

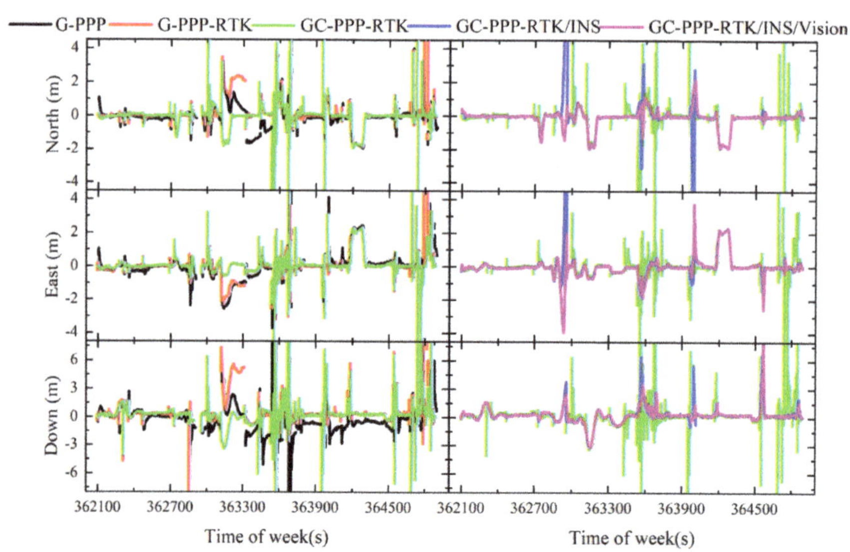


**Figure 10.** Position difference series of T02 in north, east, and down directions, respectively.

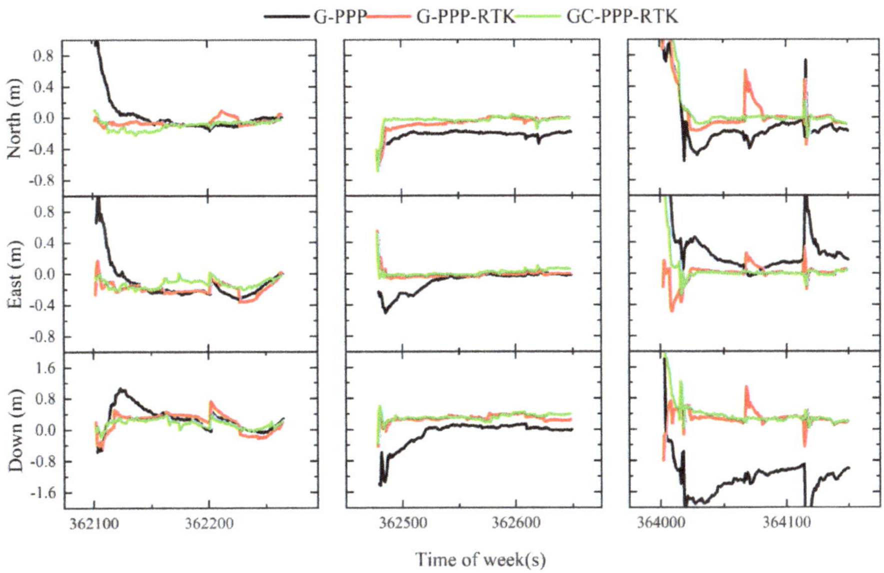


**Figure 11.** Convergence (left panel) and reconvergence (middle panel and right panel) of T02 in north, east, and down directions, respectively.

The statistics of the position difference between T01 and T02 are presented in Tables 3 and 4 to further verify the conclusion. Additionally, the number of epochs at which the position information can be obtained is denoted as A and the number of total epochs is denoted as B. Then the positioning availability can be defined as A/B, which is also included in the tables. The improvement statistics are derived in comparison with the G-PPP solution. The statistics show that the PPP-RTK performed better than PPP in both tests. The ambiguity could not be fixed in many epochs for the frequent GPS signal interruptions and disturbances in T02. Thus, the improvement brought by PPP-RTK for T02 was not as obvious as for T01. The horizontal and vertical positioning RMS of the GC-PPP-RTK/INS/vision solution for the test T01 were 0.08 m and 0.09 m, respectively. As for T02, the horizontal and vertical RMS of the GC-PPP-RTK/INS/vision solution were 0.83 m and 0.91 m, respectively. It can be seen that GC-PPP-RTK/INS/vision made significant improvements compared with the other four solutions. The improvements should have been more obvious because of the interruption in G-PPP and G-PPP-RTK. The positioning availabilities of G-PPP and G-PPP-RTK were both 90.5%. The availability increased to 95.1% with the inclusion of BDS, which means that more epochs of worse positioning were taken into consideration. Because of the accumulation errors of the INS, the statistics of GC-PPP-RTK/INS were worse than all the other solutions in the horizontal direction. The statistics of GC-PPP-RTK, GC-PPP-RTK/INS, and GC-PPP-RTK/INS/vision are shown in Table 5, in which the positioning results derived by the INS and vision are excluded, in order to better show the improvements brought by the INS and vision to the positioning performance of GC-PPP-RTK. The improvement statistics were derived in comparison with the GC-PPP-RTK solution. It can be seen that the INS improved the performance of GC-PPP-RTK by 31.4% and 37.1% in the horizontal and vertical directions, respectively. Eventually, the inclusion of vision increased the improvements to 37.1% and 42.2% in the horizontal and vertical directions, respectively.

**Table 3.** RMS of the positioning error of T01 in horizontal and vertical directions, respectively.

| Solution | Horizontal [m] | Improvement | Vertical [m] | Improvement | Availability |
|---|---|---|---|---|---|
| G-PPP | 0.37 | | 0.35 | | 100% |
| G-PPP-RTK | 0.09 | 75.7% | 0.26 | 25.7% | 100% |
| GC-PPP-RTK | 0.09 | 75.7% | 0.22 | 37.1% | 100% |
| GC-PPP-RTK/INS | 0.08 | 78.4% | 0.09 | 74.3% | 100% |
| GC-PPP-RTK/INS/Vision | 0.08 | 78.4% | 0.09 | 74.3% | 100% |

**Table 4.** RMS of the positioning error of T02 in horizontal and vertical directions, respectively.

| Solution | Horizontal [m] | Improvement | Vertical [m] | Improvement | Availability |
|---|---|---|---|---|---|
| G-PPP | 1.61 | | 2.40 | | 90.5% |
| G-PPP-RTK | 1.55 | 3.7% | 2.30 | 4.2% | 90.5% |
| GC-PPP-RTK | 0.70 | 56.5% | 1.16 | 51.7% | 95.1% |
| GC-PPP-RTK/INS | 1.84 | −14.3% | 0.98 | 59.2% | 100% |
| GC-PPP-RTK/INS/Vision | 0.83 | 48.4% | 0.91 | 62.1% | 100% |

**Table 5.** RMS of the position difference of T02 with part of the epoch excluded in horizontal and vertical directions, respectively.

| Solution | Horizontal [m] | Improvement | Vertical [m] | Improvement |
|---|---|---|---|---|
| GC-PPP-RTK | 0.70 | | 1.16 | |
| GC-PPP-RTK/INS | 0.48 | 31.4% | 0.73 | 37.1% |
| GC-PPP-RTK/INS/Vision | 0.44 | 37.1% | 0.67 | 42.2% |

Position, velocity, and attitude are of great importance for vehicle navigation in urban areas. The velocity error series of GC-PPP-RTK/INS and GC-PPP-RTK/INS/vision for T01 and T02 are shown in Figures 12 and 13, respectively. As dataset T01 was collected in a relatively open-sky environment, the error series was very stable. There was no obvious difference between GC-PPP-RTK/INS and GC-PPP-RTK/INS/vision. Because T02 was collected in a GNSS-challenged environment, there were obvious divergences in the velocity estimation around 362,900 s, 363,600 s, and 364,000 s. The inclusion of vision weakened the impact and improved the accuracy of the velocity estimation in three directions. It can be seen from the statistics in Tables 6 and 7 that vision could not bring about obvious improvements in an open-sky environment but greatly improved the velocity estimation accuracy in a GNSS-challenged environment. The error of velocity estimation in T02 was reduced by the inclusion of vision from 0.12 m/s, 0.07 m/s, and 0.07 m/s to 0.03 m/s, 0.05 m/s, and 0.05 m/s in the north, east, and down directions, respectively. The improvement in all three directions was more than 20%.

**Table 6.** RMS of the velocity error of T01 in north, east, and down directions, respectively.

| Solution | North [m/s] | East [m/s] | Down [m/s] | Improvement | | |
|---|---|---|---|---|---|---|
| GC-PPP-RTK/INS | 0.01 | 0.01 | 0.02 | | | |
| GC-PPP-RTK/INS/Vision | 0.01 | 0.01 | 0.01 | 0% | 0% | 50.0% |

**Table 7.** RMS of the velocity error of T02 in north, east, and down directions, respectively.

| Solution | North [m/s] | East [m/s] | Down [m/s] | Improvement | | |
|---|---|---|---|---|---|---|
| GC-PPP-RTK/INS | 0.12 | 0.07 | 0.07 | | | |
| GC-PPP-RTK/INS/Vision | 0.03 | 0.05 | 0.05 | 75.0% | 28.6% | 28.6% |

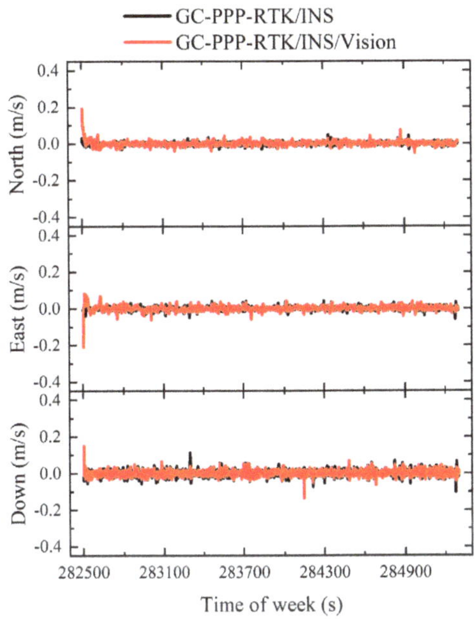


**Figure 12.** Velocity error series of GC-PPP-RTK/INS and GC-PPP-RTK/INS/vision for T01.

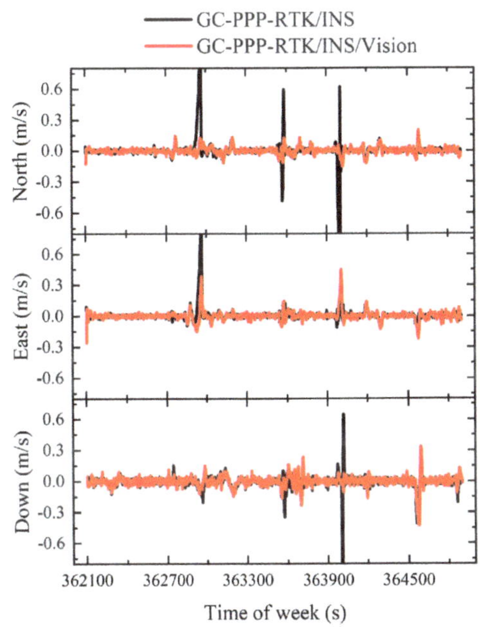


**Figure 13.** Velocity error series of GC-PPP-RTK/INS and GC-PPP-RTK/INS/vision for T02.

The attitude error series of GC-PPP-RTK/INS and GC-PPP-RTK/INS/vision for T01 and T02 are shown in Figures 14 and 15, respectively. It can be seen that there was no obvious difference between GC-PPP-RTK/INS and GC-PPP-RTK/INS/vision in the roll and pitch angles in T01. As for T02, the error of pitch and yaw angle of GC-PPP-RTK/INS accumulated when GNSS signals were blocked or interfered with. Figure 15 shows that vision helped to constrain the error divergence around 363,000 s when the GNSS signals were blocked, but there was no obvious difference in the roll and pitch angles at other parts in T02. However, the estimation accuracy of the yaw angle was significantly improved with vision aiding in both tests. The statistics of the attitude error are listed in Tables 8 and 9. The inclusion of vision reduced the error of the yaw angle from 0.24° to 0.11° and from 0.39° to 0.26° for T01 and T02, respectively. The improvement rates were more than 30% in both tests.

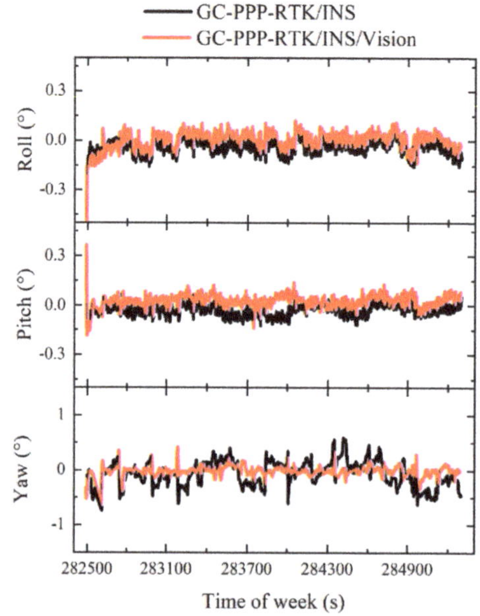


**Figure 14.** Attitude error series of GC-PPP-RTK/INS and GC-PPP-RTK/INS/vision for T01.

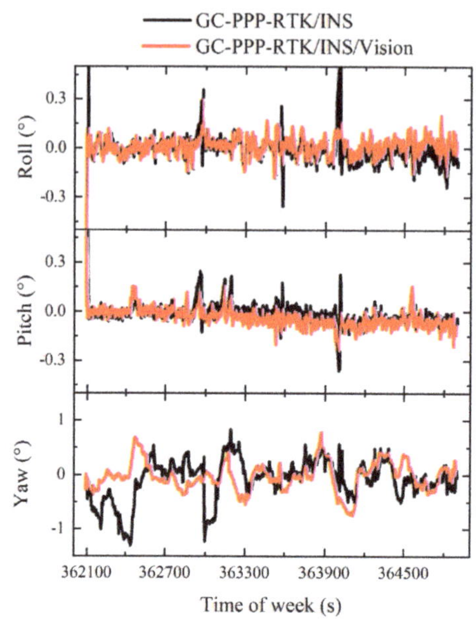


**Figure 15.** Attitude error series of GC-PPP-RTK/INS and GC-PPP-RTK/INS/vision for T02.

**Table 8.** RMS of the attitude error of T01 for GC-PPP-RTK/INS and GC-PPP-RTK/INS/vision.

| Solution | Roll [°] | Pitch [°] | Yaw [°] | Improvement | | |
|---|---|---|---|---|---|---|
| GC-PPP-RTK/INS | 0.05 | 0.04 | 0.24 | | | |
| GC-PPP-RTK/INS/Vision | 0.05 | 0.04 | 0.11 | 0% | 0% | 54.2% |

**Table 9.** RMS of the attitude error of T02 for GC-PPP-RTK/INS and GC-PPP-RTK/INS/vision.

| Solution | Roll [°] | Pitch [°] | Yaw [°] | Improvement | | |
|---|---|---|---|---|---|---|
| GC-PPP-RTK/INS | 0.07 | 0.06 | 0.39 | | | |
| GC-PPP-RTK/INS/Vision | 0.05 | 0.06 | 0.26 | 28.6% | 0% | 33.3% |

## 5. Discussion

GC-PPP-RTK/INS/vision integration with a cascading filter can provide continuous positioning information with high precision. The improvements brought about by vision are more significant in challenging environments. Vision reduces the error divergence of PPP-RTK/INS when the GNSS signals are blocked. The positioning error of PPP-RTK/INS reaches 27.16 m after the GNSS signals were lost for 60 s. The inclusion of vision reduces the positioning error to 4.88 m. Vision also helps to improve the estimation accuracy of velocity and attitude. Although the improvements are significant, the positioning accuracy still needs to be improved for vehicle navigation.

Dynamic objects can seriously affect the positioning performance of vision, so it is important to weaken the influence of dynamic objects. The dynamic object removal algorithm proposed in this paper can remove most fast-moving objects and can improve the positioning accuracy of vision. However, it is difficult to deal with slow-moving objects, which is worth further research.

Only GPS and BDS-2 were used in our test so the inclusion of BDS-3 and other systems will be the subject of further research. The integration model of PPP-RTK/INS/vision proposed in this paper is realized by a cascading filter, which can still work when one subsystem is seriously disturbed. The tightly coupled integration of PPP-RTK/INS/vision is another integration method that is worth studying. Because the update frequency of vision is higher, the tightly coupled integration of PPP-RTK/INS/vision may face a heavier computing burden. Therefore, there are still many problems worthy of study in the integration of PPP-RTK/INS/vision.

## 6. Conclusions

To improve the position, velocity, and attitude estimation performance in urban areas for vehicle navigation, a multi-GNSS PPP-RTK/INS/vision integration model with a cascading filter was developed and validated using two vehicular tests, T01 and T02, in urban areas. T01 was conducted in the suburban area of Wuhan City and T02 on the Second Ring Road of Wuhan city. The T02 test can be regarded as a typical GNSS-challenged environment. To obtain the atmospheric corrections and UPD products for PPP-RTK, observations from seven reference stations were also collected for generating those products.

A dynamic object removal model was also proposed and validated with T02. A dynamic object removal model based on position can work well in a GNSS-challenged environment and improve the positioning performance of multi-GNSS PPP-RTK/INS/vision.

PPP-RTK achieved centimeter-level positioning in the horizontal direction and decimeter-level positioning in the vertical direction under a relatively open sky environment such as T01. The performance in the vertical direction was obviously improved when BDS was included with respect to G-PPP-RTK. Moreover, it took only about 30 s for PPP-RTK convergence due to the atmospheric augmentation and ambiguity resolution. However, incorrect ambiguity resolution remained and the position performance became significantly worse in this case. The introduction of the INS weakened the influence of the incorrect ambiguity resolution and improved the positioning accuracy. The positioning error of GC-PPP-RTK/INS was 0.08 m and 0.09 m, with an improvement of 11.1% and 59.1% in the horizontal and vertical directions, respectively, in comparison with GC-PPP-RTK.

The performance of PPP-RTK degraded fast when the GNSS observation environments became complicated and challenging such as T02. G-PPP-RTK could only achieve meter-level positioning in a GNSS-challenged environment. Compared with G-PPP-RTK, the positioning availability was improved from 90.5% to 95.1% for GC PPP-RTK. The GC-PPP-RTK/INS solution, in comparison with the GC-PPP-RTK solution, contributed significantly to the improvement of the positioning accuracy in the horizontal and vertical directions and the positioning availability. The positioning availability of GC-PPP-RTK/INS increased to 100%. The positioning error of GC-PPP-RTK/INS was 0.48 m and 0.73 m in the horizontal and vertical directions, respectively, after excluding the positioning results derived by the INS. The improvements were 31.4% and 37.1% in the horizontal and vertical directions, respectively, in comparison with GC-PPP-RTK.

The two-vehicle tests showed that GC-PPP-RTK/INS could realize high-precision continuous positioning in a relatively open-sky environment but the positioning errors diverged to more than 20 m in a GNSS-challenged environment. Thus, it needed other sensors such as vision to help restrict the error divergence. Vision did not improve the positioning accuracy statistically but further reduced the fluctuation slightly in the vertical direction for T01. The results indicated that the position RMS of the GC-PPP-RTK/INS/vision tightly coupled integration were 0.08 m and 0.09 m in the horizontal and vertical directions, respectively, which could fully meet the demands of vehicle navigation in urban areas. However, the introduction of vision significantly improved the positioning performance in both the horizontal and vertical directions for T02. The RMS of GC-PPP-RTK/INS/vision reached 0.83 m and 0.91 m in the horizontal and vertical directions, respectively. It improved the positioning accuracy by 54.9% and 7.1% in the horizontal and vertical directions, respectively, compared with GC-PPP-RTK/INS. Additionally, the velocity and attitude estimation performance were also analyzed in this paper. The inclusion of vision improved the velocity performance by more than 25% in the north, east, and down directions in a GNSS-challenged environment. As for attitude, there was no obvious difference with vision in the roll and pitch angles, but GC-PPP-RTK/INS/vision performed much better in the estimation of the yaw angle. The improvements brought about by vision were more than 30% in both tests.

The results show that GC-PPP-RTK/INS/vision integration with a cascading filter performs best in the position, velocity, and attitude estimations compared with the other solutions. Multi-GNSS, INS, and vision can play their respective roles and achieve complementary advantages in vehicle navigation in urban areas. However, navigation performance in real-time still deserves further study.

**Author Contributions:** S.G., C.D. and W.F. carried out the research and the experiment. F.M. helped to solve the atmosphere augmentation products and UPD products; S.G. and C.D. analyzed the results and drafted the paper. All authors have read and agreed to the published version of the manuscript.

**Funding:** This research was funded by the National Natural Science Foundation of China, grant number 42174029.

**Data Availability Statement:** The GNSS observation, region atmosphere augmentation products, UPD products, IMU, and vision measurement data can be accessed from sftp://59.172.178.34:10016 (accessed on 10 August 2022), username: tmp-user, password: Multi-sensor@FUSING. For example, input the command "sftp -P 10016 tmp-user@59.172.178.34".

**Conflicts of Interest:** The authors declare no conflict of interest.

## References


1. Gakne, P.; O'Keefe, K. Tightly-Coupled GNSS/Vision Using a Sky-Pointing Camera for Vehicle Navigation in Urban Areas. *Sensors* **2018**, *18*, 1244. [CrossRef] [PubMed]
2. Won, D.H.; Lee, E.; Heo, M.; Lee, S.; Lee, J.; Kim, J.; Sung, S.; Lee, Y.J. Selective Integration of GNSS, Vision Sensor, and INS Using Weighted DOP Under GNSS-Challenged Environments. *IEEE Trans. Instrum. Meas.* **2014**, *63*, 2288–2298. [CrossRef]
3. Mostafa, M.M.; Moussa, A.M.; El-Sheimy, N.; Sesay, A.B. A smart hybrid vision aided inertial navigation system approach for UAVs in a GNSS denied environment. *Navigation* **2018**, *65*, 533–547. [CrossRef]
4. Yue, Z.; Lian, B.; Tang, C.; Tong, K. A novel adaptive federated filter for GNSS/INS/VO integrated navigation system. *Meas. Sci. Technol.* **2020**, *31*, 085102. [CrossRef]
5. Zumberge, J.F.; Heflin, M.B.; Jefferson, D.C.; Watkins, M.M.; Webb, F.H. Precise point positioning for the efficient and robust analysis of GPS data from large networks. *J. Geophys. Res. Solid Earth* **1997**, *102*, 5005–5017. [CrossRef]
6. Kouba, J.; Héroux, P. Precise Point Positioning Using IGS Orbit and Clock Products. *GPS Solut.* **2001**, *5*, 12–28. [CrossRef]
7. Shi, C.; Guo, S.; Gu, S.; Yang, X.; Gong, X.; Deng, Z.; Ge, M.; Schuh, H. Multi-GNSS satellite clock estimation constrained with oscillator noise model in the existence of data discontinuity. *J. Geod.* **2018**, *93*, 515–528. [CrossRef]
8. Lou, Y.; Zheng, F.; Gu, S.; Wang, C.; Feng, Y. Multi-GNSS precise point positioning with raw single-frequency and dual-frequency measurement models. *GPS Solut.* **2016**, *20*, 849–862. [CrossRef]
9. Ge, M.; Gendt, G.; Rothacher, M.; Shi, C.; Liu, J. Resolution of GPS carrier-phase ambiguities in precise point positioning (PPP) with daily observations. *J. Geod.* **2008**, *82*, 389–399. [CrossRef]
10. Laurichesse, D.; Mercier, F.; Berthias, J.P.; Broca, P.; Cerri, L. Integer ambiguity resolution on undifferenced GPS phase measurements and its application to PPP and satellite precise orbit determination. *Navigation* **2009**, *56*, 135–149. [CrossRef]
11. Collins, P.; Bisnath, S.; Lahaye, F.; Héroux, P. Undifferenced GPS ambiguity resolution using the decoupled clock model and ambiguity datum fixing. *Navigation* **2010**, *57*, 123–135. [CrossRef]
12. Geng, J.; Meng, X.; Dodson, A.H.; Teferle, F.N. Integer ambiguity resolution in precise point positioning: Method comparison. *J. Geod.* **2010**, *84*, 569–581. [CrossRef]
13. Schönemann, E.; Becker, M.; Springer, T. A new approach for GNSS analysis in a multi-GNSS and multi-signal environment. *J. Geod. Sci.* **2011**, *1*, 204–214. [CrossRef]
14. Zhang, B.; Teunissen, P.J.G.; Odijk, D. A Novel Un-differenced PPP-RTK Concept. *J. Navig.* **2011**, *64* (Suppl. 1), S180–S191. [CrossRef]
15. Gu, S.; Shi, C.; Lou, Y.; Feng, Y.; Ge, M. Generalized-Positioning for Mixed-Frequency of Mixed-GNSS and Its Preliminary Applications. In *China Satellite Navigation Conference (CSNC) 2013 Proceedings*; Springer: Berlin/Heidelberg, Germany, 2013; Volume 244, pp. 399–428.
16. Gu, S.; Shi, C.; Lou, Y.; Liu, J. Ionospheric effects in uncalibrated phase delay estimation and ambiguity-fixed PPP based on raw observable model. *J. Geod.* **2015**, *89*, 447–457. [CrossRef]
17. Gu, S.; Lou, Y.; Shi, C.; Liu, J. BeiDou phase bias estimation and its application in precise point positioning with triple-frequency observable. *J. Geod.* **2015**, *89*, 979–992. [CrossRef]
18. Zhang, B.; Chen, Y.; Yuan, Y. PPP-RTK based on undifferenced and uncombined observations: Theoretical and practical aspects. *J. Geod.* **2019**, *93*, 1011–1024. [CrossRef]
19. Wübbena, G.; Schmitz, M.; Bagge, A. PPP-RTK: Precise point positioning using state-space representation in RTK networks. In Proceedings of the 18th International Technical Meeting of the Satellite Division of the Institute of Navigation, Long Beach, CA, USA, 13–16 September 2005.
20. European GNSS Agency. *PPP-RTK Market and Technology Report*; European GNSS Agency: Prague, Czech Republic, 2019.
21. Angrisano, A.; Gaglione, S.; Gioia, C. Performance assessment of GPS/GLONASS single point positioning in an urban environment. *Acta Geod. Geophys.* **2013**, *48*, 149–161. [CrossRef]
22. Du, S.; Gao, Y. Integration of PPP GPS and low cost IMU. In Proceedings of the 2010 Canadian Geomatics Conference and Symposium of Commission I, ISPRS, Calgary, AB, Canada, 15–18 June 2010.
23. Gao, Z.; Zhang, H.; Ge, M.; Niu, X.; Shen, W.; Wickert, J.; Schuh, H. Tightly coupled integration of multi-GNSS PPP and MEMS inertial measurement unit data. *GPS Solut.* **2017**, *21*, 377–391. [CrossRef]
24. Rabbou, M.A.; El-Rabbany, A. Tightly coupled integration of GPS precise point positioning and MEMS-based inertial systems. *GPS Solut.* **2015**, *19*, 601–609. [CrossRef]
25. Liu, S.; Sun, F.; Zhang, L.; Li, W.; Zhu, X. Tight integration of ambiguity-fixed PPP and INS: Model description and initial results. *GPS Solut.* **2016**, *20*, 39–49. [CrossRef]
26. Han, H.; Wang, J. Robust GPS/BDS/INS tightly coupled integration with atmospheric constraints for long-range kinematic positioning. *GPS Solut.* **2017**, *21*, 1285–1299. [CrossRef]
27. Gu, S.; Dai, C.; Fang, W.; Zheng, F.; Wang, Y.; Zhang, Q.; Lou, Y.; Niu, X. Multi-GNSS PPP/INS tightly coupled integration with atmospheric augmentation and its application in urban vehicle navigation. *J. Geod.* **2021**, *95*, 64. [CrossRef]
28. Mourikis, A.I.; Roumeliotis, S.I. A multi-state constraint Kalman filter for vision-aided inertial navigation. In Proceedings of the IEEE International Conference on Robotics and Automation, Roma, Italy, 10–14 April 2007; pp. 3565–3572.
29. Li, M.; Mourikis, A.I. High-precision, consistent EKF-based visual-inertial odometry. *Int. J. Robot. Res.* **2013**, *32*, 690–711. [CrossRef]

30. Bloesch, M.; Omari, S.; Hutter, M.; Siegwart, R. Robust visual inertial odometry using a direct EKF-based approach. In Proceedings of the IEEE/RSJ International Conference on Intelligent Robots and Systems, Hamburg, Germany, 28 September–2 October 2015; pp. 298–304.
31. Leutenegger, S.; Lynen, S.; Bosse, M.; Siegwart, R.; Furgale, P. Keyframe-based visual–inertial odometry using nonlinear optimization. *Int. J. Robot. Res.* **2015**, *34*, 314–334. [CrossRef]
32. Qin, T.; Li, P.; Shen, S. VINS-Mono: A Robust and Versatile Monocular Visual-Inertial State Estimator. *IEEE Trans. Robot.* **2018**, *34*, 1004–1020. [CrossRef]
33. Kim, J.; Sukkarieh, S. SLAM aided GPS/INS navigation in GPS denied and unknown environments. *Positioning* **2005**, *4*, 120–128. [CrossRef]
34. Won, D.H.; Lee, E.; Heo, M.; Sung, S.; Lee, J.; Lee, Y.J. GNSS integration with vision-based navigation for low GNSS visibility conditions. *GPS Solut.* **2014**, *18*, 177–187. [CrossRef]
35. Liu, F. Tightly Coupled Integration of GNSS/INS/Stereo Vision/Map Matching System for Land Vehicle Navigation. Unpublished. Doctoral Thesis, University of Calgary, Calgary, AB, USA, 2018.
36. Li, T.; Zhang, H.; Gao, Z.; Niu, X.; El-sheimy, N. Tight Fusion of a Monocular Camera, MEMS-IMU, and Single-Frequency Multi-GNSS RTK for Precise Navigation in GNSS-Challenged Environments. *Remote Sens.* **2019**, *11*, 610. [CrossRef]
37. Liu, H.; Liu, G.; Tian, G.; Xin, S.; Ji, Z. Visual SLAM based on dynamic object removal. In Proceedings of the 2019 IEEE International Conference on Robotics and Biomimetics (ROBIO), Dali, China, 6–8 December 2019; pp. 596–601.
38. Sun, Y.; Liu, M.; Meng, M.Q.H. Improving RGB-D SLAM in dynamic environments: A motion removal approach. *Robot. Auton. Syst.* **2017**, *89*, 110–122. [CrossRef]
39. Zhao, Q.; Wang, Y.T.; Gu, S.; Zheng, F.; Shi, C.; Ge, M.; Schuh, H. Refining ionospheric delay modeling for undifferenced and uncombined GNSS data processing. *J. Geod.* **2019**, *93*, 545–560. [CrossRef]
40. Gu, S.; Wang, Y.; Zhao, Q.; Zheng, F.; Gong, X. BDS-3 differential code bias estimation with undifferenced uncombined model based on triple-frequency observation. *J. Geod.* **2020**, *94*, 45. [CrossRef]
41. Teunissen, P. The least-squares ambiguity decorrelation adjustment a method for fast GPS integer ambiguity estimation. *J. Geod.* **1995**, *70*, 65–82. [CrossRef]
42. Shin, E.H. *Estimation Techniques for Low-Cost Inertial Navigation*; UCGE Report 20219; University of Calgary: Calgary, AB, Canada, 2005.
43. Böhm, J.; Möller, G.; Schindelegger, M.; Pain, G.; Weber, R. Development of an improved empirical model for slant delays in the troposphere (GPT2w). *GPS Solut.* **2015**, *19*, 433–441. [CrossRef]
44. Shi, C.; Gu, S.; Lou, Y.; Ge, M. An improved approach to model ionospheric delays for single-frequency precise point positioning. *Adv. Space Res.* **2012**, *49*, 1698–1708. [CrossRef]

*remote sensing*

*Article*

# Assessment of Real-Time GPS/BDS-2/BDS-3 Single-Frequency PPP and INS Tight Integration Using Different RTS Products

**Jie Lv, Zhouzheng Gao *, Qiaozhuang Xu, Ruohua Lan, Cheng Yang and Junhuan Peng**

School of Land Science and Technology, China University of Geosciences Beijing, 29 Xueyuan Road, Beijing 100083, China
* Correspondence: zz.gao@cugb.edu.cn; Tel.: +86-186-2791-5172

**Abstract:** Due to the virtues of low-cost and high positioning accuracy, Single-Frequency Precise Point Positioning (SF-PPP) is becoming a prospective technique. However, SF-PPP is not as widely used as dual-frequency and triple-frequency PPP at present, owing to the effect of ionospheric delay residuals after model rectification. In recent years, with the evolution of multi-constellation Global Navigation Satellite Systems (multi-GNSS, i.e., GPS, BDS-2, and BDS-3), it has become possible to obtain credible and continuous positioning results using SF-PPP. However, such performance would be significantly degraded in challenging environments (i.e., boulevards, tunnels, and tall buildings). Under these circumstances, GNSS signals are obstructed, and it is difficult to provide sufficient observations for SF-PPP. Therefore, the Inertial Navigation System (INS) is employed to promote the positioning performance of SF-PPP. The PPP/INS integration is regarded as one of the most efficient approaches in GNSS-denied environments. To satisfy the request of supplying real-time positioning information, the Real-Time Services (RTS) of the International GNSS Service (IGS) provide real-time precise orbit and clock products for globally distributed users through the internet. In this paper, a real-time GPS/BDS-2/BDS-3 SF-PPP and INS tight integration model is proposed, and it is assessed using the data gathered by vehicle and real-time products afforded by CAS (Chinese Academy of Sciences), GFZ (Deutsche GeoForschungsZentrum), and WHU (Wuhan University). The outcomes illustrate the following: (1) GPS + BDS SF-PPP/INS can provide more accurate and continuous positioning solutions compared with those of GPS + BDS SF-PPP, with improvements of 52.8%, 31.1%, and 42.8% in the north, east, and vertical components, respectively. (2) In general, the orbit and clock products' accuracies in terms of GPS afforded by the three analysis centers are consistent with each other. For BDS, the orbit product from WHU is more accurate compared to those of CAS and GFZ. However, the accuracy of the clock product afforded by WHU is lower compared with those provided by the other two centers, especially for BDS-2 satellites. (3) The positioning accuracy in terms of Root Mean Square (RMS) values based on GFZ products are much higher than the results based on CAS and WHU products in the three directions.

**Keywords:** Single-Frequency Precise Point Positioning (SF-PPP); Inertial Navigation System (INS); multi-constellation Global Navigation Satellite Systems (multi-GNSS); real-time tight integration

**Citation:** Lv, J.; Gao, Z.; Xu, Q.; Lan, R.; Yang, C.; Peng, J. Assessment of Real-Time GPS/BDS-2/BDS-3 Single-Frequency PPP and INS Tight Integration Using Different RTS Products. *Remote Sens.* **2022**, *14*, 4367. https://doi.org/10.3390/rs14174367

Academic Editors: Chuang Shi, Shengfeng Gu, Yidong Lou and Xiaopeng Gong

Received: 16 July 2022
Accepted: 31 August 2022
Published: 2 September 2022

**Publisher's Note:** MDPI stays neutral with regard to jurisdictional claims in published maps and institutional affiliations.

**Copyright:** © 2022 by the authors. Licensee MDPI, Basel, Switzerland. This article is an open access article distributed under the terms and conditions of the Creative Commons Attribution (CC BY) license (https://creativecommons.org/licenses/by/4.0/).

## 1. Introduction

Since Precise Point Positioning (PPP) was first introduced [1,2], it has been a popular tool in numerous applications, for instance, GNSS (Global Navigation Satellite System) seismology [3,4], GNSS meteorology [5,6], deformation monitoring [7,8], etc. PPP utilizes only single-station GNSS data and orbit/clock products with high precision supplied via the International GNSS Service (IGS) centers to provide users with centimeter-level positioning solutions, and the errors contained in observations are remedied by appointed models, treated as unknown parameters, or parameterized after model correction. The research presented in [9–11] has proved the advantages of PPP, especially after applying the ambiguity fixing model [12,13] and the slant ionospheric delay and receiver Differential

Code Bias (DCB) constraint models [14,15]. However, most of the works conducted thus far mainly consider conditions of dual-frequency observations. In contrast, Single-Frequency PPP (SF-PPP) is not as widely applied as dual-/triple-frequency PPP [16]. This is chiefly owing to the fact that no effective model can be utilized in SF-PPP to decrease or eliminate the effect of the ionospheric delay. Hence, this may directly lead to low positioning accuracy and long convergence time, especially in kinematic conditions.

Recently, GNSS satellite availability has significantly improved, along with the evolution of multi-constellation GNSS, and obtaining reliable and continuous positioning solutions using SF-PPP has become possible. The performance of three SF-PPP models, namely the GRoup And PHase Ionosphere Correction (GRAPHIC) model, GRAPHIC with code observation model, and an ionosphere-constrained model were analyzed in [17]. The contribution of the QZSS (Quasi-Zenith Satellite System) to GPS/BDS/GLONASS/Galileo SF-PPP was also presented. The outcomes proved that QZSS significantly increased the positioning accuracy of BDS and GLONASS when using simulated dynamic data. Such improvement was invisible while using the GPS-only, Galileo-only, and GPS/BDS/GLONASS/Galileo data. Furthermore, in the work of [17], the SF-PPP with ionosphere constraints presented a superior convergence performance, while the positioning accuracies supplied by the three SF-PPP models were close to each other. Moreover, with the completion of the BDS-3 constellation [18] on 23 June 2020, the modeling and performance provided by BDS-2/BDS-3 SF-PPP have become a new topic. Based on data from global distribution stations, Shi et al. [18] presented a BDS-2/BDS-3 SF-PPP model and studied the improvement of the combination based on BDS-2 and BDS-3 in precise positioning. The results indicated that the positioning accuracy was relevant to the receiver type adopted in different GNSS networks. The positioning performance based on the B1I signal is superior to that based on the B1C signal from the stations of the international GNSS Monitoring and Assessment System (iGMAS), while it is opposite at the stations of the Multi-GNSS EXperiment (MGEX). Contrasted with BDS-2 SF-PPP and BDS-3 SF-PPP, the convergence performance of BDS-2 + BDS-3 SF-PPP is significantly increased. As a major error in SF-PPP, ionospheric delay must be considered carefully. For ionospheric studies, Su et al. [19] proposed a novel single-frequency ionospheric-free-half PPP technique. To validate the reliability of the method, the performances of two conventional PPP models, and the proposed method, were evaluated. Results showed that slant ionospheric observables can be extracted with submeter-level accuracy towards the proposed novel method.

However, the continuous satellite availability will decrease in most dynamic applications when the receiver passes through overbridges, tunnels, and near towering constructions. Under these circumstances, the availability of GNSS satellites will be obstructed, and it is difficult to provide sufficient observations for PPP calculation. Therefore, high positioning accuracy with SF-PPP is difficult to achieve. Fortunately, according to [20], GPS positioning performance can be improved by fusing GPS with an Inertial Navigation System (INS), especially in challenging environments. In the last few years, profiting from the rapid development of PPP, PPP/INS integration has been regarded as one of the most efficient approaches to provide positions and attitudes. However, there are just a small number of works on SF-PPP/INS tight integration. Gao et al. [21] presented a multi-GNSS tight integration technique for single-frequency measurements and IMU outputs, by which the performance supplied by the ionospheric-delay- and receiver-DCB-constrained model was ameliorated significantly. The results indicated that the proposed model can obtain more accurate, continuous, and credible solutions in open-sky and GNSS-denied environments compared with traditional SF-PPP. The work in [22] presented a GNSS/INS tight integration with augmentations on ionospheric delay and tropospheric delay. Results illustrated that tight integration has a great effect in urban environments. Regional atmospheric augmentation increases the positioning accuracy of SF-PPP significantly on account of the sensibility of single-frequency signals to the ionospheric delay.

With the development of society, the requirement for real-time positioning is becoming urgent. To satisfy this demand, the Real-Time Services (RTS) of the IGS afford precise orbit and clock products to global users through the internet. The work in [23] verified that the RTS products are, with high accuracy, comparing with the final products of the ESA (European Space Agency, Paris, France). Subsequently, the work in [24] appraised the positioning performance of BDS-2-only real-time PPP based on RTS products (CLK93) and MGEX stations. Results showed that the positioning accuracy acquired can be centimeter-/decimeter-level in static/kinematic environments. Kazmierski et al. [25] further investigated the quality of real-time products of CNES (Centre National d'Etudes Spatiales) for GPS, GLONASS, Galileo, and BDS-2. Results illustrated that the orbits and clocks of the GPS are the most accurate currently. Based on these backgrounds, this paper introduces a real-time GPS/BDS-2/BDS-3 SF-PPP/INS tight integration model. It is assessed via a set of vehicle-borne data and real-time orbit and clock products afforded by the IGS centers of CAS (Chinese Academy of Sciences), GFZ (Deutsche GeoForschungsZentrum), and WHU (Wuhan University). The paper is organized as follows: The methodology of the GPS/BDS-2/BDS-3 SF-PPP/INS tight integration model is described in Section 2. Then, the evaluations and the conclusions are presented in Sections 3 and 4.

## 2. Methodology

In this section, the methods of undifferenced and uncombined GPS/BDS-2/BDS-3 SF-PPP, recovery of real-time orbit and clock products, GPS/BDS-2/BDS-3 SF-PPP/INS tight integration, and parameter adjustments are described in detail.

### 2.1. Real-Time SF-PPP Model

The linearized observation functions for the undifferenced and uncombined pseudorange and carrier phase are expressed as [21]

$$P^s_{r,j} = \mathbf{u}^s_r \cdot \mathbf{x} + ct_r + \mathrm{Mw}^s_r \cdot \mathrm{ZWD}_r + \gamma_j \cdot I^s_{r,1} + cd_{r,j} + e^s_{r,j} \quad (1)$$

$$L^s_{r,j} = \mathbf{u}^s_r \cdot \mathbf{x} + ct_r + \mathrm{Mw}^s_r \cdot \mathrm{ZWD}_r - \gamma_j \cdot I^s_{r,1} - \lambda_j N^s_{r,j} + \varepsilon^s_{r,j} \quad (2)$$

where $P^s_{r,j}$ and $L^s_{r,j}$ refer to observed minus computed values of the original pseudorange and carrier phase observations; $s$ represents satellite, $r$ is the receiver, and $j$ is signal frequency; $\mathbf{x}$ is the receiver position increments vector; $\mathbf{u}^s_r$ is the direction cosine between receiver and satellite; $t_r$ is the receiver clock offset; $\mathrm{ZWD}_r$ and $\mathrm{Mw}^s_r$ are the residual of tropospheric delay and the relevant mapping function; $I^s_{r,1}$ is the ionospheric delay at frequency 1 when the signal is passing through the propagation path; $\gamma_j$ is the multiplier factor dependent on signal frequency; $d_{r,j}$ represents receiver hardware time delay; $c$ is the velocity of light; $\lambda_j$ is the wavelength and $N^s_{r,j}$ is integer ambiguity; $e^s_{r,j}$ and $\varepsilon^s_{r,j}$ represent multipath biases, which are unmodeled, and the observation noise from the pseudorange and carrier phase, respectively.

The ionospheric delay is a key section that influences the performance of SF-PPP. Therefore, several models, for instance, the GIM (Global Ionosphere Map) data correction model [26], the single-frequency GRAPHIC model [27], and the single-frequency ZIDE (Zenith Ionospheric Delay Estimation) PPP model [28], have been proposed to limit such influence. For example, Øvstedal [26] recommended using the GIM to amend the ionospheric delay of the pseudorange and carrier phase. However, the drawback of the GIM-correction-based SF-PPP is that the residual of each satellite's ionospheric delay and the undisposed receiver DCB will degrade the performance of SF-PPP. To overcome this drawback, Montenbruck [29] tried to use the GRAPHIC model [27] instead of GIM data to eliminate the ionospheric delay in the carrier phase. In the GRAPHIC model, the characteristic that the ionospheric delays on pseudorange and carrier phase are equal in magnitude and opposite in sign for the same satellite is utilized. Hence, a linear ionospheric-free combination between pseudorange and carrier phase is adopted to form a new ionospheric-free carrier phase observation. Even so, the SF-PPP positioning accuracy still cannot satisfy the accuracy requirement because of the large noise of the new carrier phase [30]. Therefore, based on the GRAPHIC model, Beran et al. [28,31] proposed the ZIDE model to estimate the zenith ionospheric delay of each satellite as a parameter, which has been verified to be effective to increase the performance of SF-PPP [30]. According to the works of [32–34], the position accuracy and convergence time of ZIDE SF-PPP can be ameliorated by adopting precise global and regional ionospheric models. However, the receiver DCB on pseudorange is ignored in the ZIDE model. Hence, the SF-PPP model based on Ionospheric delay and Receiver DCB Constraint (IRC) [21] is applied in this contribution. Here, to separate the receiver DCB from the ionospheric delay [14], the real-time ionospheric product provided by Wuhan University is utilized to generate a virtual external observation for each ionospheric delay. The basic observation equation can be written as

$$\begin{aligned} P^G_{r,f_1} &= \mathbf{u}^G_r \cdot \mathbf{x} + \delta t_r + \mathrm{Mw}^G_r \cdot \mathrm{ZWD}_r + I^G_{r,f_1} + d^G_{r,f_1} \\ P^B_{r,f_1} &= \mathbf{u}^B_r \cdot \mathbf{x} + \delta t_r + \mathrm{Mw}^B_r \cdot \mathrm{ZWD}_r + I^B_{r,f_1} + d^B_{r,f_1} \end{aligned} \tag{3}$$

$$\begin{aligned} L^G_{r,f_1} &= \mathbf{u}^G_r \cdot \mathbf{x} + \delta t_r + \mathrm{Mw}^G_r \cdot \mathrm{ZWD}_r - I^G_{r,f_1} - \lambda^G_{f_1} N^G_{r,f_1} \\ L^B_{r,f_1} &= \mathbf{u}^B_r \cdot \mathbf{x} + \delta t_r + \mathrm{Mw}^B_r \cdot \mathrm{ZWD}_r - I^B_{r,f_1} - \lambda^B_{f_1} N^B_{r,f_1} \end{aligned} \tag{4}$$

$$I^s_{r,f_1} = 40.28 \cdot \mathrm{VTEC} / \left( f_1^2 \cos(Z_\theta) \right) + \varepsilon_{I^s_{r,f_1}}, \varepsilon_{I^s_{r,f_1}} \sim N\left(0, \sigma^2_{\varepsilon_{I^s_{r,f_1}}}\right) \tag{5}$$

$$\begin{bmatrix} d_{DCB} \\ 0 \end{bmatrix} = \begin{bmatrix} 1 & -1 \\ \frac{f_1^2}{f_1^2 - f_2^2} & -\frac{f_2^2}{f_1^2 - f_2^2} \end{bmatrix} \begin{bmatrix} d_1 \\ d_2 \end{bmatrix} \tag{6}$$

where VTEC and $Z_\theta$ stand for vertical total electron content acquired from the WHU's real-time ionospheric product and the zenith angle at the Ionospheric Punctuation Point (IPP); $\varepsilon_{I^s_{r,f_1}}$ is the accuracy of the real-time ionospheric product with prior variance of $\sigma^2_{\varepsilon_{I^s_{r,f_1}}}$; $d_{DCB}$ is the DCB between pseudoranges $P_1$ and $P_2$ in the same GNSS system; $f_1$ and $f_2$ are signal frequencies; $d_1$ and $d_2$ are hardware time delays in frequencies 1 and 2; the meanings of other parameters are the same as those in Equations (1) and (2).

In real-time data processing, products from IGS are transmitted into the RTCM-SSR (Radio Technical Commission for Maritime Services State Space Representation) form. The real-time corrections are the differences between precise products and broadcast products. The real-time products cannot be applied in positioning directly. Therefore, the products must be recovered to satisfy the requirement of PPP. SSR corrections for real-time orbit at the present epoch $t$ can be calculated by corrections at the reference epoch $t_0$ using [35]

$$\begin{bmatrix} \Delta_r \\ \Delta_a \\ \Delta_c \end{bmatrix}_t = \begin{bmatrix} \Delta_r \\ \Delta_a \\ \Delta_c \end{bmatrix}_{t_0} + \begin{bmatrix} \dot{\Delta}_r \\ \dot{\Delta}_a \\ \dot{\Delta}_c \end{bmatrix} (t - t_0) \tag{7}$$

where $\Delta_r$, $\Delta_a$, and $\Delta_c$ are satellite position corrections in radial, along, and cross directions; $\dot{\Delta}_r$, $\dot{\Delta}_a$, and $\dot{\Delta}_c$ are the corresponding velocities.

The rotation matrix $\boldsymbol{R}$ between the orbital coordinate system and Earth-Centered Earth-Fixed (ECEF) frame can be expressed as

$$\boldsymbol{R} = \begin{bmatrix} \frac{\dot{\boldsymbol{r}}}{|\dot{\boldsymbol{r}}|} \times \frac{\boldsymbol{r} \times \dot{\boldsymbol{r}}}{|\boldsymbol{r} \times \dot{\boldsymbol{r}}|} & \frac{\dot{\boldsymbol{r}}}{|\dot{\boldsymbol{r}}|} & \frac{\boldsymbol{r} \times \dot{\boldsymbol{r}}}{|\boldsymbol{r} \times \dot{\boldsymbol{r}}|} \end{bmatrix} \tag{8}$$

where $\boldsymbol{r}$ and $\dot{\boldsymbol{r}}$ stand for position and velocity vector for satellites calculated from the broadcast ephemeris, respectively.

The orbit corrections can be transferred from the orbital coordinate system to the ECEF frame via

$$\begin{bmatrix} \Delta_x \\ \Delta_y \\ \Delta_z \end{bmatrix}_t = \boldsymbol{R} \cdot \begin{bmatrix} \Delta_r \\ \Delta_a \\ \Delta_c \end{bmatrix}_t \tag{9}$$

where $\Delta_x$, $\Delta_y$, and $\Delta_z$ are satellite orbit corrections in the ECEF frame.

Then, the precise products can be recovered by combining orbit corrections with satellite positions calculated by broadcast ephemeris

$$\begin{bmatrix} X_{pre} \\ Y_{pre} \\ Z_{pre} \end{bmatrix}_t = \begin{bmatrix} X_{brd} \\ Y_{brd} \\ Z_{brd} \end{bmatrix}_t - \begin{bmatrix} \Delta_x \\ \Delta_y \\ \Delta_z \end{bmatrix}_t \tag{10}$$

where $X_{pre}$, $Y_{pre}$, and $Z_{pre}$ are satellite coordinates in ECEF frame at the time $t$ after SSR corrections; $X_{brd}$, $Y_{brd}$, and $Z_{brd}$ are satellite coordinates calculated from broadcast ephemeris.

For real-time clock offset, the SSR corrections at the present epoch $t$ can be calculated by corrections at the reference epoch $t_0$ by [35]

$$\Delta t_C = C_0 + C_1(t - t_0) + C_2(t - t_0)^2 \tag{11}$$

where $\Delta t_C$ is the real-time clock correction in distance form calculated from SSR clock offsets. Then, the real-time precise clock can be obtained by

$$t_{pre}^{sat} = t_{brd}^{sat} - \frac{\Delta t_C}{c} \tag{12}$$

where $t_{\text{pre}}^{\text{sat}}$ is the satellite offset at epoch $t$ after SSR corrections.

The signal–noise ratio, GNSS satellite elevation, as well as GNSS measurement environment are chiefly factors to influence the observation quality. Considering these terms, the conventional satellite elevation-dependent weight function is utilized here to figure out the prior variance [36]

$$\sigma^2 = \begin{cases} \sigma_0^2 & E \geq 30^\circ \\ \sigma_0^2/(2 \cdot \sin(E))^2 & \text{else} \end{cases} \tag{13}$$

where $E$ and $\sigma_0^2$ are the elevation of satellites and the corresponding prior variance.

### 2.2. GPS/BDS-2/BDS-3 SF-PPP/INS Tight Integration Model

In the GPS/BDS-2/BDS-3 SF-PPP/INS tight integration model, the theoretical geometrical distance in SF-PPP can be replaced by the INS-predicted values. Meanwhile, the doppler observation was proved to be effective to estimate the INS sensor errors [21]. Hence, the corresponding functions can be expressed as:

$$\begin{bmatrix} \boldsymbol{P}_{GNSS,f_1} - \boldsymbol{P}_{INS,f_1} - l_p^s \\ \boldsymbol{L}_{GNSS,f_1} - \boldsymbol{L}_{INS,f_1} - l_p^s \\ \boldsymbol{D}_{GNSS,f_1} - \boldsymbol{D}_{INS,f_1} - l_v^s \\ \boldsymbol{I}_{r,f_1}^s - \boldsymbol{I}_{INS,r,f_1}^s \\ d_{r,f_1}^s - d_{INS,r,f_1}^s \end{bmatrix} = \begin{bmatrix} \mathbf{u}_r^s \cdot \mathbf{x} + \delta t_r + \mathrm{Mw}_r^s \cdot \delta \mathrm{ZWD}_r + \delta I_{r,f_1}^s + \delta d_{r,f_1}^s \\ \mathbf{u}_r^s \cdot \mathbf{x} + \delta t_r + \mathrm{Mw}_r^s \cdot \delta \mathrm{ZWD}_r - \delta I_{r,f_1}^s - \lambda_{f_1}^s \delta N_{r,f_1}^s \\ \mathbf{u}_r^s \delta v_r + \dot{\delta t}_r \\ \delta I_{r,f_1}^s \\ \delta d_{r,f_1}^s \end{bmatrix} \tag{14}$$

where $l_p^s$ and $l_v^s$ are the lever arms on position and velocity; $\boldsymbol{D}$ stands for the doppler observations with the unit of m/s.

In general, the mathematical model of INS mechanization can be written as [21]

$$\begin{bmatrix} \boldsymbol{v}_{INS,t_k}^n \\ \boldsymbol{p}_{INS,t_k}^n \\ \boldsymbol{C}_{b,t_k}^n \end{bmatrix} = \int_{t_{k-1}}^{t_k} \begin{bmatrix} \boldsymbol{f}^n - (\boldsymbol{\omega}_{ie}^n + \boldsymbol{\omega}_{in}^n) \times \mathbf{v}_{INS,t_{k-1}}^n + \boldsymbol{g}^n \\ \mathbf{v}_{INS,t_{k-1}}^n \\ (\boldsymbol{\omega}_{ib}^n \times) - (\boldsymbol{\omega}_{in}^n \times) \boldsymbol{C}_{b,t_{k-1}}^n \end{bmatrix} dt \tag{15}$$

where $\boldsymbol{\omega}_{ib}^n$ and $\boldsymbol{f}^n$ are the angular rate and specific force that measured by accelerometer and gyroscope; $\boldsymbol{p}_{INS,t_k}^n$ and $\mathbf{v}_{INS,t_k}^n$ represent the position and velocity, computed via INS; $\boldsymbol{C}_{b,t_k}^n$ is the transform matrix for attitude from body frame ($b$) to navigation frame ($n$); $\boldsymbol{C}_b^n$ is the direction cosine matrix of attitude, which can be described based on the Euler angle;

$\int_{t_{k-1}}^{t_k}()dt$ is the integral operation from epoch $t_{k-1}$ to $t_k$; $\omega_{in}^n$ and $\omega_{ie}^n$ are the rotation angular rate of the $n$ frame and ECEF frame in terms of inertial frame ($i$) projected in the $n$ frame; $g^n$ is the gravity in the $n$ frame. Then, the basic observational functions for SF-PPP/INS tight integration can be expressed as

$$\begin{bmatrix} P_{INS,r,f_1}^s \\ L_{INS,r,f_1}^s \\ D_{INS,r,f_1}^s \end{bmatrix} + \begin{bmatrix} l_p^s \\ l_p^s \\ l_p^s \\ l_v^s \end{bmatrix} = \begin{bmatrix} \left\| p^s - \left( p_{INS}^e + C_n^e C_b^n l_{INS-GNSS}^b \right) \right\| + \Delta P_{f_1}^s \\ \left\| p^s - \left( p_{INS}^e + C_n^e C_b^n l_{INS-GNSS}^b \right) \right\| - \lambda_{f_1}^s N_{r,f_1}^s + \Delta L_{f_1}^s \\ \left\| v^s - v_{INS}^e C_n^e \left( (\omega_{in}^n \times) C_b^n l_{INS-GNSS}^b + C_b^n \left( l_{INS-GNSS}^b \times \right) \omega_{ib}^b \right) \right\| + \Delta D_{f_1}^s \end{bmatrix} \tag{16}$$

where $p^s$ and $v^s$ refer to position and velocity of a satellite in the $e$ frame computed by the real-time satellite orbit and clock products; $p_{INS}^e$ and $\mathbf{v}_{INS}^e$ stand for the position at the IMU center in the geodetic coordinate system and velocity at the IMU center in $e$ frame, which can be obtained from $p_{INS}^n$ and $\mathbf{v}_{INS}^n$; $C_n^e$ is the transform matrix from the $n$ frame to the $e$ frame; $l_{INS-GNSS}^b$ is the lever arm measured from the IMU center to the GNSS receiver antenna phase center in $b$ frame; $\Delta P_{f_1}^s$, $\Delta L_{f_1}^s$, and $\Delta D_{f_1}^s$ represent the sum of the error corrections of the pseudorange, carrier phase, and doppler.

As illustrated above, the INS solutions and GNSS observations are in different frames ($n$ frame for INS and $e$ frame for GNSS). Therefore, to utilize the information on the same foundation, the following corrections on position and velocity can be introduced:

$$\begin{bmatrix} \delta p_r \\ \delta \mathbf{v}_r \end{bmatrix} = \begin{bmatrix} C_1 \delta p_{INS}^n + C_1 \left( C_b^n l_{INS-GNSS}^b \times \right) \delta\theta \\ C_2 \delta p_{INS}^n + C_n^e \delta v_{INS}^n - C_n^e \gamma_1 \delta\theta + C_n^e C_b^n \left( l_{INS-GNSS}^b \times \right) \delta\omega_{ib}^b \end{bmatrix} \tag{17}$$

where $C_1$ is the transform matrix between the geodetic coordinate frame and the $e$ frame of the position increments; $C_2$ stands for the coefficient relevant to position derived from $\delta(C_n^e v_r^n)$ [21]; $\gamma_1$ is the coefficient related to attitude; $\delta\omega_{ib}^b$ denotes the gyroscope errors (bias and scale factor). In general, this parameter can be expressed as [37]

$$\begin{bmatrix} \delta\omega_{ib}^b \\ \delta f^b \end{bmatrix} = \begin{bmatrix} S_g & 0 \\ 0 & S_a \end{bmatrix} \begin{bmatrix} \omega_{ib}^b \\ f^b \end{bmatrix} + \Delta t \begin{bmatrix} B_g \\ B_a \end{bmatrix} \tag{18}$$

where $\Delta t$ indicates the IMU observations interval.

The state vector of the multi-GNSS SF-PPP/INS model can be described as

$$X = \begin{bmatrix} \delta p_{INS}^n & \delta \mathbf{v}_{INS}^n & \delta\theta & B_g & B_a & S_g & S_a & \delta t_r & \mathrm{ZWD}_r & \delta d_{r,f_1}^s & \delta N_{r,f_1}^s & \delta I_{r,f_1}^s \end{bmatrix} \tag{19}$$

where $\delta t_r$ and $\delta d_{r,f_1}^s$ refer to the parameters related to receiver clock error and receiver DCB errors; $\delta N_{r,f_1}^s$ and $\delta I_{r,f_1}^s$ denote the ambiguities and ionospheric delays in the slant propagation path.

The PSI angle model [21] is adopted to describe the variations of position, velocity, and attitude in the temporal domain, which can be described as

$$\begin{bmatrix} \delta\dot{p}_{INS}^n \\ \delta\dot{v}_{INS}^n \\ \dot{\theta} \end{bmatrix} = \begin{bmatrix} \omega_{en}^n \times \delta p^n + \delta v_{INS}^n \\ f^n \times \theta + C_b^n \delta f^b - (2\omega_{ie}^n + \omega_{en}^n) \times \delta v_{INS}^n + \delta g^n \\ -(\omega_{ie}^n + \omega_{en}^n) \times \theta - C_b^n \delta\omega_{ib}^b \end{bmatrix} \tag{20}$$

where the meanings of the parameters are the same as those mentioned above. The variation of IMU errors can be described by the first-order Gauss–Markov process. Meanwhile, the random constant process is adopted to express the variation of float ambiguities. The random walk process is chosen as the dynamic model for the receiver-clock-related parameters and the atmosphere-related parameters.

Finally, the parameterized elements can be estimated by the Extended Kalman Filter (EKF) [38]

$$\begin{bmatrix} X_k \\ P_k \end{bmatrix} = \begin{bmatrix} \Phi_{k,k-1}X_{k-1} + K_k(Z_k - H_k\Phi_{k,k-1}X_{k-1}) \\ (\mathbf{I} - K_k)\left(\Phi_{k,k-1}P_{k-1}\Phi_{k,k-1}^T + Q_{k-1}\right)(\mathbf{I} - K_k)^T + K_kR_kK_k^T \end{bmatrix} \tag{21}$$

where $\mathbf{I}$ is the unit matrix. The elements in $\Phi_{k,k-1}$ can be acquired from the state models mentioned above; $K$ refers to the gain matrix; $Q$ is the state noise variance.

### 2.3. Implementation of SF-PPP/INS Tight Integration Model

Based on the descriptions above, the structure of the proposed real-time multi-GNSS SF-PPP/INS tight integration model is presented in Figure 1. Velocity and angular increments are provided by IMU sensors. After initializing the system, the compensated IMU outputs are processed in the INS mechanization to supply position, velocity, and attitude information. Then, the information, along with real-time GNSS products, is utilized to obtain the GNSS observation predictions (pseudorange, carrier phase, and doppler). After this, the INS-predicted observations are fused with the original observations provided by GNSS in the EKF. Finally, the estimated IMU errors are fed back to the IMU outputs before INS mechanization. Meanwhile, the navigation information is corrected.

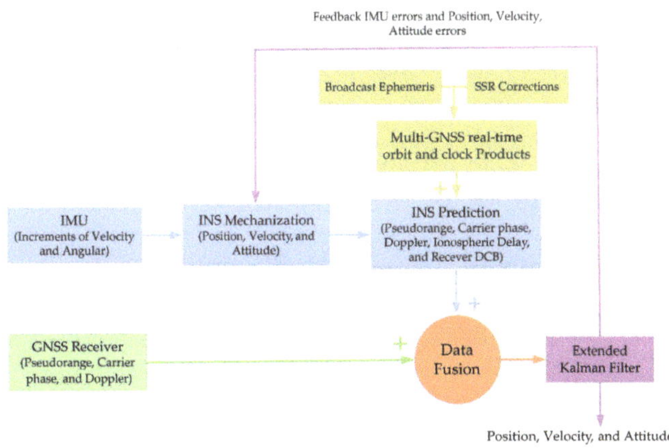


**Figure 1.** Implementation of real-time multi-GNSS SF-PPP/INS tight integration model.

## 3. Tests, Results, and Discussions

To validate the performance of the presented real-time GPS/BDS-2/BDS-3 SF-PPP/INS tight integration, the data of the GNSS original observations and IMU outputs were processed and analyzed. The first subsection demonstrates the positioning performance of GPS + BDS (G + B) SF-PPP and SF-PPP/INS tight integration based on IGS' final precise products under kinematic conditions. The second subsection is to evaluate the performance while using real-time orbit and clock products from CAS, GFZ, and WHU.

### 3.1. Data Collection

The original test data for GPS (L1) and BDS-2/BDS-3 (B1I) were gathered on 21 December 2021, at China University of Geosciences Beijing. Figure 2 (top left and top right) shows the typical scenery in data collection areas, which were mainly around large buildings and boulevards. In these environments, the GNSS signal is heavily blocked. Figure 2 (bottom left) shows the test platform and equipment. Figure 2 (bottom right) presents the mission route. The multi-GNSS multi-frequency PANDA PD318 receiver was adopted in this mission, the sampling rate of which was 1 Hz. IMU outputs were provided by POS320, and the output rate was 200 Hz. Table 1 presents the details of POS320.

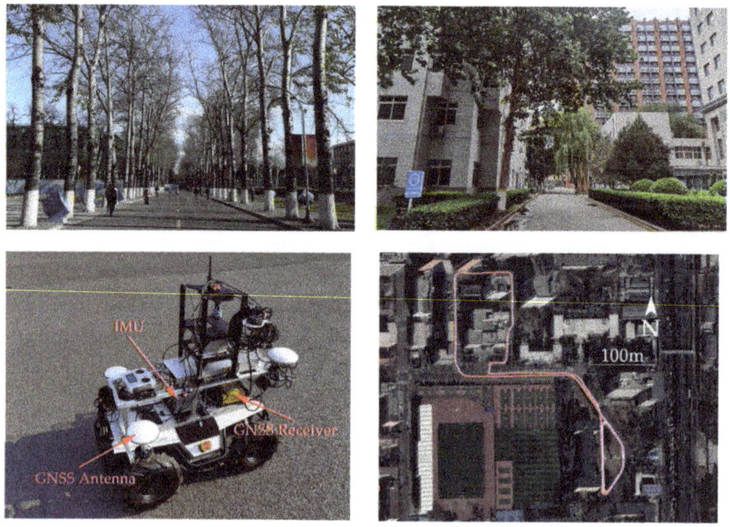


**Figure 2.** Mission details of vehicle test on 21 December 2021, in Beijing, China. Typical scenery in data collection areas (top left and top right); test platform and equipment (bottom left); and mission trajectory (bottom right).

**Table 1.** Technical parameters of POS320.

| IMU Sensor | Bias | | Random Walk | |
|---|---|---|---|---|
| | Gyro. (°/h) | Acce. (mGal) | Angular (°/$\sqrt{h}$) | Velocity (m/s/$\sqrt{h}$) |
| POS320 | 0.5 | 25 | 0.05 | 0.10 |

For data processing, the final precise orbit and clock products were afforded by Wuhan University, and the real-time products were provided by CAS, GFZ, and WHU. The DCB product was from CAS. The satellite cutoff elevation was intercalated as 10°. The satellite antenna phase center offsets were rectified by adopting igs14_2076_plus.atx. The slant ionospheric delay was corrected first using WHU's GIM data. Afterward, the residuals were parameterized as random walks. Receiver DCB, drifts and offsets of the receiver clock, and ISB were also modeled as random walk processes. Moreover, float ambiguities were parameterized as random constants. In INS data processing, coning offsets, rotational, and sculling effect engendered from inertial axes motion were rectified using the INS mechanization [39]. According to the research in [40], to reduce the impact of receiver Time Delay Bias (TDB) between BDS-2 and BDS-3, the weight ratio of BDS-2 and BDS-3 MEOs was set to 1:3. The weight ratio of BDS-2 GEOs and MEOs/IGSOs was set to 1:10. In the validation stage, the results provided by the smoothed RTK/INS tight integration were utilized as reference values.

Figure 3 describes the numbers of available satellites and relevant Position Dilution of Precision (PDOP) of GPS-only, BDS-only, and G + B modes. The numbers of satellites on average were 5.1, 12.4, and 16.2, respectively. The corresponding PDOPs on average were 3.35, 4.27, and 2.99, respectively. These results illustrate that frequent signal blocks occurred under the GPS-only mode. BDS satellites could remedy the loss of GPS satellites in most of these periods, but there was still a GNSS outage for the partially obstructed observation environment. According to the statistics, about 24 s satellite outages (from 2048 s to 2072 s) occurred during the test (the black rectangle in Figure 3); therefore, the effect of INS is mainly focused on this section.

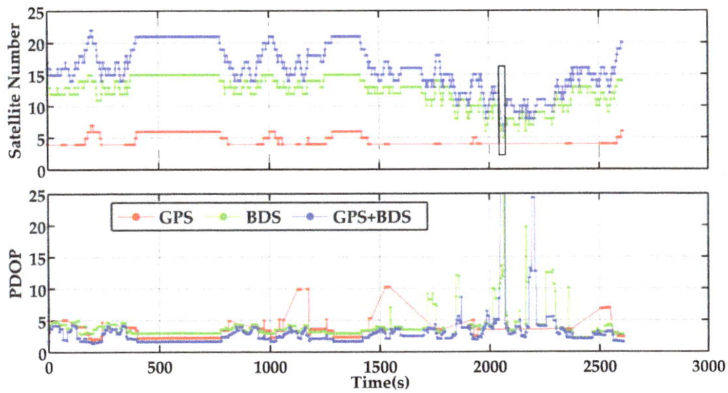


**Figure 3.** Satellite number and PDOP of GPS-only, BDS-only, and G + B (the start time is 195,453 s (GPS time)); the black rectangle represents the GNSS outage period under G + B mode.

### 3.2. *Positioning Performance of PPP and PPP/INS Tight Integration*

Figure 4 presents the position differences of G + B SF-PPP and G + B SF-PPP/INS in terms of the reference solutions, and the relevant RMS (Root Mean Square) values of the position differences are shown in Table 2. According to Figure 4, the performance of SF-PPP/INS was obviously higher than that of SF-PPP in both horizontal and vertical directions. The position RMS in the three directions is enhanced from 0.642 m, 0.649 m, and 1.331 m with SF-PPP to 0.303 m, 0.447 m, and 0.761 m with SF-PPP/INS tight integration, with improvements of 52.8%, 31.1%, and 42.8%, respectively. Figure 5 portrays the distributions of position differences of SF-PPP and SF-PPP/INS tight integration in the horizontal direction and vertical direction. Relevant statistics indicate that the horizontal position differences percentages within 0.3 m were 18.22% and 2.05% for SF-PPP/INS and SF-PPP, respectively. The vertical position differences percentages within 0.3 m were 12.90% and 2.08% for SF-PPP/INS tight integration and SF-PPP, respectively. The results also indicate that the percentage of the horizontal position differences from the SF-PPP/INS tight integration within 0.6 m reached 86.11%. However, for SF-PPP only, this percentage was 59.19%. After about 1800 s, the position accuracy of SF-PPP was strongly influenced by the surrounding observation environment. Nevertheless, the position accuracy of SF-PPP/INS tight integration was hardly influenced. Moreover, the SF-PPP/INS tight integration could still provide navigation solutions during GNSS outage periods. Therefore, INS can significantly increase position accuracy and continuity.

**Table 2.** RMS of position differences in the three directions of SF-PPP and SF-PPP/INS tight integration.

| Positioning Mode | RMS (m) | | |
|---|---|---|---|
| | North | East | Down |
| SF-PPP | 0.642 | 0.649 | 1.331 |
| SF-PPP/INS | 0.303 | 0.447 | 0.761 |

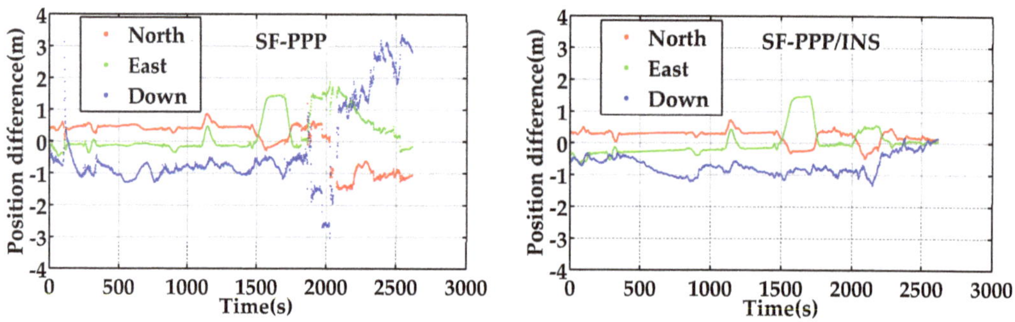


**Figure 4.** Position differences provided by G + B SF-PPP (**left**) and G + B SF-PPP/INS tight integration (**right**) in the three directions, in terms of the reference solution provided by RTK/INS tight integration (the start time is 195,453 s).

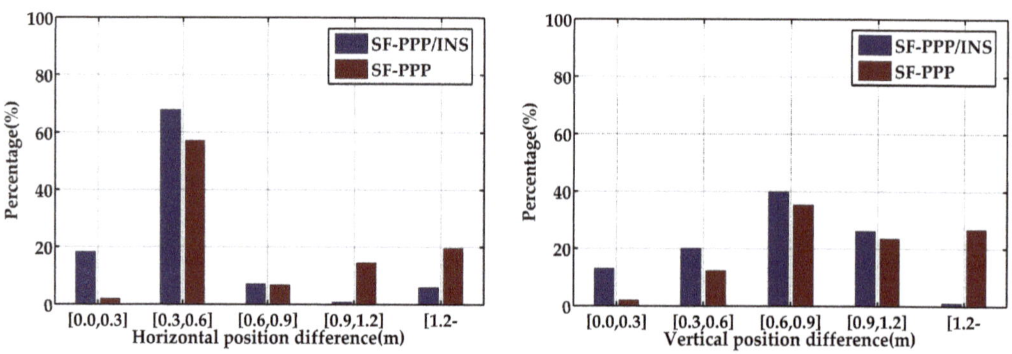


**Figure 5.** Distribution of position differences of G + B SF-PPP and G + B SF-PPP/INS tight integration in the horizontal direction (**left**) and vertical direction (**right**).

### *3.3. Evaluation of Real-Time Orbit and Clock Products*

In this section, the accuracy of the real-time orbit and clock products provided by CAS, GFZ, and WHU are evaluated. Firstly, to validate the performance of each IGS center's products, the final orbit and clock products supplied by Wuhan University were adopted as reference values. Figure 6 presents the RMS of real-time orbit products compared to reference values. Relevant mean RMS values are enumerated in Table 3. According to the statistics, the accuracy of GPS real-time orbits provided by the three analysis centers were generally consistent with each other. For the products from WHU, the orbit accuracies of GPS satellites were 1.7 cm, 4.6 cm, and 3.6 cm, with improvements of 81.3%, 65.7%, and 78.7% compared to those of BDS satellites in the three directions. For BDS satellites, the orbit accuracy of WHU was higher than that of CAS and GFZ, with improvements of 32.1%, 51.8%, and 31.9% compared to that of CAS, respectively. Such improvements were up to 78.3%, 71.1%, and 73.1% compared to that of GFZ in radial, along, and cross directions. The orbit accuracies of the BDS MEO satellites were significantly higher than those of the GEO and IGSO satellites. When the GEO and IGSO satellites were detached, the RMSs of MEO-only in the three directions obtained about 45.1%, 31.3%, and 62.1% accuracy upgradation for the products from WHU. According to the work in [35], the accuracy of final precise orbits is 2.5 cm. Therefore, the orbit accuracies of the GPS real-time products were comparable, but those of the BDS MEO orbits were marginally lower.

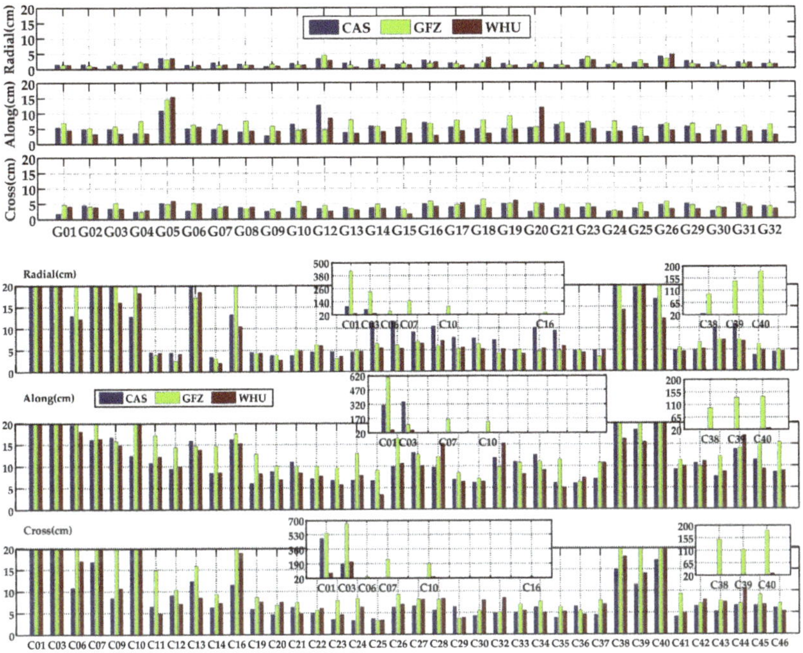


**Figure 6.** Orbit RMS of real-time products of GPS satellites (**top**) and BDS satellites (**bottom**) from the three IGS analysis centers.

**Table 3.** Mean orbit RMS of GPS and BDS satellites in radial, along, and cross directions of each analysis center.

| Analysis Center | GPS (cm) | | | BDS (GEO + IGSO + MEO) (cm) | | | BDS (MEO) (cm) | | |
|---|---|---|---|---|---|---|---|---|---|
| | Radial | Along | Cross | Radial | Along | Cross | Radial | Along | Cross |
| CAS | 2.0 | 5.5 | 3.6 | 13.4 | 27.8 | 24.8 | 6.6 | 9.0 | 5.5 |
| GFZ | 1.9 | 6.7 | 4.4 | 42.0 | 46.4 | 62.8 | 5.2 | 11.7 | 7.6 |
| WHU | 1.7 | 4.6 | 3.6 | 9.1 | 13.4 | 16.9 | 5.0 | 9.2 | 6.4 |

Figure 7 presents the RMS and STD of the real-time clock offsets from three IGS analysis centers with respect to the reference products. Table 4 illustrates the average values of RMS and STD for GPS and BDS satellite clock offsets. The results shows that the accuracies of the GPS clock offset products supported via the three centers were basically consistent with each other. The accuracies of the products provided by WHU were slightly higher than those for the other analysis centers. For the BDS satellites, the accuracy of WHU was lower compared to the others, especially for the BDS-2 satellites. The maximum clock offset RMS achieved 46.10 ns, which directly led to a larger RMS value, on average, for WHU. For the CAS and GFZ products, the accuracies of the BDS GEO and IGSO satellite clock offsets were significantly lower than those of BDS MEO satellites. When GEO and IGSO satellites were detached, the RMS of MEO-only was improved by 12.6% for the products from CAS, whereas such improvement was 43.1% for the products from GFZ.

**Table 4.** Average values of clock offset RMS and STD of GPS and BDS satellites of each analysis center.

| Analysis Center | GPS (ns) | | BDS (GEO + IGSO + MEO) (ns) | | BDS (MEO) (ns) | |
|---|---|---|---|---|---|---|
| | RMS | STD | RMS | STD | RMS | STD |
| CAS | 0.85 | 0.13 | 2.22 | 0.64 | 1.94 | 0.30 |
| GFZ | 0.84 | 0.15 | 2.83 | 1.69 | 1.61 | 0.26 |
| WHU | 0.49 | 0.10 | 14.94 | 0.59 | 6.65 | 0.26 |

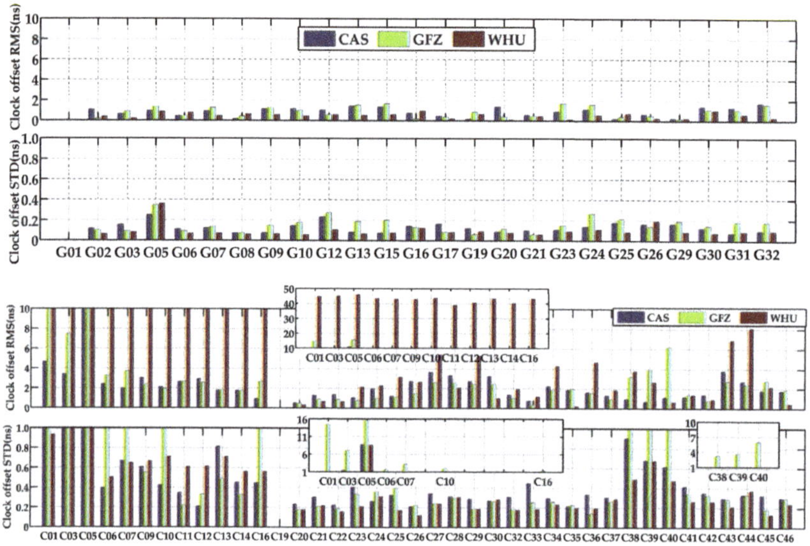


**Figure 7.** Clock offset RMS and STD of real-time products of GPS satellites (**top**) and BDS satellites (**bottom**) from each analysis center.

### *3.4. Performance of Real-Time PPP/INS Tight Integration*

To analyze the positioning and navigation performance of the presented real-time SF-PPP/INS tight integration, in addition to the real-time orbit/clock products, the real-time ionospheric data provided by Wuhan University (ftp address: igs.gnsswhu.cn), with an interval of 5 min, were also used.

Figure 8 reveals the position differences of real-time G + B PPP/INS tight integration in terms of the reference solutions provided by the smoothed RTK/INS tight integration. The RMS and STD are illustrated in Table 5. According to the results of STD, the real-time SF-PPP/INS tight integration could also provide stable and continuous positioning solutions. The position accuracies based on the real-time products from GFZ were higher in the horizontal and vertical directions compared with the results based on the products of CAS and WHU. The position RMSs of the GFZ-product-based solutions were 0.206 m, 0.542 m, and 0.368 m, which is about 65.5%, 14.9%, and 68.7% more accurate than those based on CAS products in the three directions. Compared to the solutions based on WHU products, the position improvements were more obvious, at about 50.1% and 73.1% in east and vertical directions. This is chiefly owing to the lower accuracies of the clock products from WHU, especially for BDS-2. Moreover, compared with the results based on WHU final products, the accuracies of the results based on GFZ's real-time products still present about 32.0% and 51.6% improvements in the north and vertical directions. This may be on account of a high rate of real-time products (orbit, clock offset, and ionospheric data). Figure 9 illustrates the distributions of position differences for G + B SF-PPP/INS tight integration adopting real-time products from CAS, GFZ, and WHU in horizontal and vertical directions. The statistics indicate that the horizontal position differences percentages within 0.3 m were 0.61%, 57.48%, and 0.15% for the results based on the products of CAS, GFZ, and WHU, respectively. The corresponding percentages of the vertical position differences within 0.3 m were 0.00%, 52.62%, and 3.36%, respectively. Accordingly, the positioning accuracy was principally affected by the real-time product quality of BDS MEO. The analysis of positioning performance and the accuracies of real-time products infer that the positioning accuracy is greatly impacted by the quality of the clock products under the premise of consistent orbit accuracy. When using higher-accuracy clock products, it will be easier to obtain positioning solutions with high accuracy.

**Table 5.** Position RMS and STD in the three directions of SF-PPP/INS tight integration using real-time products from CAS, GFZ, and WHU.

| Analysis Center | RMS (m) | | | STD (m) | | |
|---|---|---|---|---|---|---|
| | North | East | Down | North | East | Down |
| CAS | 0.597 | 0.637 | 1.177 | 0.209 | 0.463 | 0.319 |
| GFZ | 0.206 | 0.542 | 0.368 | 0.205 | 0.471 | 0.367 |
| WHU | 0.296 | 1.086 | 1.369 | 0.275 | 0.460 | 0.458 |

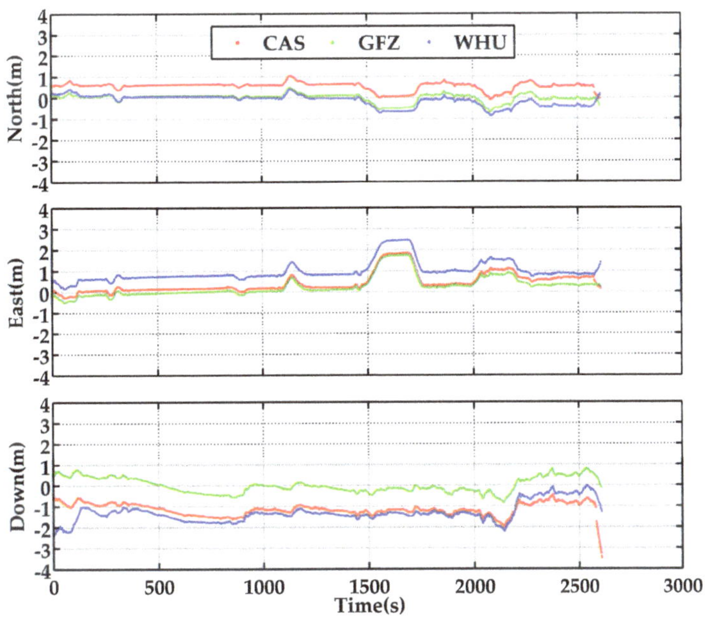


**Figure 8.** Position differences of G + B SF-PPP/INS tight integration in the three directions using the real-time products from CAS, GFZ, and WHU (the start time is 195,453 s).

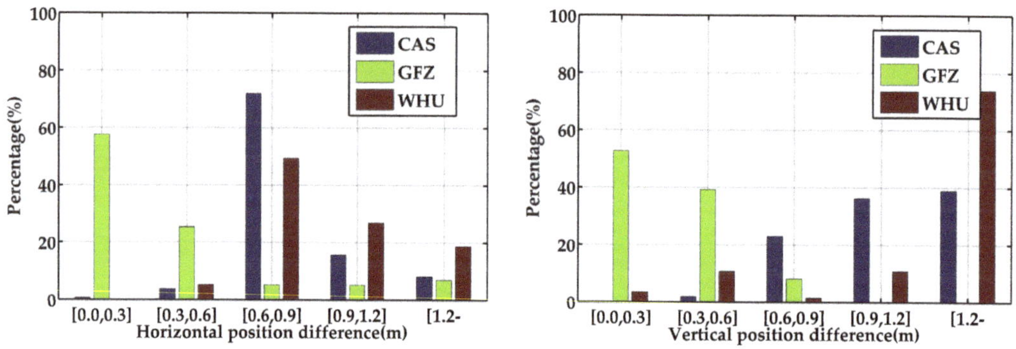


**Figure 9.** Distribution of position differences of G + B SF-PPP/INS tight integration in the horizontal direction (**left**) and vertical direction (**right**) using real-time products from CAS, GFZ, and WHU.

Other than the positioning solutions, the G + B SF-PPP/INS tight integration can also support real-time attitudes with regard to roll, pitch, and heading. The attitude offsets via comparing the attitude solutions with the reference values are portrayed in Figure 10. In the light of the RMSs enumerated in Table 6, attitude accuracy also illustrates a certain relationship to the accuracy of orbit/clock products. The attitude RMSs based on WHU's real-time products were 0.020°, 0.019°, and 0.523° in roll, pitch, and heading, with improvements of 37.5%, 40.6%, and 46.4% compared to the solutions based on CAS's products. Similarly, there were about 25.9% and 29.6% enhancements in roll and pitch compared to the results supplied by GFZ's products. Figure 11 presents the distributions of attitude offsets of the G + B SF-PPP/INS tight integration using the real-time products from CAS, GFZ, and WHU in roll, pitch, and heading. Relevant statistics indicate that the percentages of roll attitude offsets within 0.03° were 65.98%, 69.42%, and 91.20% for the results based on the products of CAS, GFZ, and WHU, respectively. The percentages of pitch attitude offsets within 0.03° were 62.23%, 66.93%, and 92.34%, respectively. In general, the offset distribution in roll and pitch was consistent for solutions based on CAS and GFZ. Additionally the percentages of heading attitude offsets within 0.5° were 32.99%, 66.32%, and 83.16%, respectively.

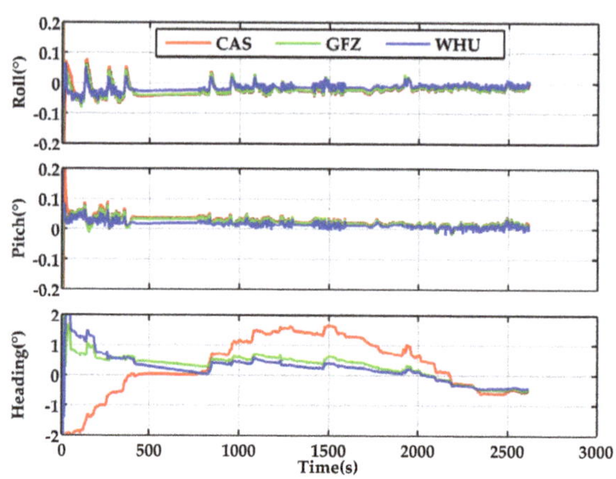


**Figure 10.** Attitude offsets of G + B SF-PPP/INS tight integration in roll, pitch, and heading directions using the real-time products from CAS, GFZ, and WHU, in terms of reference solutions (the start time is 195,453 s).

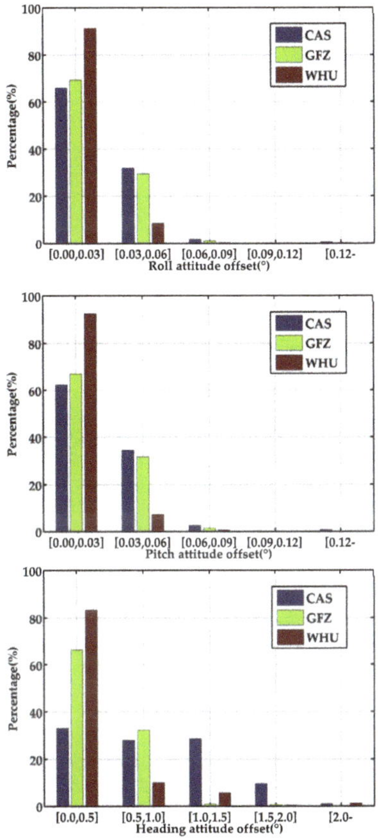


**Figure 11.** Distribution of the offsets in roll (**top**), pitch (**middle**), and heading (**bottom**) of the G + B SF-PPP/INS tight integration using the real-time products from CAS, GFZ, and WHU.

**Table 6.** RMS of attitude and velocity offsets of SF-PPP/INS tight integration using the real-time products from CAS, GFZ, and WHU.

| Analysis Center | Attitude (°) | | | Velocity (m/s) | | |
|---|---|---|---|---|---|---|
| | Roll | Pitch | Heading | North | East | Down |
| CAS | 0.032 | 0.032 | 0.975 | 0.033 | 0.033 | 0.010 |
| GFZ | 0.027 | 0.027 | 0.489 | 0.033 | 0.033 | 0.008 |
| WHU | 0.020 | 0.019 | 0.523 | 0.033 | 0.032 | 0.007 |

Theoretically, in GNSS/INS tight integration, attitudes are discernible when the vehicle makes maneuvers. In this condition, the accuracy of attitude estimation will be marginally affected by the positioning information change. In addition, the estimation accuracy of other parameters in the Kalman filtering could also slightly influence the estimation accuracy of attitude [21]. Therefore, the GNSS/INS tight integration based on different real-time products will make the attitude solutions a little different. Figure 12 illustrates the velocity differences of G + B SF-PPP/INS tight integration. The distributions of velocity offsets of the G + B SF-PPP/INS tight integration using real-time products from CAS, GFZ, and WHU in horizontal and vertical directions are presented in Figure 13. The statistics indicate that percentages of the horizontal velocity offsets within 0.03 m/s were 61.23%, 62.04%, and 62.11% for the results based on the products of CAS, GFZ, and WHU, respectively.

The corresponding percentages of the vertical velocity differences within 0.03 m/s were 98.09%, 99.89%, and 99.85%, respectively. In general, the distribution of velocity offsets based on the real-time products from three centers were consistent in horizontal and vertical directions. In terms of the statistics in Table 6, the velocity offset RMS was almost equivalent when adopting the real-time products from different centers (CAS, GFZ, and WHU). This is because (1) the velocity accuracy of GNSS/INS tight integration chiefly relies on the quality of doppler observations and the performance of IMU sensors, and (2) it has a weak relationship with positioning accuracy. Plotted in Figure 14 are the estimated biases and scale factors of the accelerometer and gyroscope of POS320 in the body frame. Additionally, the estimations present a visible relationship with the quality of real-time satellite orbit/clock products.

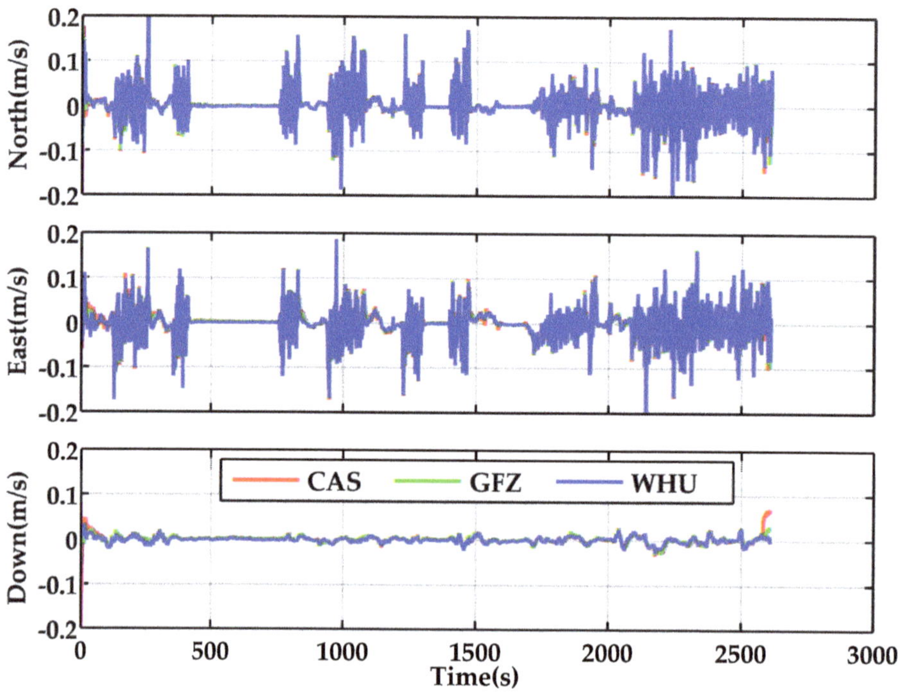


**Figure 12.** Velocity offsets of G + B SF-PPP/INS tight integration in the three directions using the real-time products from CAS, GFZ, and WHU, in terms of reference solutions (the start time is 195,453 s).

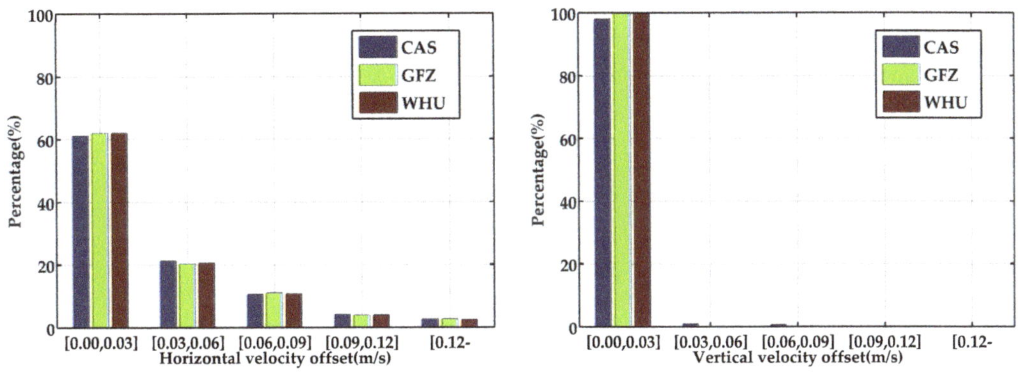


**Figure 13.** Distribution of velocity offsets in horizontal (**left**) and vertical (**right**) directions of G + B SF-PPP/INS tight integration using the real-time products from CAS, GFZ, and WHU.

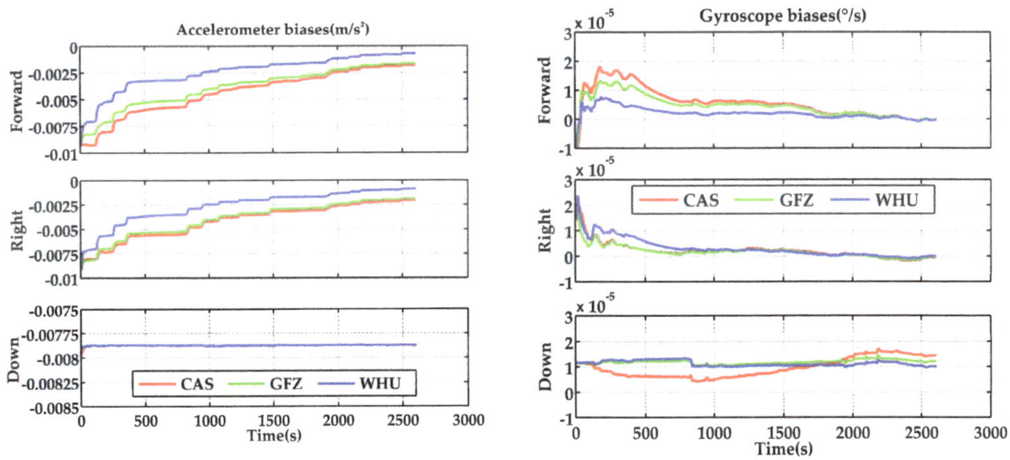


**Figure 14.** Accelerometer (**left**) and gyroscope (**right**) biases and scale factors in the body frame (Forward–Right–Down) by utilizing real-time products from CAS, GFZ, and WHU (the start time is 195,453 s).

## 4. Conclusions

In this contribution, a real-time GPS/BDS-2/BDS-3 SF-PPP/INS tight integration model is introduced. To assess its performance, we firstly dissect the positioning performance of SF-PPP and SF-PPP/INS tight integration. Then, the accuracy of real-time products afforded by CAS, GFZ, and WHU are evaluated. To present the performance using orbit and clock products from different analysis centers, a set of vehicle-borne data was processed using the same strategy. According to the results, it can be concluded that: (1) GPS + BDS SF-PPP/INS tight integration can obtain more accurate and continuous positioning solutions, especially during GNSS outage, compared to the solutions of the GPS + BDS SF-PPP. (2) The accuracy of GPS orbit and clock products provided by the three analysis centers are consistent with each other. For BDS satellites, the accuracy of orbits provided by WHU is higher, but the accuracy of the clocks is lower than the others, especially for BDS-2 satellites. The accuracy of BDS MEO satellite products is significantly superior to that of GEO and IGSO satellites. (3) The positioning RMS values based on GFZ's real-time products are better than those based on the products of CAS and WHU. (4) Owing to the high rate of real-time products, the positioning accuracies in north and vertical directions based on GFZ real-time products are even higher compared to the solutions based on the IGS final products. In addition, the accuracy of attitude and the convergence of IMU sensor errors also present a visible relationship with the real-time orbit/clock products' accuracy.

With the development of multi-sensor fusion, other sensors such as LiDAR, camera, odometer, and Ultra Wide Band (UWB) can be utilized for positioning and navigation in a GNSS challenging environment.

**Author Contributions:** Conceptualization, J.L. and Z.G.; data curation, J.L., Q.X. and R.L.; funding acquisition, C.Y. and J.P.; investigation, J.L. and Z.G.; software, Z.G.; visualization, C.Y. and J.P.; writing—original draft preparation, J.L.; writing—review and editing, J.L. and Z.G. All authors have read and agreed to the published version of the manuscript.

**Funding:** This research was partly supported by the National Key Research and Development Program of China (Grant No. 2020YFB0505802) and the National Natural Science Foundation of China (NSFC) (Grants No. 42074004).

**Data Availability Statement:** The datasets adopted in this paper are managed by the School of Land Science and Technology, China University of Geosciences, Beijing, and are available on request from the corresponding author.

**Acknowledgments:** The authors would like to thank anonymous reviewers who provided valuable suggestions that have helped to improve the quality of the manuscript.

**Conflicts of Interest:** The authors declare no conflict of interest.

## References

1. Zumberge, J.F.; Heflin, M.B.; Jefferson, D.C.; Watkins, M.M.; Webb, F.H. Precise point positioning for the efficient and robust analysis of GPS data from large networks. *J. Geophys. Res. Solid Earth* **1997**, *102*, 5005–5017. [CrossRef]
2. Malys, S.; Jensen, P.A. Geodetic point positioning with GPS carrier beat phase data from the CASA UNO experiment. *Geophys. Res. Lett.* **1990**, *17*, 651–654. [CrossRef]
3. Larson, K.M.; Bodin, P.; Gomberg, J. Using 1-Hz GPS data to measure deformations caused by the Denali fault earthquake. *Science* **2003**, *300*, 1421–1424. [CrossRef] [PubMed]
4. Li, X.; Ge, M.; Lu, C.; Zhang, Y.; Wang, R.; Wickert, J.; Schuh, H. High-Rate GPS Seismology Using Real-Time Precise Point Positioning with Ambiguity Resolution. *IEEE Trans. Geosci. Remote Sens.* **2014**, *52*, 6165–6180.
5. Yao, Y.; Liu, C.; Xu, C. A New GNSS-Derived Water Vapor Tomography Method Based on Optimized Voxel for Large GNSS Network. *Remote Sens.* **2020**, *12*, 2306. [CrossRef]
6. Li, X.; Zhang, X.; Guo, F. Predicting atmospheric delays for rapid ambiguity resolution in precise point positioning. *Adv. Space Res.* **2014**, *54*, 840–850. [CrossRef]
7. Zhang, K.; Liang, S.; Gan, W. Crustal strain rates of southeastern Tibetan Plateau derived from GPS measurements and implications to lithospheric deformation of the Shan-Thai terrane. *Earth Planet. Phys.* **2019**, *3*, 45–52. [CrossRef]
8. Zhang, K.; Wang, Y.; Gan, W.; Liang, S. Impacts of Local Effects and Surface Loads on the Common Mode Error Filtering in Continuous GPS Measurements in the Northwest of Yunnan Province, China. *Sensors* **2020**, *20*, 5408. [CrossRef]
9. Lou, Y.; Zheng, F.; Gu, S.; Wang, C.; Guo, H.; Feng, Y. Multi-GNSS precise point positioning with raw single-frequency and dual-frequency measurement models. *GPS Solut.* **2016**, *20*, 849–862. [CrossRef]
10. Ge, Y.; Dai, P.; Qin, W.; Yang, X.; Zhou, F.; Wang, S.; Zhao, X. Performance of Multi-GNSS Precise Point Positioning Time and Frequency Transfer with Clock Modeling. *Remote Sens.* **2019**, *11*, 347. [CrossRef]
11. Lv, J.; Gao, Z.; Kan, J.; Lan, R.; Li, Y.; Lou, Y.; Yang, H.; Peng, J. Modeling and assessment of multi-frequency GPS/BDS-2/BDS-3 kinematic precise point positioning based on vehicle-borne data. *Measurement* **2022**, *189*, 110453. [CrossRef]
12. Li, X.; Li, X.; Ma, F.; Yuan, Y.; Zhang, K.; Zhou, F.; Zhang, X. Improved PPP Ambiguity Resolution with the Assistance of Multiple LEO Constellations and Signals. *Remote Sens.* **2019**, *11*, 408. [CrossRef]
13. Geng, J.; Meng, X.; Dodson, A.H.; Ge, M.; Teferle, F.N. Rapid re-convergences to ambiguity-fixed solutions in precise point positioning. *J. Geod.* **2010**, *84*, 705–714. [CrossRef]
14. Zhang, H.; Gao, Z.; Ge, M.; Niu, X.; Huang, L.; Tu, R.; Li, X. On the convergence of ionospheric constrained precise point positioning (IC-PPP) based on undifferential uncombined raw GNSS observations. *Sensors* **2013**, *13*, 15708–15725. [CrossRef]
15. Zhang, B.; Teunissen, P.J.G.; Yuan, Y.; Zhang, X.; Li, M. A modified carrier-to-code leveling method for retrieving ionospheric observables and detecting short-term temporal variability of receiver differential code biases. *J. Geod.* **2019**, *93*, 19–28. [CrossRef]
16. Guo, F.; Zhang, X.; Wang, J.; Ren, X. Modeling and assessment of triple-frequency BDS precise point positioning. *J. Geod.* **2016**, *90*, 1223–1235. [CrossRef]

17. Hong, J.; Tu, R.; Zhang, R.; Fan, L.; Zhang, P.; Han, J. Contribution analysis of QZSS to single-frequency PPP of GPS/BDS/GLONASS/Galileo. *Adv. Space Res.* **2020**, *65*, 1803–1817. [CrossRef]
18. Shi, C.; Wu, X.; Zheng, F.; Wang, X.; Wang, J. Modeling of BDS-2/BDS-3 single-frequency PPP with B1I and B1C signals and positioning performance analysis. *Measurement* **2021**, *178*, 109355. [CrossRef]
19. Su, K.; Jin, S. A novel GNSS single-frequency PPP approach to estimate the ionospheric TEC and satellite pseudorange observable-specific signal bias. *IEEE Trans. Geosci. Remote Sens.* **2022**, *60*, 5801712. [CrossRef]
20. Cox, D.B. Integration of GPS with Inertial Navigation Systems. *Navig. J. Inst. Navig.* **1978**, *25*, 236–245. [CrossRef]
21. Gao, Z.; Ge, M.; Shen, W.; Zhang, H.; Niu, X. Ionospheric and receiver DCB-constrained multi-GNSS single-frequency PPP integrated with MEMS inertial measurements. *J. Geod.* **2017**, *91*, 1351–1366. [CrossRef]
22. Gu, S.; Dai, C.; Fang, W.; Zheng, F.; Wang, Y.; Zhang, Q.; Lou, Y.; Niu, X. Multi-GNSS PPP/INS tightly coupled integration with atmospheric augmentation and its application in urban vehicle navigation. *J. Geod.* **2021**, *95*, 1–15. [CrossRef]
23. Hadas, T.; Bosy, J. IGS RTS precise orbits and clocks verification and quality degradation over time. *GPS Solut.* **2015**, *19*, 93–105. [CrossRef]
24. Wang, L.; Li, Z.; Ge, M.; Neitzel, F.; Wang, X.; Yuan, H. Investigation of the performance of real-time BDS-only precise point positioning using the IGS real-time service. *GPS Solut.* **2019**, *23*, 1–12. [CrossRef]
25. Kazmierski, K.; Zajdel, R.; Sośnica, K. Evolution of orbit and clock quality for real-time multi-GNSS solutions. *GPS Solut.* **2020**, *24*, 1–12. [CrossRef]
26. Øvstedal, O. Absolute positioning with single-frequency GPS receivers. *GPS Solut.* **2002**, *5*, 33–44. [CrossRef]
27. Yunck, T.P. Orbit determination. In *Global Positioning System-Theory and Applications*; Parkinson, B.W., Spilker, J.J., Eds.; AIAA: Washington, DC, USA, 1996.
28. Beran, T.; Kim, D.; Langley, R.B. High-precision single-frequency GPS point positioning. In Proceedings of the 16th International Technical Meeting of the Satellite Division of The Institute of Navigation (ION GPS/GNSS 2003), Portland, OR, USA, 9–12 September 2003; pp. 1192–1200.
29. Montenbruck, O. Kinematic GPS positioning of LEO satellites using ionosphere-free single frequency measurements. *Aerosp. Sci. Technol.* **2003**, *7*, 396–405. [CrossRef]
30. Gao, Y.; Zhang, Y.; Chen, K. Development of a real-time single frequency precise point positioning system and test results. In Proceedings of the 19th International Technical Meeting of the Satellite Division of The Institute of Navigation (ION GNSS 2006), Fort Worth, TX, USA, 26–29 September 2006; pp. 2297–2303.
31. Beran, T.; Bisnath, S.B.; Langley, R.B. Evaluation of high-precision, single-frequency GPS point positioning models. In Proceedings of the 17th International Technical Meeting of the Satellite Division of The Institute of Navigation (ION GNSS 2004), Long Beach, CA, USA, 21–24 September 2004; pp. 1893–1901.
32. Le, A.Q.; Tiberius, C.C.J.M.; Van der Marel, H.; Jakowski, N. Use of global and regional ionosphere maps for single-frequency precise point positioning. In *Observing Our Changing Earth*; Springer: Berlin, Germany, 2009; pp. 759–769.
33. Shi, C.; Gu, S.; Lou, Y.; Ge, M. An improved approach to model ionospheric delays for single-frequency precise point positioning. *Adv. Space Res.* **2012**, *49*, 1698–1708. [CrossRef]
34. Yao, Y.; Zhang, R.; Song, W.; Shi, C.; Lou, Y. An improved approach to model regional ionosphere and accelerate convergence for precise point positioning. *Adv. Space Res.* **2013**, *52*, 1406–1415. [CrossRef]
35. Di, M.; Zhang, A.; Guo, B.; Zhang, J.; Liu, R.; Li, M. Evaluation of Real-Time PPP-Based Tide Measurement Using IGS Real-Time Service. *Sensors* **2020**, *20*, 2968. [CrossRef]
36. Gendt, G.; Dick, G.; Reigber, C.H.; Tomassini, M.; Liu, Y. Demonstration of NRT GPS water vapor monitoring for numerical weather prediction in Germany. *J. Meteorol. Soc. Jpn.* **2003**, *82*, 360–370.
37. Niu, X.; Goodall, C.; Nassar, S.; El-Sheimy, N. An efficient method for evaluating the performance of MEMS IMUs. In Proceedings of the Position Location and Navigation Symposium, 2006 IEEE/ION, San Diego, CA, USA, 25–27 April 2006; pp. 766–771.
38. Brown, R.G.; Hwang, P.Y.C. *Introduction to Random Signals and Applied Kalman Filtering*; Willey: New York, NY, USA, 1992.
39. Shin, E.H. *Estimation Techniques for Low-Cost Inertial Navigation*; Library and Archives Canada: Ottawa, ON, Canada, 2006.
40. Su, K.; Jin, S. Analysis and comparisons of the BDS/Galileo quad-frequency PPP models performances. *Acta Geod. Cartogr. Sin.* **2020**, *49*, 1189–1201.

*remote sensing*

MDPI

*Technical Note*

# Preliminary Analysis of Intersystem Biases in BDS-2/BDS-3 Precise Time and Frequency Transfer

**Pengfei Zhang [1,2,3,4], Rui Tu [1,3,5,*], Linlin Tao [1,3], Bing Wang [1,3], Yuping Gao [1,2,3] and Xiaochun Lu [1,3,5]**

1 National Time Service Center, Chinese Academy of Sciences, Xi'an 710600, China
2 Key Laboratory of Time and Frequency Primary Standards, Chinese Academy of Sciences, Xi'an 710600, China
3 University of Chinese Academy of Sciences, Yu Quan Road, Beijing 100049, China
4 State Key Laboratory of Geo-information Engineering, Xi'an Research Institute of Surveying and Mapping, Xi'an 710054, China
5 Key Laboratory of Precision Navigation Positioning and Timing Technology, Chinese Academy of Sciences, Xi'an 710600, China
* Correspondence: turui@ntsc.ac.cn; Tel.: +86-029-8389-0246

**Abstract:** The Chinese BeiDou global satellite system (BDS-3) and regional system (BDS-2) are predicted to coexist over the next decade. Intersystem biases (ISBs) in BDS-2/BDS-3 play a key role in maintaining the consistency and continuity from the BDS-2 to BDS-3 time transfer. Here, we discuss the temporal characteristics, parameter composition, generation mechanism, and the effect of ISBs in BDS-2/BDS-3 on time and frequency transfer. The satellite orbits and clock products from three international GNSS service analysis centers, namely Wuhan University (WUM, China), GeoForschungsZentrum Potsdam (GFZ, Germany), and the Center for Orbit Determination in Europe (CODE), were employed to investigate the time-transfer stability of ISBs when BDS-2 and BDS-3 were used in combination. We analyzed the intrinsic characteristics of ISBs, the receiver types, antennas, and frequency standards. Our first results showed that ISBs are stable for different analysis center products, although the mean values of daily results differed markedly for the three analysis centers. With respect to the relationship between station attribution and ISB difference for a time link, the receiver type, antenna, and frequency standard influence the ISB differences in time and frequency transfer. The effect of three ISB stochastic models was evaluated with respect to time and frequency transfer. The "walk" and "constant" schemes were slightly superior to "noise", with the improvement in their frequency stability being approximately 5% compared with that of "noise".

**Keywords:** time and frequency transfer; precise point positioning; BDS-2; BDS-3; intersystem biases

**Citation:** Zhang, P.; Tu, R.; Tao, L.; Wang, B.; Gao, Y.; Lu, X. Preliminary Analysis of Intersystem Biases in BDS-2/BDS-3 Precise Time and Frequency Transfer. *Remote Sens.* **2022**, *14*, 4594. https://doi.org/10.3390/rs14184594

Academic Editors: Shengfeng Gu, Xiaopeng Gong, Yidong Lou and Chuang Shi

Received: 19 August 2022
Accepted: 9 September 2022
Published: 14 September 2022

**Publisher's Note:** MDPI stays neutral with regard to jurisdictional claims in published maps and institutional affiliations.

**Copyright:** © 2022 by the authors. Licensee MDPI, Basel, Switzerland. This article is an open access article distributed under the terms and conditions of the Creative Commons Attribution (CC BY) license (https://creativecommons.org/licenses/by/4.0/).

## 1. Introduction

The BeiDou Navigation Satellite System (BDS) provides positioning, navigation, and timing (PNT) information to global users. The system was developed in three phases. The first is the demonstration system (BDS-1), developed in 2003 and mainly providing services through two first-generation experimental geostationary Earth orbit (GEO) satellites. The second phase is a regional system (BDS-2) for the Asia–Pacific region, operational since 25 October 2012. This phase comprises five GEO, five inclined geosynchronous orbit (IGSO), and four medium-altitude Earth orbit (MEO) satellites. The third phase is the global system (BDS-3), operational since July 2020, which comprises 30 satellites, including 3 GEO satellites, 3 IGSO satellites, and 24 MEO satellites [1,2]. The BDS-2 is expected to remain in service for at least another decade [3,4], although BDS-3 is already fully operational. In combination, the system provides exceptional potential for PNT users as it employs more BDS satellite resources.

Multi-GNSS constellations have known benefits of time and frequency transfer with respect to precision, integrity, and availability, because of the increased number of available satellites [5–10]. In particular, the multi-GNSS carrier phase technique (CP) has been proven to outperform a single GNSS with respect to remote time and frequency transfer with precision at the nanosecond level [11,12]. Accordingly, this technique is applied widely by international time laboratories for the campaign of time and frequency transfer. However, multi-GNSS precise time and frequency transfer is not free of challenges. In this regard, the unification of coordinate frames and time benchmarks and the differences in receiver hardware delays related to using signals from different systems should be considered [13–15]. Such biases are called intersystem biases (ISBs) and they affect data processing when combined multi-GNSS are employed for time and frequency transfer [16].

Abundant satellite sources are provided by BDS-3 and BDS-2, with common coordinate frames and time benchmarks, namely Geodetic Coordinate System 2000 (CGCS2000) and BDS Time (BDT). The combination of BDS-3 and BDS-2 signals has aroused the interest of numerous scholars. For instance, Li et al. analyzed the ISB between the BDS-2 and BDS-3 experimental systems using the International GNSS Monitoring and Assessment System (iGMAS) and the Multi-GNSS Experiment (MGEX) observations [17]. These authors point out that there a systematic bias between BDS3 and BDS2, and the systematic biases are different for stations in different networks. Pan et al. evaluated the multi-GNSS positioning performance with a priori ISB constraint. According to these authors, the ISBs in BDS-2/BDS-3 could not be ignored [18]. Zhao et al. showed that the ISB values of stations with the same type of receiver were similar, while a substantial difference existed for different receiver types [19]. Fu et al. introduced the ISB parameter between BDS-2 and BDS-3 and improved the standard deviation (STD) of all satellite clocks, even using overlapping B1I/B3I measurements [20]. Although the performance of BDS-2/BDS-3 time and frequency transfer has been investigated thoroughly [21–25], few studies have focused on the characteristics and mechanism of ISBs in BDS-2/BDS-3.

It is well-known that the GNSS CP technique relies heavily on precise satellite products for time and frequency transfer. However, it is not clear whether ISB characteristics also depend on satellite clock products. Accordingly, estimating the ISBs in BDS-2/BDS-3 and determining their influence on the performance would facilitate optimal utilization of the BDS system.

In this study, we discuss the temporal characteristics, parameter composition, generation mechanism, and the effect of ISBs in BDS-2/BDS-3 on time and frequency transfer. This study starts with a brief description of mathematical models of BDS-2/BDS-3 time and frequency transfer with the CP technique, after which ISB estimation methods are discussed. Subsequently, the data sources with several types of receivers, antennas, external time and frequency references, and data-processing strategies are introduced. In addition, the temporal characteristics of ISB based on satellite products from different analysis centers and the reasons for changes are analyzed. Furthermore, the influence of different ISB estimation methods on the time and frequency transfer results is evaluated. Finally, we present our summary and conclusions.

## 2. Materials and Methods of Multi-GNSS CP Technique

The classical CP time transfer comprises two components for the pseudorange with time information and carrier phase measurements with millimeter-scale noise levels. The ionosphere-free (IF) linear combination with dual-frequency observation is used to eliminate the effect of the first-order ionosphere. The observation model can be expressed as follows, taking the single BDS as an example [12]:

$$\begin{cases} P_{IF}^{i} = \rho_r^i + c\cdot\left(dt_r^i - dt^{s,i}\right) + T_{trop} + b_{r,IF}^{i} - b_{IF}^{s,i} + e_{IF}^{i} \\ L_{IF}^{i} = \rho_r^i + c\cdot\left(dt_r^i - dt^{s,i}\right) + T_{trop} + \lambda_{IF}\cdot\left(N_{IF}^{i} + B_{r,IF}^{i} - B_{IF}^{s,i}\right) \\ \qquad +\varepsilon_{IF}^{i} \end{cases} \tag{1}$$

where the superscript $i$ denotes the BDS system (i.e., BDS-2, BDS-3). $P_{IF}^i$ and $L_{IF}^i$ are the IF combination of pseudorange and carrier phase observation, respectively; $\rho_r^i$ is the geometric distance between the phase center of the satellite and receiver antenna; $c$ is the speed of light in vacuum; $dt_r^i$ and $dt^{s,i}$ represent the receiver and satellite clock offsets, respectively. $T_{trop}$ is the tropospheric delay; $b_{r,IF}^i$ and $B_{r,IF}^i$ are the IF combination of reciever pseudorange and phase hardware delay, respectively; $b_{IF}^{s,i}$ and $B_{IF}^{s,i}$ are the IF combination of satellite pseudorange and phase hardware delay. $\lambda_{IF}$ is the wavelength of the IF combination; $N_{IF}^i$ is the phase ambiguity of the IF combination; and $e_{IF}^i$ and $\varepsilon_{IF}^i$ are measurement noise for the pseudorange and carrier phase observation, respectively. The GNSS satellite and receiver phase center offset and variation, phase wind-up, solid tide, ocean load, pole tide, and relativistic delay should also be considered, although these terms are not listed in Equation (1).

In precise time and frequency transfer employing the CP technique, the satellite orbit and clock products provided by the International GNSS Service (IGS) are used to reduce the orbit and clock errors. The IF combination is used in the data processing of the IGS to determine the satellite orbit and clock parameter. The satellite hardware delay $b_{IF}^{s,i}$ is absorbed in the satellite clock offset, providing a reference for the receiver clock offset. Therefore, the receiver IF combination pseudorange hardware delay $b_{r,IF}^i$ is assimilated into the receiver clock offset. The CP delays $B_{r,IF}^i$ and $B_{IF}^{s,i}$ at the receiver and satellite are related closely to the ambiguity parameter $N_{IF}^i$ and lumped with the estimated ambiguity parameter. Therefore, Equation (1) can be written further as:

$$\begin{cases} \overline{P}_{IF}^i = \rho_r^i + c \cdot \overline{dt}_r^i + T_{trop} + e_{IF}^i \\ \overline{L}_{IF}^i = \rho_r^i + c \cdot \overline{dt}_r^i + T_{trop} + \lambda_{IF} \cdot \overline{N}_{IF}^i + \varepsilon_{IF}^i \end{cases} \tag{2}$$

where $\overline{P}_{IF}^i$ and $\overline{L}_{IF}^i$ are the actual pseudorange and CP observations when using the IGS precise satellite orbit and clock products. Therefore,

$$\overline{P}_{IF}^i = P_{IF}^i + c \cdot dt^{s,i} + b_{IF}^{s,i} \tag{3}$$

$$\overline{L}_{IF}^i = L_{IF}^i + c \cdot dt^{s,i} + b_{IF}^{s,i} \tag{4}$$

$$\overline{dt}_r^i = dt_r^i + b_{r,IF}^i \tag{5}$$

$$\overline{N}_{IF}^i = N_{IF}^i + B_{r,IF}^i - \frac{b_{r,IF}^i}{\lambda_{IF}} - B_{IF}^{s,i} + \frac{b_{IF}^{s,i}}{\lambda_{IF}} \tag{6}$$

where $\overline{dt}_r^i$ and $\overline{N}_{IF}^i$ are the new receiver clock offset and ambiguity parameter, lumped with the corresponding hardware delays.

When BDS-2 and BDS-3 are combined for precise time and frequency transfer, an additional ISB parameter ($ISB^{C3,C2}$) between the two systems is introduced to obtain a common receiver clock offset that references a unique system time scale [26–28]. Therefore, the BDS-2/BDS-3 time and frequency transfer model can be written as:

$$\begin{cases} \overline{P}_{IF}^{C2} = \rho_r^{C2} + c \cdot \overline{dt}_r^{C2} + T_{\text{trop}} + e_{IF}^{C2} \\ \overline{L}_{IF}^{C2} = \rho_r^{C2} + c \cdot \overline{dt}_r^{C2} + T_{\text{trop}} + \lambda_{IF} \cdot \overline{N}_{IF}^{C2} + \varepsilon_{IF}^{C2} \\ \overline{P}_{IF}^{C3} = \rho_r^{C3} + c \cdot \overline{dt}_r^{C2} + ISB^{C3,C2} + T_{\text{trop}} + e_{IF}^{C3} \\ \overline{L}_{IF}^{C3} = \rho_r^{C3} + c \cdot \overline{dt}_r^{C2} + ISB^{C3,C2} + T_{\text{trop}} + \lambda_{IF} \cdot \overline{N}_{IF}^{C3} + \varepsilon_{IF}^{C3} \end{cases} \tag{7}$$

where superscript C2 and C3 denote the BDS-2 and BDS-3 system. Among the parameters, the unique receiver clock offset $\overline{dt}_r^{C2}$ is the most interesting parameter for precise time transfer, which is determined jointly by the BDS-2 and BDS-3 observations, although it is denoted simply as the BDS-2 system. Two stations, A and B, located at different places on

Earth, are equipped with their corresponding time and frequency references. The operation of time transfer between the two references can be obtained using the following expression:

$$\begin{aligned}\Delta T_{A,B} &= \overline{dt_r^{C2}}(A) - \overline{dt_r^{C2}}(B) \\ &= \left(t_A^{C2} - BDT + b_{r,IF}^{C2}(A)\right) - \left(t_B^{C2} - BDT + b_{r,IF}^{C2}(B)\right) \\ &= t_A^{C2} - t_B^{C2} + b_{r,IF}^{C2}(A) - b_{r,IF}^{C2}(B) = \Delta t_{A,B}^{C2} + \Delta b_{r,IF}^{C2}(AB)\end{aligned} \tag{8}$$

where $t^{C2}$ is the external time and frequency reference when BDS-2 observation is used. The term BDT is the BeiDou time scale, which uses the international system of units (SI) second without leap seconds, connects with universal time coordinated (UTC) through UTC (NTSC, national time service center), and the deviation of BDT to UTC is maintained within 50 nanoseconds. The initial epoch of BDT is 00:00:00 on January 1, 2006, of UTC (BeiDou Navigation Satellite System Open Service Performance Standard, Version 3.0, May 2021). The $\Delta t_{A,B}^{C2}$ is the clock difference between two external time and frequency references at stations A and B; $\Delta b_{r,IF}^{C2}(AB)$ is the delay difference of receiver pseudorange between stations A and B, usually calibrated using the time-transfer link calibration or receiver calibration approaches [29,30]. After the combined observation equation (Equation (7)) has been transformed and linearized, the unknown parameter vector X can be summarized as:

$$X = \left[x, y, z, \overline{dt_r^{C2}}, ISB^{C3,C2}, T_{trop}, \overline{N_{IF}^{C2}}, \overline{N_{IF}^{C3}}\right] \tag{9}$$

where $(x, y, z)$ is the station coordinate parameter.

In order to further clarify the origin of ISB, referring to Equations (5), (7), and (8), the ISB can be written further as:

$$\begin{aligned}ISB^{C3,C2} &= c{\cdot}\overline{dt_r^{C3}} - c{\cdot}\overline{dt_r^{C2}} = c{\cdot}dt_r^{C3} - c{\cdot}dt_r^{C2} + b_{r,IF}^{C3} - b_{r,IF}^{C2} \\ &= c{\cdot}\Delta dt_r^{C3,C2} + \Delta b_{r,IF}^{C3,C2}\end{aligned} \tag{10}$$

The $ISB$ theoretically comprises two components, namely the time difference of two receiver clock offsets with different GNSS observations, and the difference in the receiver hardware delays for two GNSS systems [31]. For the former, $\Delta dt_r^{C3,C2}$ is a function of the receiver clock offset, which is the difference between the external time and frequency reference and GNSST (GNSS time, GNSST), as discussed previously. Unlike the combination of different GNSSs, such as BDS, GPS, Galileo, and GLONASS, with their individual system time scales, BDT (BeiDou system time, BDT), GPST (GPS system scale, GPST), GST (Galileo time scale, GST), and UTC (SU) include the component of time deviation between different GNSSTs for the term. If $\Delta dt_r^{C3,C2}$ does not contain this term, the formula can be expressed as follows:

$$\Delta dt_r^{C3,C2} = \left(t^{C3} - BDT\right) - \left(t^{C2} - BDT\right) = t^{C3} - t^{C2} \tag{11}$$

where $t^{C3}$ is the external time and frequency references when using BDS-3 observation and is equivalent to $t^{C2}$ when using a multimode BDS receiver. Considering the occurrence of unknown errors and unmodeled deviation in ISB estimation, we introduced parameter $\tau_{ISB}$ to represent these errors. Equation (10) can be written further as:

$$ISB^{C3,C2} = \Delta b_{r,IF}^{C3,C2} + \tau_{ISB} \tag{12}$$

As shown by Equations (7)–(9) and (12), the ISB parameter is important when determining the receiver clock and further carrying out the time and frequency transfer.

## 3. ISB Stochastic Models in Multi-GNSS Time Transfer

With respect to the ISB parameter, it usually performs three stochastic models, namely white noise, random constant, and random walk process. Although the previous research shows that the stochastic model is closely related to the ISB performance in the multi-GNSS positioning and time transfer [32,33], the performance in BDS-2/BDS-3 is still unclear.

Regarding the white noise process, the ISB is assumed to be uncorrelated between the different epochs. The white noise process is applied widely when the characteristics of a parameter are not known. The model is expressed as

$$Q_{ISB}(k) \sim N\left(0,\sigma^2\right) \tag{13}$$

where $Q$ denotes a covariance; $k$ is the epoch index.

For the random constant process, it is estimated as a piecewise mode. As for the entire data block, it is usually divided into sub-blocks; the mathematical model in the sub-blocks is expressed as

$$Q_{ISB}(k+1) = Q_{ISB}(k) \tag{14}$$

where $k$ is the epoch index.

The random walk process can be formulated as follows:

$$Q_{ISB}(k+1) = Q_{ISB}(k) + \omega_{ISB}, \quad \omega_{ISB} \sim N\left(0,\sigma^2_{\omega_{ISB}}\right) \tag{15}$$

where $\omega$ is the process noise of a random walk.

## 4. Results

To explicitly investigate the temporal characteristics of ISBs in BDS-2/BDS-3 time transfer, we collected data from the Multi-GNSS Experiment (MGEX), which is piloted by the IGS to collect all available GNSS observations of new signals since 2013. The MGEX network has expanded to more than 300 stations, providing an excellent opportunity to track multi-GNSS constellations and to conduct tracking data analysis. More than 200 stations are tracking BDS satellites. However, most stations only track the dual-frequency BDS-2 signals, but single-frequency BDS-3 observations or the data received are only for a short period during a day. Moreover, the external time and frequency of atomic clocks are not equipped in most stations. Consequently, a limited number of available BDS-2 and BDS-3 stations are available for analyzing ISB variation. Ten stations that track common dual-frequency (B1I and B3I) signals for BDS-2 and BDS-2, equipped with atomic clocks, and that have a relatively complete receiver period in a day, are collected. The geographical distribution of collected stations is shown in Figure 1. The experiment was conducted from day of year (DOY) 100–120, 2021. Detailed information on these stations, e.g., type of receiver, antenna, and frequency standard, is presented in Table 1.

**Table 1.** Information of utilized BDS stations in the experiment.

| Site | Receiver Type | Antenna | Frequency Standard | Country | Location |
|---|---|---|---|---|---|
| GOP6 | SEPT POLARX5 | SEPCHOKE_B3E6 | PASSIVE HYD | Czechia | 49.9°N, 14.8°E |
| HOB2 | SEPT POLARX5 | AOAD/M_T | H-MASER | Australia | 42.8°S, 147.4°E |
| KIRU | SEPT POLARX5 | SEPCHOKE_B3E6 | CESIUM | Sweden | 67.9°N, 20.9°E |
| TID1 | SEPT POLARX5 | AOAD/M_T | H-MASER | Australia | 35.4°S, 148.9°E |
| ONSA | SEPT POLARX5TR | AOAD/M_B | H-MASER | Sweden | 57.4°N, 11.9°E |
| PARK | SEPT POLARX5TR | ASH701945C_M | H-MASER | Australia | 33.0°S, 148.3°E |
| SPT0 | SEPT POLARX5TR | TRM59800.00 | H-MASER | Sweden | 57.7°N, 12.9°E |
| BOR1 | TRIMBLE NETR9 | TRM59800.00 | H-MASER | Poland | 52.3°N, 17.1°E |
| NYAL | TRIMBLE NETR9 | AOAD/M_B | H-MASER | Norway | 78.9°N, 11.9°E |
| TRO1 | TRIMBLE NETR9 | TRM59800.00 | RUBIDIUM | Norway | 69.7°N, 18.9°E |

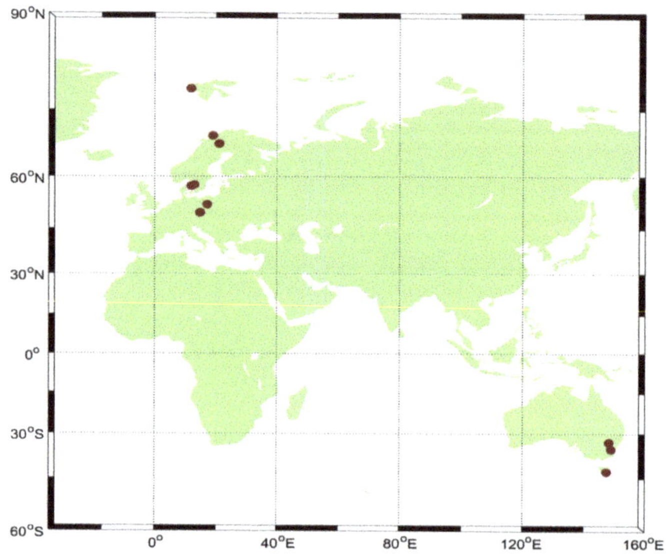


**Figure 1.** Geographical distribution of collected stations in the experiment.

During the data processing, the precise time-transfer solution (PTTSol) was used [34]. The BDS satellite orbit and clock products from three IGS analysis centers, WUM, GFZ, and CODE, were employed for further research of the stability of ISBs when combining BDS-2 and BDS-3 for time transfer. The Wuhan University analysis center has provided BDS-3 satellite orbit and clock products since GPS week 2034, with 15 min and 5 min updates, using "wum" ID. GeoForschungsZentrum Potsdam has provided them since GPS week 2081, with 5 min and 30 s intervals, using "gbm" ID. Fortunately, starting from GPS week 2148, the CODE satellite solution has included BDS-3 (apart from GEO satellites), with a 5 min orbit and 30 s clock, using "com" ID. The pseudorange and CP measurements for B1, B3 of BDS-2, and BDS-3 dual-frequency observations are used to alleviate the ionosphere effect. In preprocessing, both the geometry-free combination and the Melbourne–Wübbena combination were used for cycle slip detection. Tropospheric wet delay is typically modeled as the sum of the Saastamoinen model and a random walk process. The receiver clock offset parameter was estimated as a white stochastic noise process. We also considered the wind-up effects on phase measurements. The data-processing strategies in this study are summarized in Table 2.

**Table 2.** Data-processing strategies in this study.

| Item | Models and Strategies |
|---|---|
| Observation | Pseudorange and carrier phase measurement |
| Signal Frequency | B1I (1561.098 MHz), B3I (1268.52 MHz) |
| Sampling rate | 30 s |
| Elevation cut-off | 7° |
| Satellite orbit and clock | Using precise satellite products from three analysis centers |
| Satellite antenna phase center | Corrected, using MGEX value |
| Ionosphere | Eliminated by ionosphere-free combination |
| Tropospheric delay | Saastamoinen + random walk process |
| Mapping function | Neill mapping function (NMF) |
| Estimator | Least-squares solution in sequential mode |
| Receiver clock offset | Estimated with white noise |
| Phase wind-up effect | Model corrected |

### 4.1. Characterization of ISB over Different Time Periods

As the research was conducted over numerous days, we randomly selected the result on DOY 100, 110, and 120 of 2021 to analyze ISB variation. Figure 2 shows that the ISBs are stable with the different analysis center products on the three days. The average variation over the experimental period is within 0.08m for ISB_com and 0.07 m for ISB_gbm and ISB_wum. The variations on DOY 100 and 120 are within 0.06 m; however, the range on DOY 110 at 0.09 m is much larger than that for the other days. The difference between the minimum and maximum values is approximately 0.28 m for the ISB_com solution, 0.26 m for ISB_wum, and 0.25 m for ISB_gbm. Notably, the mean values of the results from three analysis centers differ markedly for one daily result. Figure 3 shows the mean value character of the results of different analysis centers of 10 stations. Although obvious systematic bias does exist among the com, gbm, and wum results, it is relatively stable among the ten stations. The systematic bias values between the com and wum results are −0.46 m, −0.50 m, and −0.33 m for DOY 100, 110, and 120, respectively, whereas for com and gbm the values are −1.65 m, −1.75 m, and −0.78 m, respectively, i.e., larger than the former. Systematic bias is caused mainly by the different data-processing strategies of the three analysis centers when determining the BDS-2 and BDS-3 satellite orbit and clock products. Figures 2 and 3 show that the ISB trend is generally stable. The stability of ISB_gbm and ISB_wum is slightly superior to that of ISB_com. Remarkably, the mean value of the former four stations equipped with SEPT POLARX5 receivers is not more stable than that of the middle three stations (SEPT POLARX5TR) or the latter three stations (TRIMBLE NETR9). Relevant details on this aspect are presented in the following sections.

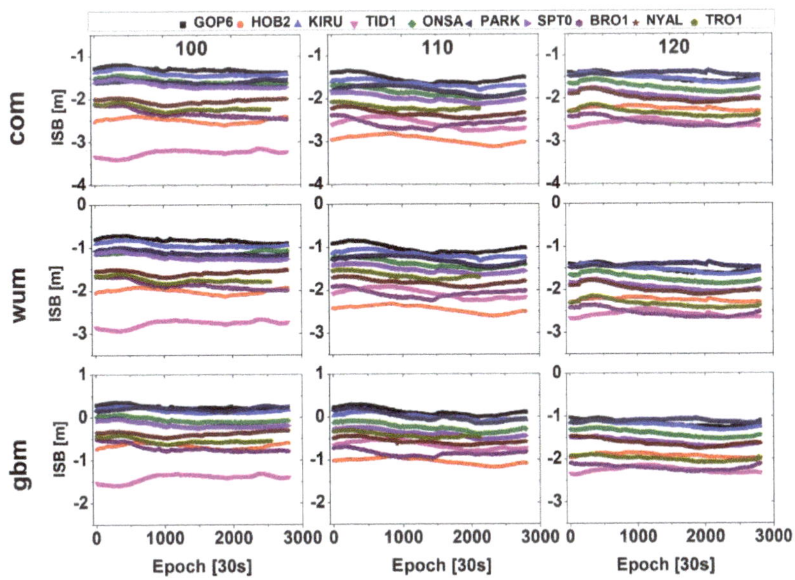


**Figure 2.** Daily variations in ISB between BDS-3 and BDS-2 for 10 stations. The panels from top to bottom show the results of the com, wum, and gbm precise products.

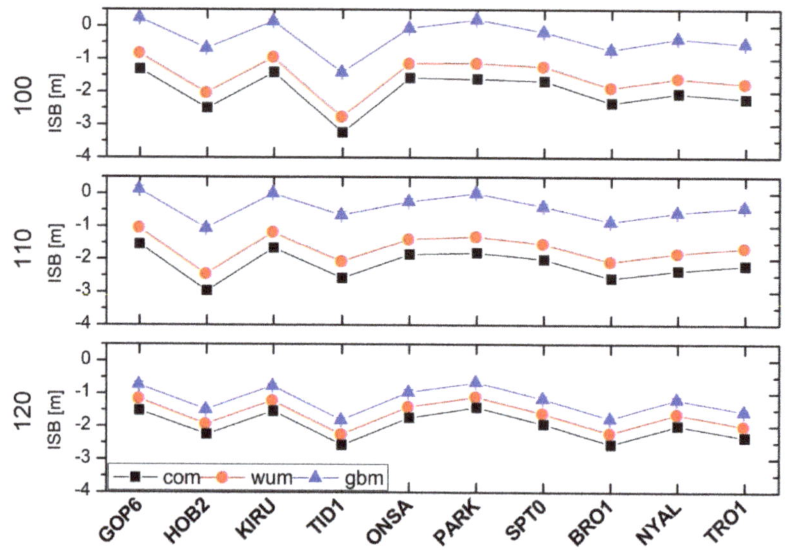


**Figure 3.** Average of ISB between BDS-3 and BDS-2 for 10 stations for the com, wum, and gbm solutions.

### 4.2. Analysis of ISB for Different Station Attributes

From the Equation (10), we know that the ISB contains two components, the time difference of two receiver clock offsets with different GNSS observations, and the difference in the receiver hardware delays for two GNSS systems. For the latter, the receiver type, antenna, and frequency standard are the important factors to affect the receiver hardware delays. For further analyses of the relationship between the station attribute and ISB stability, the stations shown in Table 3 were regrouped into five comparative schemes according to three indicators, namely receiver type, antenna, and frequency standard. Table 3 shows the comparative strategies and corresponding stations, where • means the same, and ○ means different. Further, to alleviate the effect of systematic bias in the results from different stations of three analysis centers on ISB stability analysis, the ISB difference of stations is more focused, which is also related to the calibration of the time-transfer links.

**Table 3.** Comparative schemes and corresponding stations.

| ID | Receiver Type | Antenna | Frequency Standard | Stations |
|---|---|---|---|---|
| Scheme 1 | • | • | • | HOB2, TID1 |
| Scheme 2 | ○ | • | • | NYAL, ONSA, SPT0, BOR1 |
| Scheme 3 | • | ○ | • | PARK, SPT0, |
| Scheme 4 | • | • | ○ | BOR1, TRO1, KIRU, GOP6 |
| Scheme 5 | ○ | ○ | ○ | HOB2, TRO1 |

As 10 stations were involved in the 20 d experimental period, we randomly used the results of DOY 101, 106, 111, 115, and 119 of 2021 in the five comparison schemes, as shown in Figures 3–7. To clearly plot and compare the receiver type, antennas, and frequency standards in the figures, we used abbreviations, namely "Rec", "Ant", and "Fre".

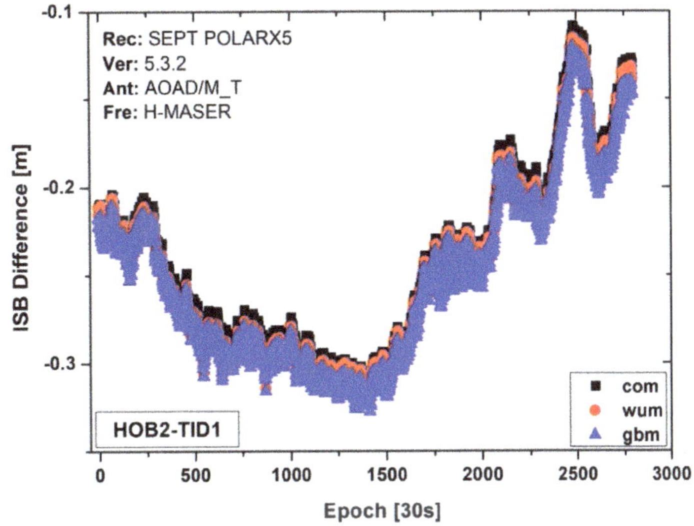


**Figure 4.** ISB difference series for scheme 1 on DOY 101, 2021.

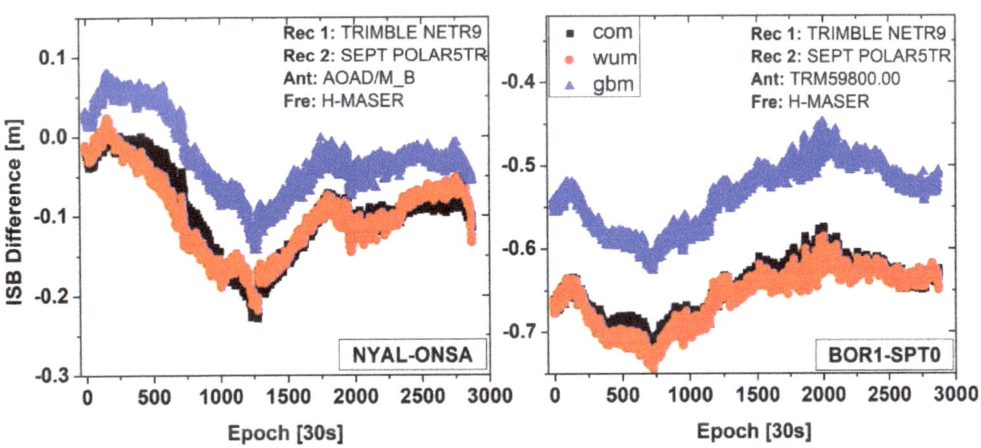


**Figure 5.** ISB difference series for scheme 2 on DOY 106, 2021.

Figure 4 shows the ISB difference series for scheme 1, using the same receiver type, antenna, and frequency standard for SEPT POLARX5, AOAD/M_T, and H-MASER, respectively. One can see that the variation trends of ISB_com, ISB_wum, and ISB_gbm agree very well. The standard deviation (STD) values are all 0.05 m for the three analysis center products. The ranges between the minimum and maximum are 0.2 m.

Figure 5 shows the ISB difference series for scheme 2, which uses the same type of antenna and frequency standard, but different receiver types for one time-transfer link. The left panel shows the time link of station NYAL-ONSA, with the same type of antenna AOAD/M_B and frequency standard H-MASER. The right panel shows station BOR1-SPT0, with the same type of antenna TRM59800.00 and frequency standard H-MASER. The ISB difference series of $ISB_{com}$ and $ISB_{wum}$ agree relatively well, whereas the $ISB_{gbm}$ difference series shows obvious bias compared with that of $ISB_{com}$ and $ISB_{wum}$ for the two time-transfer links. The bias for NYAL-ONSA is approximately 0.07 m and that for BOR1-SPT0 is approximately 0.12 m. The standard deviation of divergence is approximately 0.014 m and 0.013 m, respectively. The above analyses show that the ISB difference series has a close relationship with the type of receiver of the different analysis center products.

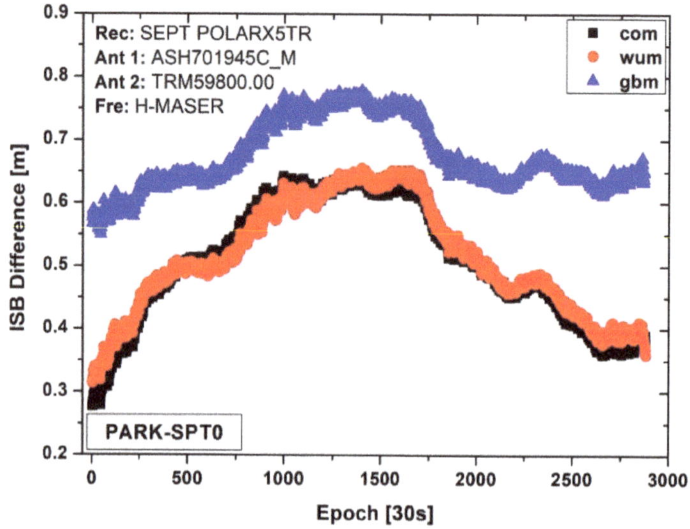


**Figure 6.** ISB difference series for scheme 3 on DOY 111, 2021.

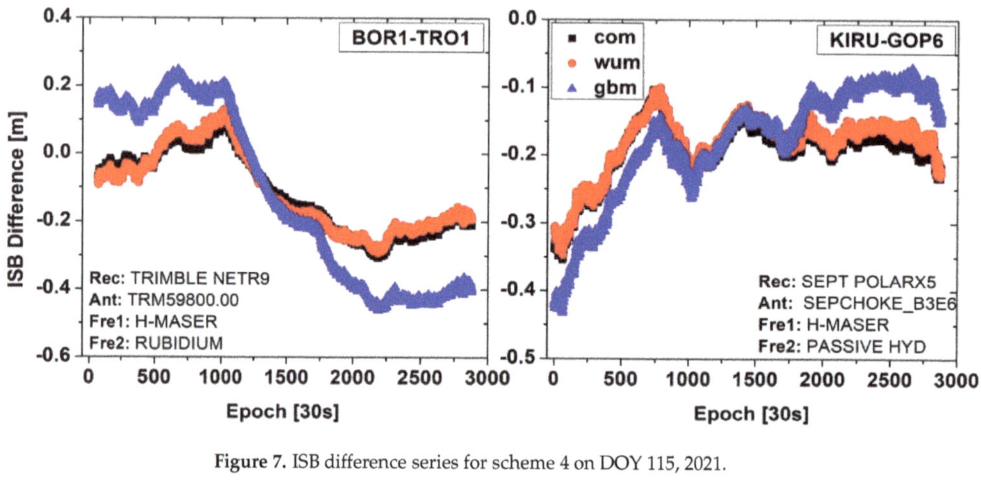


**Figure 7.** ISB difference series for scheme 4 on DOY 115, 2021.

Figure 6 shows the ISB difference series for scheme 3, which uses the same type of receiver and frequency standard, but a different antenna type. This scheme is similar to scheme 2, and the difference series of $ISB_{com}$ and $ISB_{wum}$ show good agreement, although the antenna type differs. Obvious bias exists between $ISB_{gbm}$, $ISB_{com}$, and $ISB_{wum}$, with the corresponding values ranging from 0.1 m to 0.25 m. This finding indicates that the ISB difference series has a certain relationship with the antenna type for the three analysis centers' satellite orbits and clock products. Compared with $ISB_{com}$, the mean value of divergence is 0.010 m for $ISB_{wum}$ and 0.167 m for $ISB_{gbm}$, and the standard deviations are 0.013 m and 0.050 m.

Figure 7 shows the ISB difference series for scheme 4, which uses the same type of receiver and antenna, but a different frequency standard. The left panel is the time link of station BOR1-TRO1, with the same type of receiver TRIMBLE NETR9 and antenna

TRM59800.00. The right panel is station KIRU-GOP6, with the same type of receiver SEPT POLARX5 and antenna SEPCHOKE_B3E6. The variations in the ISB difference series for the $ISB_{com}$ and $ISB_{wum}$ solutions are in extremely good agreement. Although the general trend of $ISB_{gbm}$ is somewhat similar, the divergence between $ISB_{wum}$ and the former two solutions does not show constant bias, but indicates significant variation for the two time-transfer links.

Figure 8 shows the ISB difference series for scheme 5, which uses different receivers, antennas, and frequency standards. Although the general trend is somewhat similar, the divergence among ISB_com, ISB_wum, and ISB_gbm shows not only obvious bias but also the variation term. As indicated by the analyses and discussions of the five schemes, the ISB difference of ISB_com and ISB_wum agrees well. The mean value of divergence is 0.023 m for ISB_wum and 0.030 m for ISB_gbm, and the standard deviations are 0.030 m and 0.036 m compared with those of ISB_com.

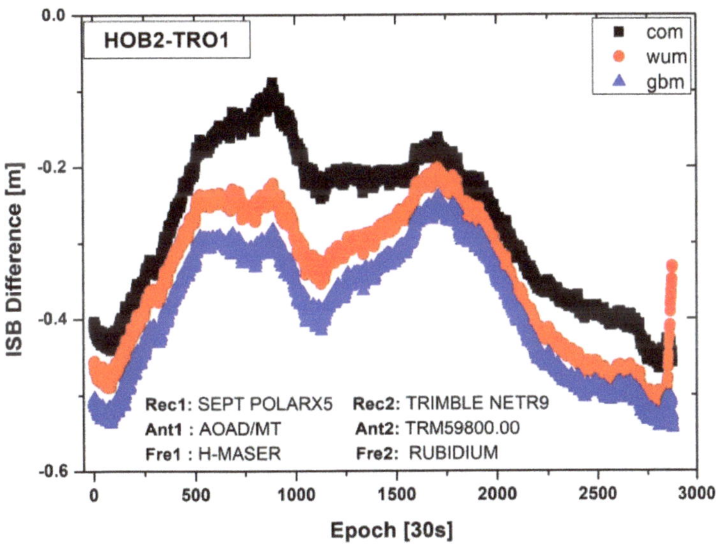


**Figure 8.** ISB difference series for scheme 5 on DOY 119, 2021.

### *4.3. Influence of Different ISB Stochastic Models on Time and Frequency Transfer*

To assess the characteristics of different ISB stochastic models in the time and frequency transfer, the previous three modes are applied in the experiment. Then, the ISB were modeled as constants for one hour in a model of random constant process, marked as "constant." For the random walk process, it was defined as "walk" and the power density was 1 mm/$s^{0.5}$, whereas the white noise process was marked as "noise." Figure 9 shows the results of time and frequency with three ISB stochastic models on the time link PTBB–WTZZ, using GFZ precise products. The variations in the clock difference agree well for the three ISB stochastic models. Figure 10 shows a comparison of Allan deviations of time-transfer results for the three ISB stochastic models at the PTBB–WTZZ time link. It is clear that the "constant" and "walk" schemes show slightly superior performances for frequency stability compared with that of "noise" at different time intervals. The average values within 10,000 s among the solutions of the three models are $1.08 \times 10^{-13}$ for "noise", $1.00 \times 10^{-13}$ for "constant," and $9.95 \times 10^{-13}$ for "walk," with the improvements being 5.08% and 5.67%, respectively.

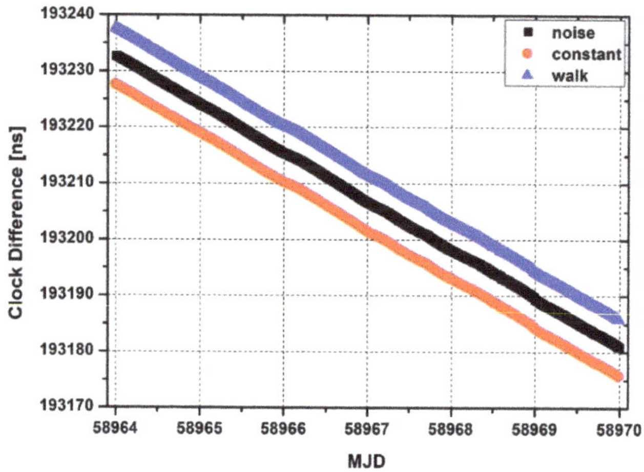


**Figure 9.** Result of time and frequency transfer with three ISB stochastic models on time link PTBB–WTZZ. For plotting purposes, the overall values for "walk" results were translated up to 5 ns, whereas those for results using the constant were translated down to 5 ns.

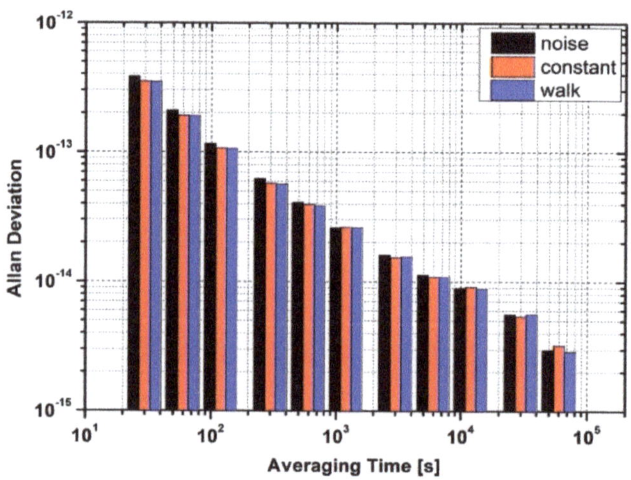


**Figure 10.** Comparison of Allan deviations in time-transfer results for the three ISB stochastic models at the PTBB–WTZZ time link.

## 5. Discussion

The multi-GNSS time and frequency transfer is essential for UTC comparison and traceability services, particularly for the existing BDS-2 regional and BDS-3 global system. However, the character of intersystem biases in the BDS-2/BDS-3 GNSS time and frequency transfer is still unclear. Therefore, the spatiotemporal characterization, different station attributes, and stochastic models of ISB were focused.

One can note that the daily ISB in BDS-2/BDS-3 is relatively stable, but exhibits obvious discrepancies among the three IGS analysis centers. Considering that the current BDS daily products have an obvious day-boundary jump, the daily stability of ISB helps to precisely estimate parameters. From the results of ISB for different station attributes, it can be see that common receiver type, antenna, and frequency standard can contribute to improving consistency for the three different analysis centers, these mainly being caused by the relationship between the attribute of the station (receiver DCB [35], receiver calibration, type of frequency standard) and the strategies of satellite products (i.e., sample interval, data-processing strategy, used stations, and so on). In addition, the three ISB stochastic models are compared in the time and frequency transfer, which agree with the results in previous research [32].

Of course, this study proposes only the first step of this research, and several topics still require further investigation in our near future work; for example, how to use a functional model to improve the estimation precision of ISB, and how to calibrate the ISB delay in the time link based on multisystem time and frequency transfer.

## 6. Conclusions

To maintain the consistency and continuity from BDS-2 to BDS-3 time transfer for one time link, we analyzed the ISBs in BDS-2/BDS-3. We deduced the mathematical model of BDS-2/BDS-3 time and frequency transfer, including observation and stochastic models. The temporal characteristics of ISB for different types of receivers, antennas, and frequency standards, with different IGS analysis center products were discussed. The three stochastic models of ISB were evaluated using one time link.

Our results indicated that the ISB series exhibit obvious discrepancies among the three analysis centers, but relatively stable characteristics. The mean values of the daily results of differ markedly for the three analysis centers. The receiver type, antenna, and frequency standard have a certain influence on the ISB difference in time and frequency transfer. The receiver type, antenna, and frequency standard are different for the two ends of the time link; the obvious system bias exists among the com, gbm, and wum analysis centers. As the only different receiver type scheme, the ISB difference series of ISB_com and ISB_wum agree relatively well, whereas the ISB_gbm series shows obvious bias compared with ISB_com and ISB_wum for the two time-transfer links. The bias differs for the two time links. The bias for station NYAL-ONSA is approximately 0.07 m, and that for station BOR1-SPT0 is approximately 0.12 m. The ISB difference series of ISB_com and ISB_wum agree relatively well for the only different antenna type scheme, whereas the ISB_gbm series shows obvious bias compared with ISB_com and ISB_wum for the two time-transfer links. It should be noted that the bias is not a constant but varies with time. As the only different-frequency-standard scheme, the general trend of ISB_gbm is somewhat similar; the divergence between ISB_wum and the other two solutions is not constant but shows significant variations for the two time-transfer links. The effect of the three different ISB stochastic models was assessed with respect to time and frequency transfer. The "walk" and "constant" schemes were slightly superior to the "noise", with the improvements in frequency stability being approximately 5.08% and 5.67%, respectively, compared with that of "noise".

This study proposes only the first step of this research, and several topics still require further investigation in our near-future work; for example, how to use a functional model to improve the estimation precision of ISB, and how to calibrate the ISB delay in the time link based on multisystem time and frequency transfer.

**Author Contributions:** P.Z. and R.T. conceived and designed the experiments; P.Z. performed the experiments, analyzed the data and wrote the paper. L.T., B.W., Y.G. and X.L. contributed to discussions and revisions. All authors have read and agreed to the published version of the manuscript.

**Funding:** This research was funded by National Natural Science Foundation of China (Grant No: 11903040, 41674034, 41974032, 42030105) and Chinese Academy of Sciences (CAS) programs of "Youth Innovation Promotion Association" (Grant No: 2022414), "Western young scholars" (Grant No.: XAB2019B21), China Postdoctoral Science Foundation (Grant No: 2020M683763).

**Data Availability Statement:** The datasets analyzed in this study are managed by the MGEX and National Time Service Center, Chinese Academy of Sciences, which can be available on request from the corresponding author.

**Acknowledgments:** Many thanks go to the IGS MGEX for providing multi-GNSS ground tracking data, precise orbit, and clock products.

**Conflicts of Interest:** The authors declare no conflict of interest.

## References

1. Wang, M.; Wang, J.; Dong, D.; Meng, L.; Chen, J.; Wang, A.; Cui, H. Performance of BDS-3: Satellite visibility and dilution of precision. *GPS Solut.* **2019**, *23*, 56. [CrossRef]
2. Yang, Y.; Mao, Y.; Sun, B. Basic performance and future developments of BeiDou global navigation satellite system. *Satell. Navig.* **2020**, *1*, 1. [CrossRef]
3. Yang, Y.; Gao, W.; Guo, S.; Mao, Y.; Yang, Y. Introduction to BeiDou-3 navigation satellite system. *Navigation* **2019**, *66*, 7–18. [CrossRef]
4. Mi, X.; Sheng, C.; El-Mowafy, A.; Zhang, B. Characteristics of receiver-related biases between BDS-3 and BDS-2 for five frequencies including inter-system biases, differential code biases, and differential phase biases. *GPS Solut.* **2021**, *25*, 113. [CrossRef]
5. Harmegnies, A.; Defraigne, P.; Petit, G. Combining GPS and GLONASS in all-in-view for time transfer. *Metrologia* **2013**, *50*, 277–287. [CrossRef]
6. Zhang, P.; Tu, R.; Gao, Y.; Zhang, R.; Liu, N. Improving the Performance of Multi-GNSS Time and Frequency Transfer Using Robust Helmert Variance Component Estimation. *Sensors* **2018**, *18*, 2878. [CrossRef] [PubMed]
7. Dow, J.; Neilan, R.; Rizos, C. The International GNSS Service in a changing landscape of Global Navigation Satellite Systems. *J. Geod.* **2009**, *83*, 191–198. [CrossRef]
8. Montenbruck, O.; Steigenberger, P.; Prange, L.; Deng, Z.; Zhao, Q.; Perosanz, F.; Schaer, S. The Multi-GNSS Experiment (MGEX) of the International GNSS Service (IGS)–Achievements, prospects and challenges. *Adv. Sp. Res.* **2017**, *59*, 1671–1697. [CrossRef]
9. Li, X.; Ge, M.; Dai, X.; Ren, X.; Fritsche, M.; Wickert, J.; Schuh, H. Accuracy and reliability of multi-GNSS real-time precise positioning: GPS, GLONASS, BeiDou, and Galileo. *J. Geod.* **2015**, *89*, 607–635. [CrossRef]
10. Jin, S.; Wang, Q.; Dardanelli, G. A Review on Multi-GNSS for Earth Observation and Emerging Applications. *Remote Sens.* **2022**, *14*, 3930. [CrossRef]
11. Jiang, Z.; Lewandowski, W. Use of GLONASS for UTC time transfer. *Metrologia* **2012**, *49*, 57–61. [CrossRef]
12. Petit, G.; Defraigne, P. The performance of GPS time and frequency transfer: Comment on A detailed comparison of two continuous GPS carrier-phase time transfer techniques. *Metrologia* **2016**, *53*, 1003–1008. [CrossRef]
13. Yi, H.; Wang, H.; Zhang, S.; Wang, H.; Shi, F.; Wang, X. Remote time and frequency transfer experiment based on BeiDou Common View. In Proceedings of the European Frequency and Time Forum (EFTF), York, UK, 26 May 2016; pp. 1–4. [CrossRef]
14. Guang, W.; Dong, S.; Wu, W.; Zhang, J.; Yuan, Y.; Zhang, S. Progress of BeiDou time transfer at NTSC. *Metrologia* **2018**, *55*, 175–187. [CrossRef]
15. Wu, M.; Liu, W.; Wang, W.; Zhang, X. Differential Inter-System Biases Estimation and Initial Assessment of Instantaneous Tightly Combined RTK with BDS-3, GPS, and Galileo. *Remote Sens.* **2019**, *11*, 1430. [CrossRef]
16. Defraigne, P.; Aerts, W.; Harmegnies, A.; Petit, G.; Rovera, D.; Uhrich, P. Advances in multi-GNSS time transfer. In Proceedings of the European Frequency and Time Forum and International Frequency Control Symposium, Prague, Czech Republic, 21–25 July 2013; pp. 508–512.
17. Li, X.; Xie, W.; Huang, J.; Ma, T.; Zhang, X.; Yuan, Y. Estimation and analysis of differential code biases for BDS3/BDS2 using iGMAS and MGEX observations. *J. Geod.* **2019**, *93*, 419–435. [CrossRef]
18. Pan, L.; Zhang, Z.; Yu, W.; Dai, W. Intersystem Bias in GPS, GLONASS, Galileo, BDS-3, and BDS-2 Integrated SPP: Characteristics and Performance Enhancement as a Priori Constraints. *Remote Sens.* **2021**, *13*, 4650. [CrossRef]
19. Zhao, W.; Chen, H.; Gao, Y.; Jiang, W.; Liu, X. Evaluation of Inter-System Bias between BDS-2 and BDS-3 Satellites and Its Impact on Precise Point Positioning. *Remote Sens.* **2020**, *12*, 2185. [CrossRef]
20. Fu, W.; Wang, L.; Li, T.; Chen, R.; Han, Y.; Zhou, H.; Li, T. Combined BDS-2/BDS-3 real-time satellite clock estimation with the overlapping B1I/B3I signals. *Adv. Sp. Res.* **2021**, *66*, 4470–4483. [CrossRef]
21. Dai, P.; Yang, X.; Qin, W.; Wang, R.; Zhang, Z. Analysis of BDS-2+BDS-3 Combination Real-Time Time Transfer Based on iGMAS Station. In Proceedings of the 10th China Satellite Navigation Conference (CSNC), Beijing, China, 22–25 May 2019.
22. Su, K.; Jin, S. Triple-frequency carrier phase precise time and frequency transfer models for BDS-3. *GPS Solut.* **2019**, *23*, 86. [CrossRef]
23. Zhang, P.; Tu, R.; Zhang, R.; Liu, N.; Gao, Y. Time and frequency transfer using BDS-2 and BDS-3 carrier-phase observations. *IET Radar Sonar Navig.* **2019**, *13*, 1249–1255. [CrossRef]
24. Jiao, G.; Song, S.; Chen, Q.; Huang, C.; Su, K.; Wang, Z.; Cheng, N. Modeling and Analysis of BDS-2 and BDS-3 Combined Precise Time and Frequency Transfer Considering Stochastic Models of Inter-System Bias. *Remote Sens.* **2021**, *13*, 793. [CrossRef]
25. Gong, X.; Zheng, F.; Gu, S.; Zheng, F.; Lou, Y. The long-term characteristics of GNSS signal distortion biases and their empirical corrections. *GPS Solut.* **2022**, *26*, 52. [CrossRef]
26. Nicolini, L.; Caporali, A. Investigation on Reference Frames and Time Systems in Multi-GNSS. *Remote Sens.* **2018**, *10*, 80. [CrossRef]
27. Defraigne, P.; Baire, Q. Combining GPS and GLONASS for time and frequency transfer. *Adv. Space Res.* **2011**, *47*, 265–275. [CrossRef]

28. Gong, X.; Gu, S.; Lou, Y.; Zheng, F.; Yang, X.; Wang, Z.; Liu, J. Research on empirical correction models of GPS Block IIF and BDS satellite inter-frequency clock bias. *J. Geod.* **2020**, *94*, 36. [CrossRef]
29. Rovera, G.D.; Torre, J.M.; Sherwood, R.; Abgrall, M.; Courde, C.; Laas-Bourez, M.; Uhrich, P. Link calibration against receiver calibration: An assessment of GPS time transfer uncertainties. *Metrologia* **2014**, *51*, 476–490. [CrossRef]
30. Zhang, P.; Tu, R.; Gao, Y.; Guang, W.; Zhang, R.; Cai, H. Study of time link calibration based on GPS carrier phase observation. *IET Radar Sonar Navig.* **2018**, *12*, 1330–1335. [CrossRef]
31. Lou, Y.; Gong, X.; Gu, S.; Zheng, F.; Feng, Y. Assessment of code bias variations of BDS triple-frequency signals and their impacts on ambiguity resolution for long baselines. *GPS Solut.* **2017**, *21*, 177–186. [CrossRef]
32. Zhang, P.; Tu, R.; Han, J.; Zhang, R.; Gao, Y.; Lu, X. Characterization of biases between BDS-3 and BDS-2, GPS, Galileo and GLONASS observations and their effect on precise time and frequency transfer. *Meas. Sci. Technol.* **2021**, *32*, 035006. [CrossRef]
33. Zhou, F.; Dong, D.; Li, P.; Li, X.; Schuh, H. Influence of stochastic modeling for inter-system biases on multi-GNSS undifferenced and uncombined precise point positioning. *GPS Solut.* **2019**, *23*, 59. [CrossRef]
34. Zhang, P.; Tu, R.; Zhang, R.; Gao, Y.; Cai, H. Combining GPS, BeiDou, and Galileo Satellite Systems for Time and Frequency Transfer Based on Carrier Phase Observations. *Remote Sens.* **2018**, *10*, 324. [CrossRef]
35. Choi, B.; Lee, S. The influence of grounding on GPS receiver differential code biases. *Adv. Sp. Res.* **2018**, *62*, 457–463. [CrossRef]

*remote sensing*

*Article*

# Analyses of GLONASS and GPS+GLONASS Precise Positioning Performance in Different Latitude Regions

**Yanli Zheng [1,*], Fu Zheng [2], Cheng Yang [1], Guigen Nie [3] and Shuhui Li [1]**

1 School of Land Science and Technology, China University of Geosciences, Beijing 100083, China
2 Research Institute for Frontier Science, Beihang University, Beijing 100191, China
3 GNSS Research Center, Wuhan University, Wuhan 430079, China
* Correspondence: ylzheng@cugb.edu.cn

**Abstract:** The orbital inclination angle of the GLONASS constellation is about 10° larger than that of GPS, Galileo, and BDS. Theoretically, the higher orbital inclination angle could provide better observation geometry in high latitude regions. A wealth of research has investigated the positioning accuracy of GLONASS and its impact on multi-GNSS, but rarely considered the contribution of the GLONASS constellation's large orbit inclination angle. The performance of GLONASS in different latitude regions is evaluated in both stand-alone mode and integration with GPS in this paper. The performance of GPS is also presented for comparison. Three international GNSS service (IGS) networks located in high, middle, and low latitudes are selected for the current study. Multi-GNSS data between January 2021 and June 2021 are used for the assessment. The data quality check shows that the GLONASS data integrity is significantly lower than that of GPS. The constellation visibility analysis indicates that GLONASS has a much better elevation distribution than GPS in high latitude regions. Both daily double-difference network solutions and daily static Precise Point Positioning (PPP) solutions are evaluated. The statistical analysis of coordinate estimates indicates that, in high latitude regions, GLONASS has a comparable or even better accuracy than that of GPS, and GPS+GLONASS presents the best estimate accuracy; in middle latitude regions, GPS stand-alone constellation provides the best positioning accuracy; in low latitude regions, GLONASS offers the worst accuracy, but the positioning accuracy of GPS+GLONASS is better than that of GPS. The tropospheric estimates of GLONASS do not present a resemblance regional advantage as coordinate estimates, which is worse than that of GPS in all three networks. The PPP processing with combined GPS and GLONASS observations reduces the convergence time and improves the accuracy of tropospheric estimates in all three networks.

**Keywords:** GLONASS; GPS; double-differenced; static PPP

**Citation:** Zheng, Y.; Zheng, F.; Yang, C.; Nie, G.; Li, S. Analyses of GLONASS and GPS+GLONASS Precise Positioning Performance in Different Latitude Regions. *Remote Sens.* **2022**, *14*, 4640. https://doi.org/10.3390/rs14184640

Academic Editor: Xiaoli Deng

Received: 3 August 2022
Accepted: 12 September 2022
Published: 16 September 2022

**Publisher's Note:** MDPI stays neutral with regard to jurisdictional claims in published maps and institutional affiliations.

**Copyright:** © 2022 by the authors. Licensee MDPI, Basel, Switzerland. This article is an open access article distributed under the terms and conditions of the Creative Commons Attribution (CC BY) license (https://creativecommons.org/licenses/by/4.0/).

## 1. Introduction

Currently, four satellite navigation systems with global coverage have been developed: GPS, GLONASS, BDS, and Galileo. A notable design difference among the different constellations is the satellite orbital inclination angle. The inclination angle is 55° for GPS, 56° for Galileo, 55° for BDS, and 64.8° for GLONASS. Among the four constellations, the GLONASS has the largest orbit inclination angle, which is about 10° larger than other systems, to provide the availability of the high-latitude of the Soviet Union.

The first Final Operational Capability (FOC) of GLONASS was achieved in 1995. However, due to the short satellite service life and the budget decrease, the GLONASS constellation dropped to 7 satellites by 2002 [1]. During 2001–2011, the GLONASS program progressed steadily, and by late 2011, GLONASS declared FOC again.

During the period of several satellites, many studies have been performed to investigate the advantages and disadvantages of combining GPS and GLONASS [2,3]. Bruyninx [4] concluded that using the GLONASS constellation of 13 satellites does not significantly improve the precision of the double-difference (DD) network solutions, and similar results have been obtained with IGS and CODE (Center of Orbit Determination in Europe) orbits. Habrich [5] obtained similar results with 16 GLONASS satellites. Additionally, Cai and Gao [6] indicated that adding GLONASS satellites in Precise Point Positioning (PPP) would reduce the convergence time and improve the positioning accuracy.

As GLONASS was gradually restored, in terms of DD network processing, Alcay et al. [7] illustrated that the GLONASS stand-alone baseline solutions are inconsistent compared to that of GPS and that solutions using the combined GPS and GLONASS constellations do not provide any superiority over stand-alone GPS. Nardo et al. [8] presented that the additional GLONASS observations add little improvement to the estimates of the coordinates when compared to GPS-only processing. Zheng et al. [9] concluded that the repeatability of GLONASS coordinates is slightly worse than that of GPS. The research on GLONASS PPP increased as the usage of PPP increased. Cai and Gao [10] indicated that integrating the GLONASS with GPS could not significantly improve the PPP accuracy if the stand-alone GPS has adequate visible satellites with good observation geometry. Yigit et al. [11] also revealed that the static PPP performance among GPS, GLONASS, and GPS+GLONASS with long observation periods was similar. Choy et al. [12] further demonstrated that the benefits of combining GLONASS with GPS in daily static PPP are negligible. Mohammed et al. [13] assessed the static PPP performance of GPS, GLONASS, and GPS+GLONASS, and concluded that the GLONASS PPP could achieve similar coordinate estimate accuracy as GPS and GPS+GLONASS in daily solutions. However, Malik [14] provided a different conclusion that the accuracy of undifferenced ionosphere-free dual-frequency PPP with GPS and GLONASS observations is better than that of GPS. Hamed et al. [15] obtained similar results with single-frequency PPP. The analysis of PPP convergence time indicates that the combination of GPS and GLONASS significantly shortened the convergence time of static PPP solutions [10,16]. Li and Zhang [17] studied the combination of GPS and GLONASS and illustrated that the convergence time of ambiguity-float static PPP could be reduced by 45.9% compared to GPS.

There are also many studies concerning the contribution of GLONASS to three or more GNSS systems' combined constellations [18–20]. However, the previous research rarely considers the constellation characteristics of GLONASS, especially the effect of the large orbit inclination angle of the GLONASS constellation, which benefits the positioning performance in high latitude regions. Therefore, this paper aims to evaluate the performance of GLONASS and its contribution to GPS+GLONASS processing in different latitude regions in terms of satellite visibility and positioning performance. Three networks located in high, middle, and low latitude regions are employed. The performance of both daily DD network solutions and daily static PPP solutions is used for the study.

The structure of this article is arranged as follows. Section 2 describes the methods of data quality evaluation, the positioning strategies, and the evaluation indicators. Section 3 describes the data and data selection factors. Section 4 presents the data quality results, the constellation visibility of different systems, as well as the analysis and discussion of the performance of the DD network and PPP solutions. Finally, the main conclusions and findings are shown in Section 5.

## 2. Evaluation Methods

This section describes the methods and metrics of data quality check, the analysis indicators for satellite visibility of different systems, and the positioning strategies of DD network and PPP processing, as well as the evaluation indicator for DD network solutions and PPP solutions.

### 2.1. Data Quality Check

The measurement quality assessment aimed to detect the GLONASS and GPS poor observations, which further affect the positioning performance of different combinations. TEQC [21] toolkit was employed to perform the quality assessment of GLONASS and GPS

L1 and L2 signals, which are used for performance assessment. The quality check was conducted using the following indicators:

1. Data Integrity (DI). Data integrity rate is the recorded valid observation data divided by the receivable observation data calculated by ephemeris and the station location.
2. Signal-to-Noise Ratio (SNR). SNR is the ratio of signal power to noise power within a given bandwidth. It is usually expressed in the unit of decibels.
3. Pseudorange Multipath (MP). Pseudorange multipath indicators are computed using the linear combination of pseudorange and carrier phase observations:

$$MP1 = P_1 - \left(1 + \frac{2}{\alpha - 1}\right)\varnothing_1 + \left(1 + \frac{2}{\alpha - 1}\right)\varnothing_1 \quad (1)$$

$$MP2 = P_2 - \left(\frac{2\alpha}{\alpha - 1}\right)\varnothing_1 + \left(\frac{2\alpha}{\alpha - 1} - 1\right)\varnothing_2 \quad (2)$$

where MP1 denotes the multipath effect on L1 frequency and MP2 indicates the multipath effect on L2 frequency; $P_1$ and $P_2$ denote the pseudorange observations at L1 and L2 frequencies, respectively; $\varnothing_1$ and $\varnothing_2$ denote the carrier phase observations of L1 and L2 frequencies, respectively; $\alpha = \left(\frac{f_1}{f_2}\right)^2$, where $f_1$ denotes the L1 frequency, and $f_2$ denotes the L2 frequency.

### *2.2. Constellation Visibility Analysis*

The visibility of GPS, GLONASS, and GPS+GLONASS in different latitude networks was analyzed and evaluated using the following criteria:

1. The number of visible satellites. The mean number of visible satellites in each network at each epoch is computed, and the observed probability of different positioning combinations is also analyzed.
2. The elevation angle of visible satellites. The mean elevation angle of visible satellites in each network at each epoch is computed, and the occurrence probability corresponding to different degrees is evaluated.
3. The Position Dilution of Precision (PDOP). The mean PDOP of each network at each epoch is calculated.

As the GLONASS constellation geometry repeats about every 8 sidereal days, we used observation data from 28 March 2021 (day of year (DOY) 087) to 4 April 2021 (DOY 094) to evaluate the observation quality. The elevation cutoff angle was set to 3°. The study employed the same data set as the data quality check for visibility assessment.

### *2.3. DD Network Processing Strategy*

The DD network processing was conducted using the Bernese GNSS Software, Version 5.2. The software is developed at the Astronomical Institute of the University of Bern (AIUB), Bern, Switzerland. A daily batch processing scheme is used for the data processing. The final precise orbits from CODE were adopted, containing consistent orbits for GPS and GLONASS. The different code biases (DCB) files and the Earth Rotation Parameters (ERP) of CODE were also used for consistency. The PCC model used was igs14.atx. The ocean tides model used was FES2004 (Finite Element Solutions). The elevation mask for data preprocessing was set to 3°. The baselines were defined with the OBS-MAX strategy. An attempt to fix the GPS and GLONASS integer value ambiguities was attempted with the Quasi Ionospheric Free (QIF) strategy [22]. The VMF1 (Vienna Mapping Function) [23] grid file [24] and NET WET model were used for the tropospheric estimate. Zenith Tropospheric Delay (ZTD) parameter was estimated per hour. The datum definition was realized with the minimum-constraint solution by a set of reference stations of IGS14. The processing scheme is displayed in Figure 1.

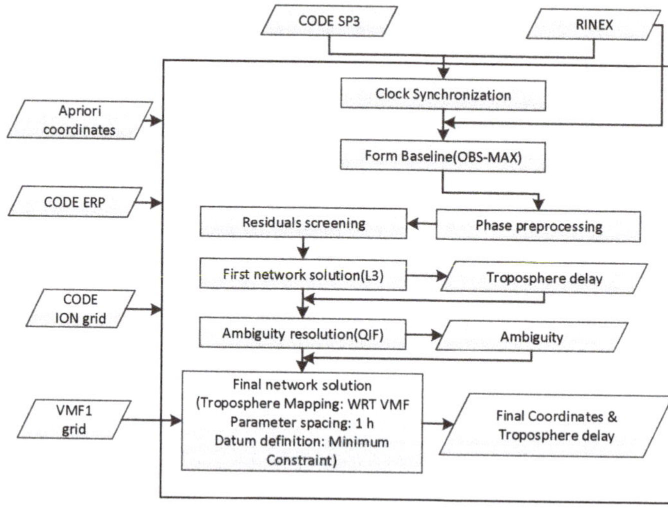


**Figure 1.** DD network processing scheme.

### 2.4. Static PPP Processing Strategy

The static PPP processing was carried out by FUSING (FUSing IN GNSS) software [25], Version 2.0, developed by Wuhan University, Wuhan, China. The Ionosphere-free (IF) linear combination with L1 and L2 was employed. The elevation mask, the precise products, and the ocean tides model, as well as the PCC model, were the same as the DD processing strategy. The GPT2 (Global Pressure and Temperature) [26] model and VMF1 [23] model were used for the tropospheric estimate. PPP static in 24 h window was processed with a forward extended Kalman filter. The processing strategies of PPP are summarized in Table 1.

**Table 1.** The processing strategies of PPP.

| Item | Processing Strategies |
|---|---|
| Signal selection | L1 and L2 |
| Sampling rate | 30 s |
| Elevation mask | 3° |
| Precise products | CODE final precise products |
| Weight for observations | Elevation-dependent weighting |
| Receiver clock | Estimated as white noise |
| Ionosphere | IF combination |
| Troposphere | GPT2, VMF1 |
| Ocean tidal loading | FES2004 |
| DCB | CODE DCB monthly files |
| Antenna center offset and variation | IGS14.atx |
| Processing mode | PPP static in 24 h window |
| Strategy | Forward extended Kalman filter |

DD network and static PPP processing were carried out for different positioning combination modes: GPS stand-alone positioning mode, GLONASS stand-alone positioning mode, and GPS+GLONASS combined positioning mode.

### 2.5. Accuracy Assessment

The station coordinates and ZTDs provided by the IGS were used as references to assess the accuracy of DD network solutions and PPP solutions. The Root Mean Square Error (RMSE) of daily coordinate estimates was used as the accuracy assessment indicator,

$$\mathrm{RMSE}_{\mathrm{COORDINATE}} = \sqrt{\frac{(\mathrm{COORDINATE}_{\mathrm{estimated}} - \mathrm{COORDINATE}_{\mathrm{IGS}})^2}{\mathrm{n}}} \quad (3)$$

where n is the total number of daily coordinate estimates.

The tropospheric products from IGS are sampled every 300 s, while, in this paper, the ZTDs estimated interval by DD strategy was one hour, and 30 s by PPP strategy. Thus, the tropospheric estimates or the IGS products needed to be resampled to match the sampling intervals. The current study resampled the IGS tropospheric products to 1 h and the PPP tropospheric to 300 s to coincide with IGS products. The RMSE of ZTD estimates is,

$$\mathrm{RMSE}_{\mathrm{ZTD}} = \sqrt{\frac{(\mathrm{ZTD}_{\mathrm{estimated}} - \mathrm{ZTD}_{\mathrm{IGS}})^2}{\mathrm{n}}} \quad (4)$$

where n is the total number of available ZTD estimates after resampling.

## 3. Data Selection

To comprehensively study the GLONASS performance, three networks located in high, middle, and low latitude regions were employed. In addition to the differences in latitude, the following three factors were also considered in the selection of the IGS station:

1. To comprehensively evaluate and compare the performance of GLONASS stand-alone mode, GPS stand-alone mode, and GPS+GLONASS combined mode, the station's receiver should receive both GPS and GLONASS observations. The receivers employed in the three networks are listed in Tables 2–4, respectively.
2. Using the GPS antenna PCC model for GLONASS will introduce systematic bias [9,27,28]. To avoid this bias, the station's antenna and radome types should have GPS and GLONASS-specific PCC models in the IGS antenna files. The antenna and radome types used in the three networks are also given in Tables 2–4, respectively.
3. The baseline accuracies are related to the length of the baseline [29]. To precisely assess the performance of GLONASS in terms of DD network processing, the mean baseline lengths of the networks should be similar.

**Table 2.** The GNSS receivers and the antenna + radome types of the high latitude stations.

| Station Name | Receiver Type | Antenna + Radome Type |
|---|---|---|
| KIRU | SEPT POLARX5 | SEPCHOKE_B3E6 SPKE |
| MAR7 | TRIMBLE ALLOY | LEIAR25.R3 LEIT |
| METG | SEPT POLARX5 | TRM59800.00 SCIS |
| NYA1 | TRIMBLE NETR8 | ASH701073.1 SNOW |
| SOD3 | JAVAD TRE_3 DELTA | JAVRINGANT_DM SCIS |
| SVTL | JAVAD TRE_3 DELTA | JAVRINGANT_DM JVDM |
| TRO1 | TRIMBLE NETR9 | TRM59800.00 SCIS |

**Table 3.** The GNSS receivers and the antenna + radome types of the middle latitude stations.

| Station Name | Receiver Type | Antenna + Radome Type |
|---|---|---|
| AJAC | SEPT POLARX5 | TRM115000.00 NONE |
| HERT | LEICA GRX1200GGPRO | LEIAT504GG NONE |
| JOZE | SEPT POLARX5 | SEPCHOKE_B3E6 NONE |
| MATG | LEICA GR10 | LEIAR25 NONE |
| TLSG | SEPT POLARX5TR | TRM59800.00 NONE |
| WARN | JAVAD TRE_3 DELTA | LEIAR25.R4 LEIT |
| WTZR | LEICA GR50 | LEIAR25.R3 LEIT |

**Table 4.** The GNSS receivers and the antenna + radome types of the low latitude stations.

| Station Name | Receiver Type | Antenna + Radome Type |
|---|---|---|
| BRAZ | TRIMBLE NETR9 | TRM57971.00 NONE |
| CHPI | SEPT POLARX5 | TPSCR.G3 NONE |
| SALU | TRIMBLE NETR9 | TRM115000.00 NONE |
| SAVO | TRIMBLE NETR9 | TRM115000.00 NONE |
| SPTU | TRIMBLE NETR9 | TRM57971.00 NONE |
| TOPL | TRIMBLE NETR9 | TRM115000.00 NONE |
| UFPR | TRIMBLE NETR9 | TRM115000.00 NONE |

According to the above station select criterion, 21 IGS stations were selected and formed 3 networks, distributed in high, middle, and low latitude regions, as shown in Figure 2. The high latitude network locates between 60°N and the North Pole, the middle latitude network lies between 30°N and 60°N, and the low latitude network situates between the equator and 30°S. The baseline lengths of the three networks are approximately 660 km, 880 km, and 778 km, respectively. The performance evaluation period was from 1 January 2021 (DOY 001) to 30 June 2021 (DOY 181). The GPS and GLONASS observations were downloaded from NASA CDDIS [30].

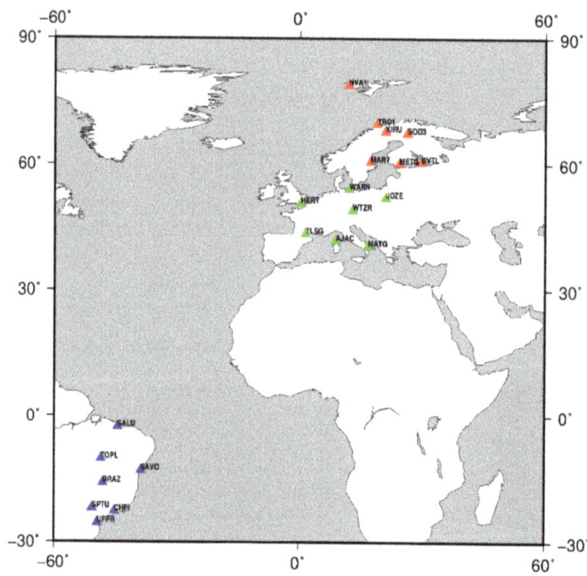


**Figure 2.** The IGS tracking stations of the high (red triangle), middle (green triangle), and low latitude networks (blue triangle).

## 4. Results and Discussion

The data quality of GPS and GLONASS, the constellation visibility of GPS, GLONASS, and GPS+GLONASS, and the performance of the DD network solutions and PPP solutions, are shown and discussed in this section.

### *4.1. Data Quality*

The DI rate, MP1, MP2, SN1 (the SNR of L1), and SN2 (the SNR of L2) of GPS and GLONASS observations for each station are calculated and presented in Figure 3. The mean value of each indicator is listed in Table 5, where G and R denote GPS and GLONASS, respectively.

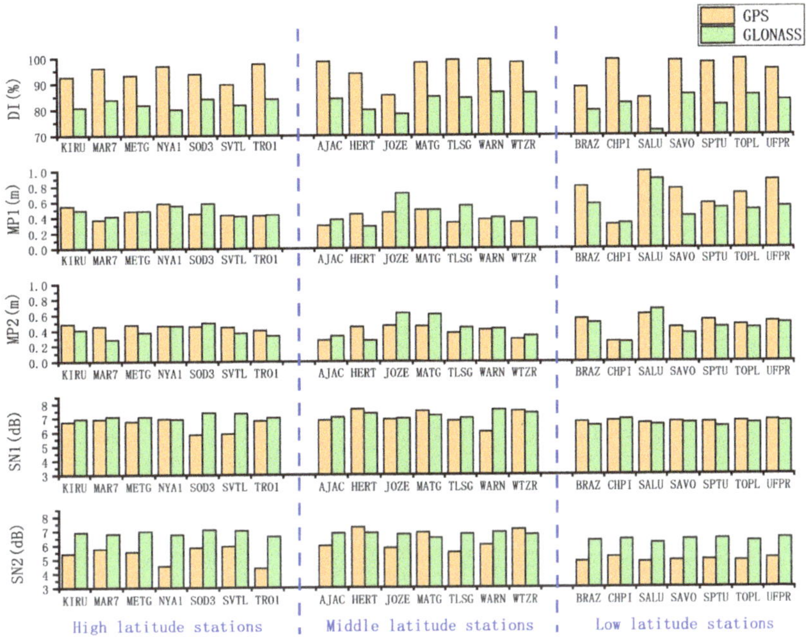


**Figure 3.** The results of the data quality check.

**Table 5.** The mean value of each data quality indicator in the three networks.

| Region \ Indicator | DI (%) | | MP1 (m) | | MP2 (m) | | SN1 (dB) | | SN2 (dB) | |
|---|---|---|---|---|---|---|---|---|---|---|
| | G | R | G | R | G | R | G | R | G | R |
| High latitude network | 94.33 | 82.37 | 0.47 | 0.48 | 0.45 | 0.39 | 6.56 | 7.12 | 5.33 | 6.88 |
| Middle latitude network | 96.11 | 83.46 | 0.39 | 0.46 | 0.39 | 0.43 | 7.02 | 7.20 | 6.35 | 6.76 |
| Low latitude network | 94.81 | 81.33 | 0.72 | 0.53 | 0.48 | 0.44 | 6.72 | 6.59 | 4.92 | 6.32 |

The data integrity rate of GLONASS in 21 stations of the 3 networks is lower than that of GPS, as is evident in Figure 3. As shown in Table 5, the calculated average data integrity rates of GLONASS in the three networks are 82.37%, 83.46%, and 81.33%, respectively, significantly lower than those of GPS with 94.33%, 96.11%, and 94.81%, respectively.

The MP1 and MP2 of GPS presented similar performance to that of GLONASS in the high latitude network. However, the MP1 and MP2 of GPS in the middle latitude network are smaller than those of GLONASS in most stations, except station HERT. By contrast, the MP1 and MP2 of GLONASS in the low latitude network are smaller than those of GPS. The calculated average value of MP1 and MP2 in Table 5 indicates similar results, but the differences in MP1 and MP2 between GPS and GLONASS are insignificant.

The difference in SN1 between GPS and GLONASS is minor in the three networks, except SOD3 and SVTL in the high latitude network and WARN in the middle latitude network. The SN2 of GLONASS is significantly better than that of GPS in the high and low latitude networks and similar to that of GPS in the middle latitude network.

Among the 21 tracking stations, however, the data quality of SALU in the low latitude network is significantly worse than other stations. The DI rate of SALU is 84.3% for GPS and 71.6% for GLONASS. The MP1 of SALU is up to 0.99 m and 0.89 m for GPS and GLONASS observations, respectively, much larger than those calculated average values, 0.39 m and 0.46 m for GPS and GLONASS. The MP2 of SALU presented similar results as that of MP1.

### 4.2. Constellation Visibility

The average number of visible satellites, the average PDOP value, and the satellite elevation distribution of the three tracking networks were analyzed with numerical comparison and statistical study. The number of visible satellites and their statistical property is shown in Figures 4 and 5, respectively.

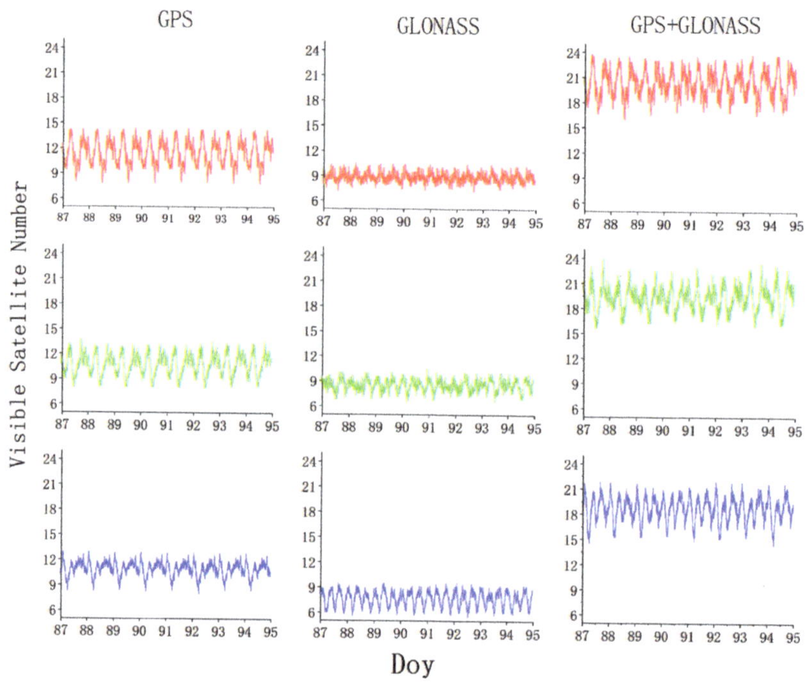


**Figure 4.** The average number of visible satellites of GPS, GLONASS, and GPS+GLONASS in the high (red), middle (green), and low (blue) latitude networks.

From Figure 5, during the 8 days test period, both constellations have the largest average number of visible satellites in the high latitude network among the three networks. The average visible number of GPS in high latitude regions is significantly larger than in the middle and low latitude regions. The average number of visible satellites of GLONASS in the high and the middle latitude regions is similar but significantly larger than that of the low latitude region.

From the statistical point of view, the GLONASS visible satellites have the smallest standard deviation in the high and middle latitude networks. The percentages of most observed 9 and 8 GLONASS satellites in the high latitude network are 52.37% and 27.63%, respectively. The percentages of most observed 9 and 8 GLONASS satellites in the middle latitude network are 34.20% and 33.64%, respectively. However, the number of GLONASS visible satellites reduces significantly, and the standard deviation increases significantly in the low latitude network. Moreover, the standard deviation of GPS+GLONASS is relatively larger than GPS or GLONASS, and the average number of visible satellites is also larger than those of GPS and GLONASS. The probability of observing satellites less than 12 in three networks of combined GPS and GLONASS is almost zeros, indicating that the combined constellations provide more than 12 visible satellites in most cases.

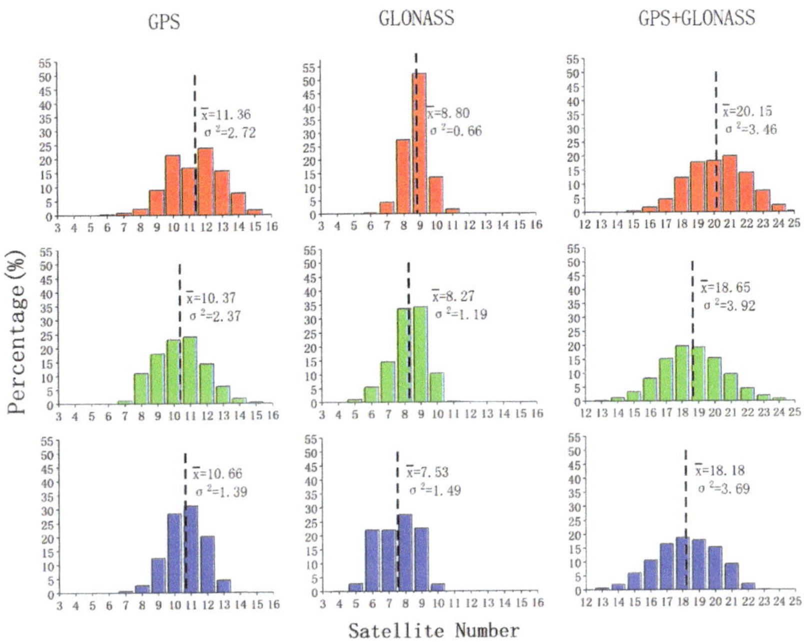


**Figure 5.** The histogram of visible satellites in the high (red), middle (green), and low (blue) latitude networks. The position of the dashed line and the value $x^{-}$ indicate the average number of visible satellites. The $\sigma^2$ represents the variance of the visible satellites' distribution. Please note that the horizontal axis of GPS and GLONASS visible satellites differs from that of GPS+GLONASS.

The histogram of the elevation distribution related to GLONASS, GPS, and GLONASS+GPS is also presented in Figure 6 to further evaluate the quality of the observed satellites.

From Figure 6, GLONASS has a much better elevation distribution than GPS in the high latitude network. The most observed GLONASS satellites' elevation angles in the high latitude network are between 13–33 degrees, while those of GPS are between 6–26 degrees, which are obviously lower than that of GLONASS. In addition, the mean elevation angle of GLONASS is 35.01°, which is 3.97° higher than that of GPS. In the middle latitude network, the elevation angle distribution and average elevation angle between GPS and GLONASS have inconspicuous disparity. In the low latitude network, the percentage peak of GLONASS corresponds to an obviously lower elevation angle than GPS. The elevation distribution of GPS+GLONASS lies between GPS and GLONASS.

The PDOP of GPS, GLONASS, and GPS+GLONASS during the testing period is illustrated in Figure 7.

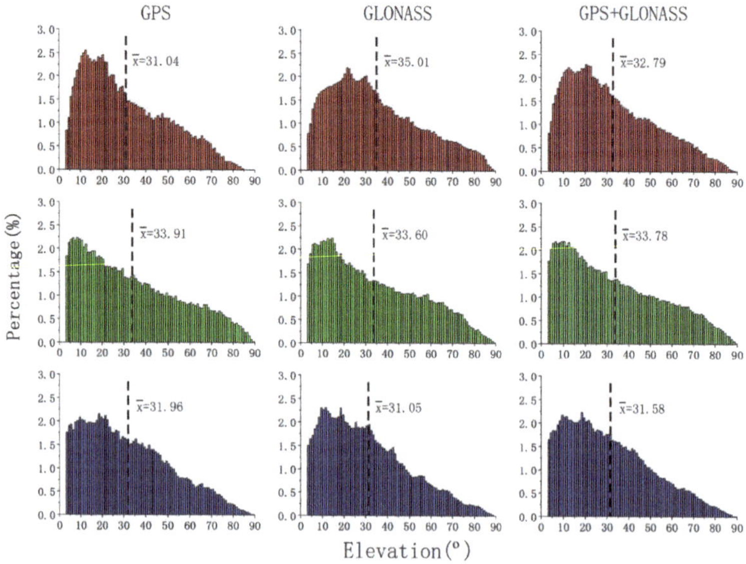


**Figure 6.** The histogram of elevation distribution of the high (red), middle (green), and low (blue) latitude networks, the position of the dashed line and the value x⁻ indicate the mean elevation.

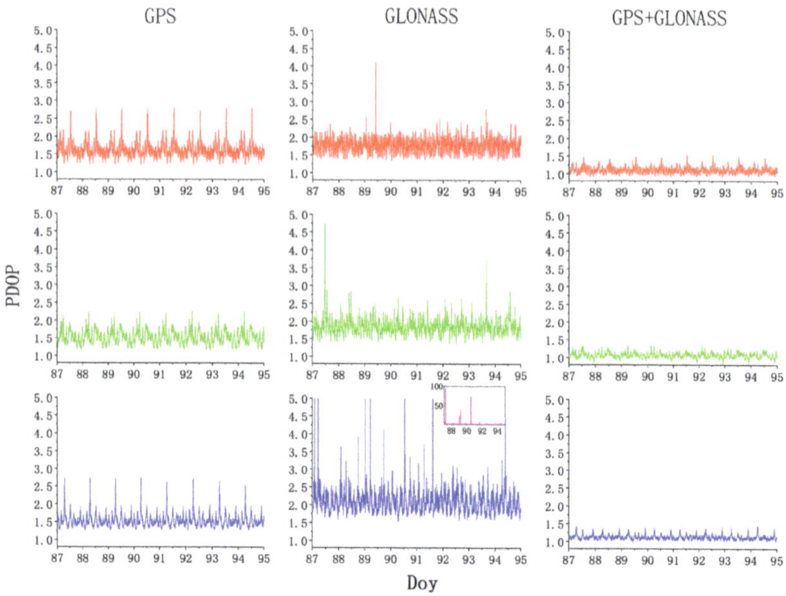


**Figure 7.** The PDOP of GPS, GLONASS, and GPS+GLONASS in the high (red), middle (green), and low (blue) latitude networks (The GLONASS PDOP for the low latitude network exceed the coordinate threshold, and the small picture in the upper right corner with the rose thread shows the full view of the PDOP).

Figure 7 shows that the PDOP values of GPS are less than 3.0 in all three regions, and the most stable PDOP values appear in middle latitude regions, indicating that the observation geometry of GPS in that region is the best among the three regions. Moreover, the PDOP values of GPS exist daily period in all three regions. GLONASS has the most stable PDOP in high latitude regions, which is more stable than GPS in the same region. However, the PDOP values of GLONASS increase obviously as the latitude decreases. In addition, the PDOP values of GLONASS at low latitude regions show some abnormally large values, up to 92.68, as shown in the upper right corner of the subfigure for low latitude GLONASS PDOP. The PDOP values of GPS+GLONASS exhibit better performance than both GLONASS and GPS stand-alone systems. Although some periods exist in high latitude regions where the PDOP of GPS is up to 2.80, the PDOP values of GPS+GLONASS are quite stable and less than 1.55. Although the PDOP of GPS is up to 2.75 in some periods in low latitudes and the PDOP of GLONASS is very large at certain epochs, the PDOP of GPS+GLONASS remains stable and below 1.44. The combined GPS and GLONASS constellations have the most obvious improvement in observation geometry at low latitudes.

### *4.3. DD Network Solutions*

The accuracy of coordinates, the ambiguity fixing rate, and the accuracy of ZTD estimates with the DD network processing strategy were analyzed.

#### 4.3.1. Accuracy of Coordinates

Taking IGS daily coordinates as the reference value, the differences between the estimated coordinates and the IGS daily coordinates were calculated. The stations SOD3 (in the high latitude network), WTZR (in the middle latitude network), and SPTU (in the low latitude network) were employed to illustrate the positioning performance of GPS, GLONASS, and GPS+GLONASS. The other stations present results similar to these three stations. The coordinate error in the North (N), East (E), and Up (U) components of the three selected stations are presented in Figure 8.

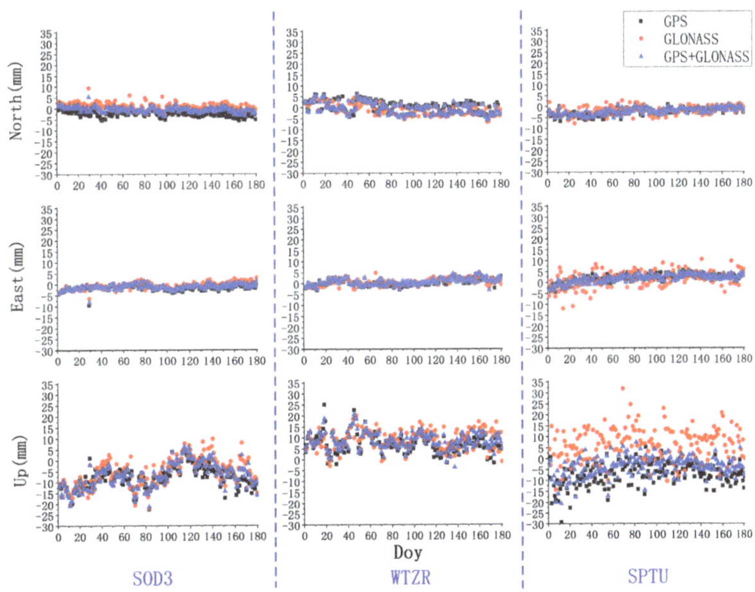


**Figure 8.** The coordinate error series of station SOD3, WTZR, and SPTU estimated with DD processing.

As shown in Figure 8, the GLONASS coordinate error components of station SOD3 and WTZR, located in the high and middle latitude networks, respectively, are as steady as GPS and GPS+GLONASS results. The GLONASS error components are more fluctuated than that of GPS and GPS+GLONASS results for the low latitude station SPTU.

The RMSE of N, E, and U components, together with the three-dimensions (3D) RMSE for all the stations, were calculated and are presented in Figure 9. Figure 9 shows that in the low latitude network, the GLONASS positioning accuracy is obviously worse than that of GPS and GPS+GLONASS. The coordinates of the SALU station exhibit the worst accuracy, and the station also has poor data quality, as shown in Figure 3. Similar results can be found from the mean RMSE of coordinate estimates of each network, as shown in Table 6.

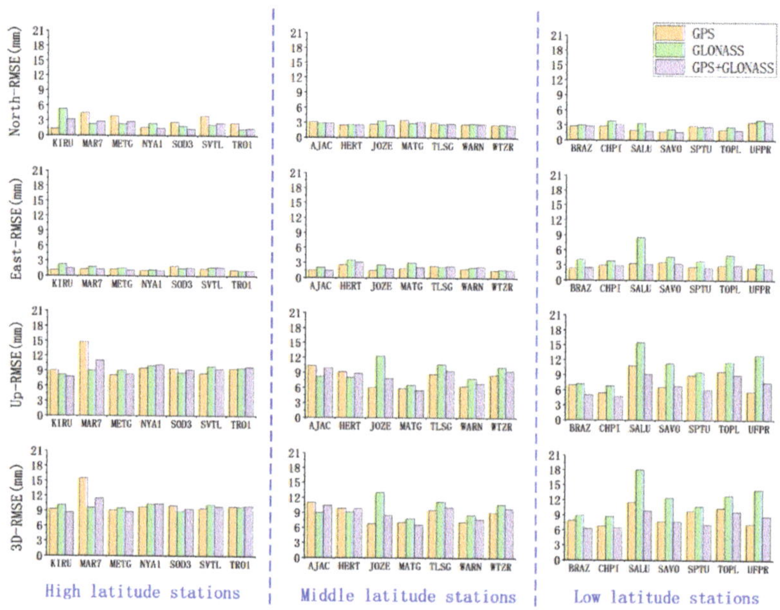


**Figure 9.** The RMSE of coordinates estimated with DD processing.

**Table 6.** The mean RMSE of coordinates of each network estimated with DD processing and their comparison among different processing modes (where R/G indicates the accuracy comparison of GLONASS and GPS results, (G + R)/G indicates the accuracy comparison of GPS+GLONASS results and GPS estimates. The positive red and negative green values indicate the percentage increment and reduction of accuracy, respectively).

| Component \ System | G (mm) | R (mm) | G + R (mm) | R/G (%) | (G + R)/G (%) |
|---|---|---|---|---|---|
| N | 2.99 | 2.58 | 2.27 | +13.79 | +24.29 |
| E | 1.38 | 1.60 | 1.38 | −16.17 | +0.00 |
| U | 9.87 | 9.24 | 9.41 | +6.35 | +4.66 |
| 3D | 10.40 | 9.73 | 9.77 | +6.50 | +6.05 |
| N | 2.91 | 2.88 | 2.78 | +1.14 | +4.48 |
| E | 1.90 | 2.47 | 2.12 | −30.17 | −11.77 |
| U | 7.89 | 9.14 | 8.25 | −15.79 | −4.59 |
| 3D | 8.62 | 9.89 | 8.96 | −14.74 | −3.96 |
| N | 2.60 | 3.18 | 2.57 | −22.13 | +1.37 |
| E | 2.94 | 4.79 | 2.87 | −62.86 | +2.39 |
| U | 7.77 | 10.74 | 6.93 | −38.24 | +10.89 |
| 3D | 8.71 | 12.18 | 7.92 | −39.93 | +9.00 |

The RMSE results clearly show that the accuracies of coordinates estimated with GLONASS are 13.79% and 6.35% better than that of GPS on N and U components in the high latitude network. On the E components, however, the GLONASS positioning accuracy decreased by 16.17% compared to GPS. Therefore, the 3D accuracy of GLONASS is 6.50% better than that of GPS. In addition, the GPS+GLONASS combined mode presents the best results among the three constellations, the accuracy improvements on N and U components are 24.29% and 4.66% compared to GPS, respectively, and the 3D accuracy is 6.05% better than that of GPS. The positioning performance is consistent with the analysis of PDOP in Section 4.2. The stable and good PDOP enables the high positioning accuracy of GLONASS. The PDOP of the combined constellations significantly improved over GPS; therefore, GPS+GLONASS shows the highest accuracy in the high latitude network.

The positioning accuracy of the middle latitude network reveals that GPS presents the best positioning results in E and U, as well as 3D components. GLONASS and GPS+GLONASS exhibit a slightly better positioning accuracy of 1.14% and 4.48% than GPS on the N component, respectively. The positioning accuracies of GLONASS on E and U components are 30.17% and 15.79% worse than that of GPS. In addition, the positioning accuracies of GPS+GLONASS on E and U components are 11.77% and 4.59% worse than that of GPS. Furthermore, the 3D positioning accuracy of GLONASS and GPS+GLONASS is reduced by 14.74% and 3.96% than that of GPS, respectively. As can be seen from Figure 7, GPS has the best and the most stable PDOP values in the middle latitudes when compared with the PDOP of high and low latitudes. The improvement of the GPS+GLONASS combined constellations' PDOP over GPS is weaker when compared to high and low latitudes. Besides, currently, the accuracy of GLONASS satellite ephemerides is about 3 cm, a bit lower than that of GPS, with a 2.5 cm accuracy. Hence, it can be inferred that when the stand-alone GPS has adequate visible satellites with good observation geometry, the addition of GLONASS observations shows no positive contribution to the accuracy of the coordinates. This is consistent with the conclusion of Cai and Gao [10].

The positioning accuracy of the middle latitude network shows that GLONASS presented the worst positioning accuracy among the three modes. The positioning accuracies of GLONASS on N, E, and U components are 22.13%, 62.86%, and 38.24% worse than that of GPS, and the 3D positioning accuracy is 39.93% lower than that of GPS. The positioning accuracy of GPS+GLONASS, however, is better than that of GPS, which is 1.37%, 2.39%, and 10.89% on N, E, and U components, respectively, and the 3D accuracy increases by 9.00%. The poor positioning results of GLONASS and the improvement of GPS+GLONASS positioning performance by introducing GLONASS observations are consistent with the analysis of PDOP.

4.3.2. Ambiguity Fixing Rate

The ambiguity fixing rate of GPS, GLONASS, and GPS+GLONASS modes of the three networks is listed in Table 7. GPS has the lowest ambiguity fixing rate in the middle latitude network but the highest ambiguity fixing rate in the low latitude network. As there is no obvious difference in the GPS positioning accuracy of the three networks, we analyzed the types of receivers used. The receivers of the high latitude network were of 3 brands and 5 models, the receivers of the middle latitude network were of 3 brands and 6 types, while in the low latitude network, there were only 2 brands and 2 models receivers. This suggests that the different levels of ambiguity fixing rate could be related to the number of receiver types used in different networks. The different signal distortion biases between inhomogeneous receivers affect the GNSS data processing [31,32]. Despite having the best receiver homogeneity, the ambiguity fixing rate of GLONASS in the low latitude network was only 26.4%, but 74.7% for GPS.

**Table 7.** The ambiguity fixing rate of DD processing (G, R, and G + R in black body denote the data processing mode, respectively. No bold G, R, and G + R represent the ambiguity fixing rate of GPS+GLONASS processing mode, respectively).

| System / Region | **G (%)** | **R (%)** | **G + R** | | |
|---|---|---|---|---|---|
| | | | **G (%)** | **R (%)** | **G + R (%)** |
| High latitude network | 70.3 | 63.9 | 69 | 62.8 | 66.5 |
| Middle latitude network | 53.1 | 56 | 51.5 | 54.8 | 53.0 |
| Low latitude network | 74.7 | 26.4 | 74.6 | 33.9 | 57.5 |

4.3.3. Tropospheric Estimates

Taking IGS tropospheric products as the reference value, the differences between the estimated ZTDs and the IGS products were calculated. The stations SOD3, WTZR, and SPTU were also used to illustrate the estimation accuracy of three different positioning modes. The ZTD errors are presented in Figure 10. The data gaps in Figure 10 were caused by missing observations or the reference data. It can be seen that the ZTD estimates present wider discrepancies as the latitude decreases.

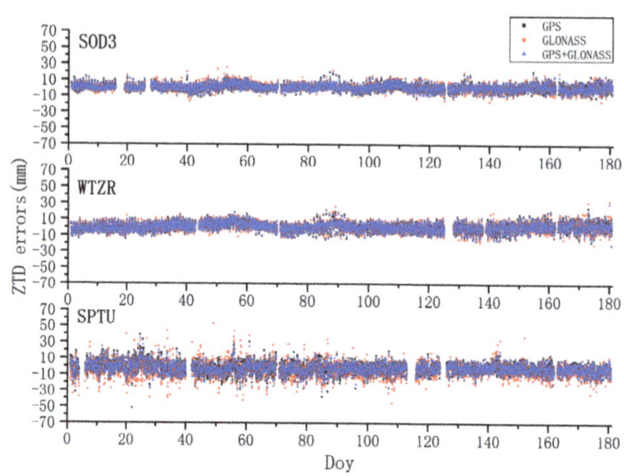


**Figure 10.** The ZTD error series of station SOD3, WTZR, and SPTU estimated with DD processing.

The RMSE of all the stations is shown in Figure 11. The mean RMSE of each network and their comparison between different processing modes are shown in Table 8.

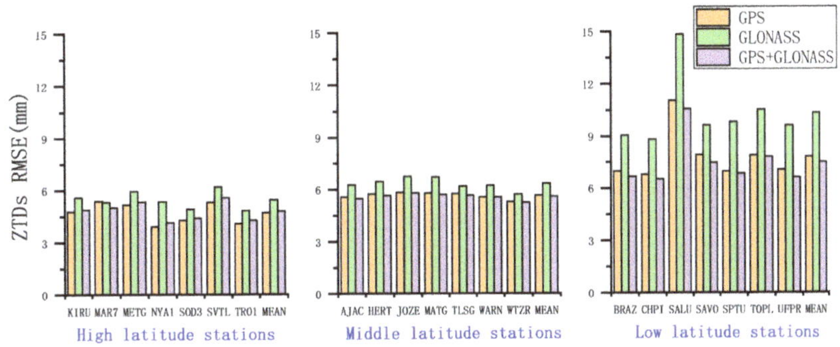


**Figure 11.** The RMSE of ZTDs estimated with DD processing.

**Table 8.** The mean RMSE of ZTDs estimated with DD processing for each network and their comparison between different processing modes (The positive red and negative green values indicate the percentage increment and reduction of accuracy, respectively).

| System / Region | G (mm) | R (mm) | G + R (mm) | R/G (%) | G + R/G (%) |
|---|---|---|---|---|---|
| High latitude network | 4.70 | 5.45 | 4.80 | −15.85 | −2.08 |
| Middle latitude network | 5.63 | 6.32 | 5.57 | −12.26 | +1.06 |
| Low latitude network | 7.81 | 10.33 | 7.50 | −32.24 | +4.00 |

The statistical results show that, in all the three networks, the ZTD estimates of GLONASS are obviously worse than that of GPS, which are 15.85%, 12.26%, and 32.24% lower in the high, middle, and low latitude networks, respectively. The accuracy of GPS+GLONASS in the high latitude network is 2.08% worse than that of GPS but slightly better in the middle and low latitude networks, about 1.06% and 4.00%, respectively. The consistency between the estimated results and the IGS products decreases as the latitude decreases for GPS, GLONASS, and GPS+GLONASS.

### *4.4. Static PPP Results*

The positioning accuracy, the convergence time, and the accuracy of ZTD estimates with the PPP processing strategy are analyzed.

#### 4.4.1. Positioning Accuracy

Taking IGS daily coordinates as the reference value, the coordinate difference between the estimated coordinates with PPP and the IGS daily coordinates was calculated. The coordinate error series of GPS, GLONASS, and GPS+GLONASS modes on N, E, and U components of SOD3, WTZR, and SPTU are shown in Figure 12.

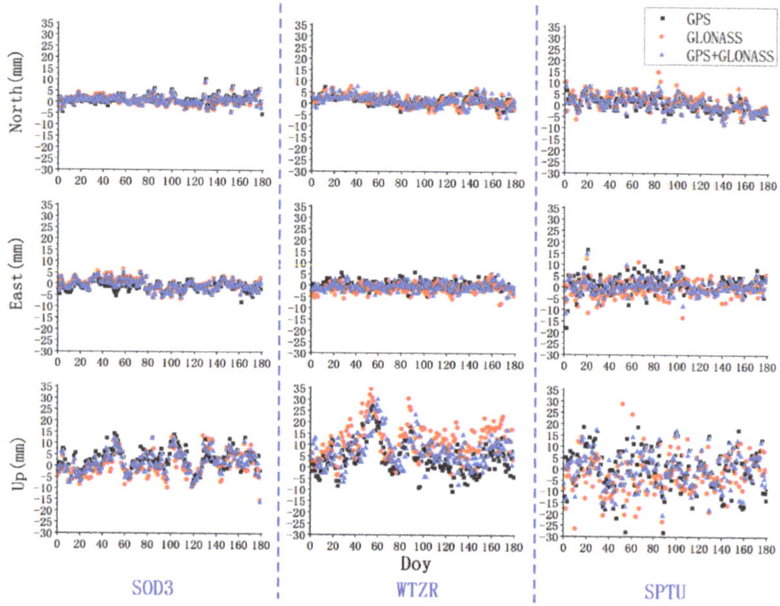


**Figure 12.** The coordinate error series of station SOD3, WTZR, and SPTU estimated with PPP.

As shown in Figure 12, the GLONASS coordinate error series of SOD3 is as steady as that of GPS, while the coordinate error of WTZR and SPTU estimated with GLONASS observations is much more dispersed than other estimate modes. In addition, the magnitude of the error series on U components of three positioning modes at the SPTU station is larger than at the other stations.

The RMSE of N, E, and U components, together with the 3D RMSE for all the stations, are shown in Figure 13. The mean RMSE of N, E, and U components, together with the 3D RMSE for each network, are calculated and listed in Table 9.

**Table 9.** The mean RMSE of coordinates for each network estimated with PPP and their comparison between different processing modes (The positive red and negative green values indicate the percentage increment and reduction of accuracy, respectively).

| System / Component | G (mm) | R (mm) | G + R (mm) | R/G (%) | G + R/G (%) |
|---|---|---|---|---|---|
| N | 2.66 | 2.30 | 2.16 | +13.62 | +18.78 |
| E | 3.16 | 3.13 | 2.83 | +0.87 | +10.37 |
| U | 6.67 | 6.86 | 6.15 | −2.83 | +7.87 |
| 3D | 7.92 | 8.03 | 7.18 | −1.37 | +9.35 |
| N | 2.97 | 3.56 | 3.14 | −19.83 | −5.63 |
| E | 2.64 | 3.21 | 2.56 | −21.77 | +2.99 |
| U | 7.54 | 9.08 | 7.62 | −20.40 | −1.03 |
| 3D | 8.56 | 10.39 | 8.69 | −21.43 | −1.52 |
| N | 3.44 | 3.86 | 3.48 | −12.06 | −0.96 |
| E | 3.91 | 4.09 | 3.14 | −4.54 | +19.73 |
| U | 8.19 | 8.87 | 7.23 | −8.30 | +11.79 |
| 3D | 9.78 | 10.52 | 8.66 | −7.58 | +11.40 |

**Figure 13.** The RMSE of coordinates of all the stations estimated with PPP.

Figure 13 and Table 9 illustrate the RMSE of coordinates estimated with PPP. The GLONASS estimated coordinates have comparable accuracy with GPS in the high latitude network. The accuracies on N and E components are 13.62% and 0.87% better than that of GPS, but the accuracies on U and 3D components are 2.83% and 1.37% worse than that of GPS. GPS+GLONASS presents the best coordinate estimates in the high latitude network. The percentages of improvement over GPS on N, E, and U components are 18.78%, 10.37% and 7.87%, respectively, and the 3D positioning accuracy is 9.35% better than that of GPS. GPS stand-alone mode presents the best coordinate estimates except for the E component in the middle latitude network. The coordinate accuracy on the E component of combined GPS+GLONASS constellations has an advantage of 2.99% over GPS. The accuracies of GLONASS on N, E, and U components are worse than that of GPS, at 19.83%, 21.77%, and 20.40%, respectively. The 3D positioning accuracy of GPS+GLONASS is 1.52% lower than that of GPS. GLONASS shows the worst coordinate estimates among the three positioning modes in the low latitude network. The 3D positioning accuracy is 7.58% lower than that of GPS; however, the 3D positioning accuracy improvement of GPS+GLONASS is 11.40% over GPS. The positioning performance of PPP with different modes in different latitudes is basically consistent with that of the DD network solutions.

#### 4.4.2. Convergence Time

The convergence time performance of GPS, GLONASS, and GPS+GLONASS processing modes was studied. The criterion of convergence is achieving a positioning error of less than 1 decimeter on N, E, and U components. The mean convergence time for each station is shown in Figure 14. The mean convergence time for each network is listed in Table 10. The results clearly show that the convergence time of GLONASS PPP is distinctly longer than that of GPS, and the increased percentages are 51.90%, 45.28%, and 105.30% in the high, middle, and low latitude networks, respectively. Compared with GPS PPP, GPS+GLONASS processing reduces the convergence time, and the shortened percentages are 6.11%, 16.57%, and 14.60% in the high, middle, and low latitude networks, respectively. The convergence time gets longer as the latitude decreases for GPS, GLONASS, and

GPS+GLONASS positioning modes, and the convergence time of GLONASS in the low latitude network is obviously longer, up to 49.83 minutes.

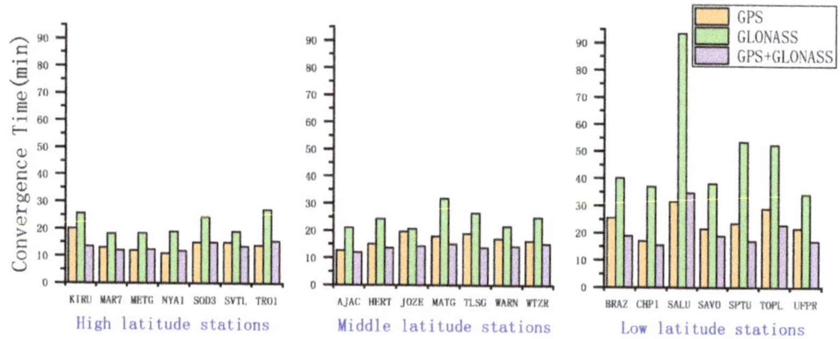


**Figure 14.** The convergence time of PPP.

**Table 10.** The mean convergence time of PPP for each network and their comparison between different processing modes (The red values indicate the percentage reduction of convergence time and the green values indicate the percentage increment of convergence time).

| System / Region | G (min) | R (min) | G + R (min) | R/G (%) | G + R/G (%) |
|---|---|---|---|---|---|
| High latitude network | 14.26 | 21.66 | 13.39 | 51.90 | 6.11 |
| Middle latitude network | 16.81 | 24.43 | 14.03 | 45.28 | 16.57 |
| Low latitude network | 24.27 | 49.83 | 20.73 | 105.30 | 14.60 |

4.4.3. Tropospheric Estimates

The differences between the estimated ZTDs with PPP after convergence and the IGS products were calculated to obtain the time series and the RMSE of ZTDs. Figure 15 shows the ZTD error series of station SOD3, WTZR, and SPTU. The RMSE of all the stations is shown in Figure 16. The mean RMSE of ZTDs for each network and their comparison between different processing modes are shown in Table 11.

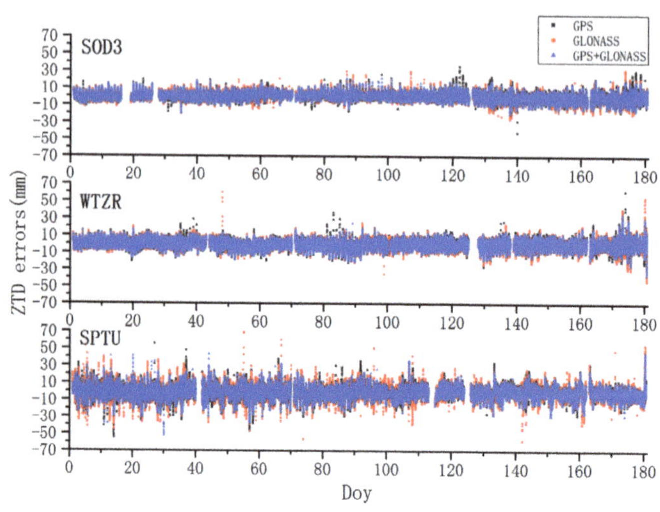


**Figure 15.** The ZTD error series of station SOD3, WTZR, and SPTU estimated with PPP.

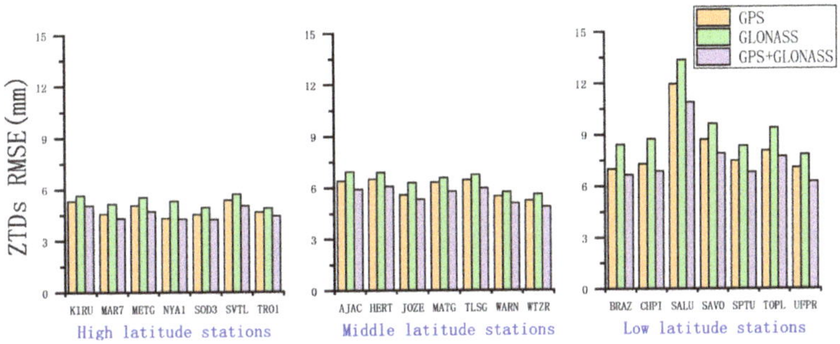


**Figure 16.** The RMSE of ZTDs estimated with PPP.

**Table 11.** The mean RMSE of ZTDs estimated with the PPP of each network and their comparison between different processing modes (The positive red and negative green values indicate the percentage increment and reduction of accuracy, respectively).

| System / Region | G (mm) | R (mm) | G + R (mm) | R/G (%) | G + R/G (%) |
|---|---|---|---|---|---|
| High latitude network | 4.84 | 5.33 | 4.59 | −10.10 | +5.19 |
| Middle latitude network | 6.00 | 6.41 | 5.58 | −6.82 | +6.98 |
| Low latitude network | 8.23 | 9.39 | 7.59 | −14.08 | +7.79 |

Figure 15 shows that the ZTD estimates are dispersed as the latitude decreases. From Figure 16 and Table 11, we can see that the ZTD estimates of GLONASS are less accurate than that of GPS in all three networks, which are 10.10%, 6.82%, and 14.08% lower in the high, middle, and low latitude networks, respectively. The addition of GLONASS will improve the accuracy of tropospheric estimates, and the accuracy improvements are 5.19%, 6.98%, and 7.79% in the high, middle, and low latitude networks, respectively. The accuracy of estimated ZTDs decreases as the latitude decreases for GPS, GLONASS, and GPS+GLONASS.

## 5. Conclusions

It has been more than 10 years since GLONASS restored its full constellation. A solid understanding of the positioning performance of GLONASS and its contribution to the multi-GNSS is important in muti-GNSS positioning applications. This paper presents the performance of GPS, GLONASS, and GPS+GLONASS with both DD network and PPP processing strategies in different latitude regions. According to the statistical analysis results of the current study, the following conclusions can be drawn:

1. The data integrity rate of GLONASS is lower than that of GPS;
2. Both GPS and GLONASS have the mean maximum number of visible satellites in high latitudes; however, the mean elevation angle of GLONASS is higher than that of GPS;
3. GLONASS has a comparable or even better positioning accuracy than GPS in high latitude regions, and the coordinates of GPS+GLONASS show the best accuracy;
4. GPS stand-alone mode gets the best positioning accuracy in middle latitude regions, and the additional GLONASS observations show no positive impact on GPS+GLONASS processing;
5. GLONASS shows the worst accuracy in low latitude regions, but the adding of GLONASS can improve the positioning accuracy of GPS+GLONASS processing mode when compared to GPS-only processing mode;
6. The addition of GLONASS will reduce the convergence time and improve the accuracy of ZTDs for PPP processing in high, medium, and low latitude regions.

Overall, for high-precision positioning users, GLONASS can be used as a stand-alone solution in high latitude regions. The inclusion of GLONASS as a combined multi-GNSS component in high and low latitude areas is advised. The addition of GLONASS observations is always recommended for PPP users of meteorology information.

**Author Contributions:** Conceptualization, Y.Z.; Data curation, Y.Z.; Investigation, Y.Z.; Software, F.Z.; Resources, G.N.; Validation, Y.Z. and F.Z.; Formal analysis, Y.Z. and C.Y.; Visualization, Y.Z.; Writing—original draft preparation, Y.Z.; Writing—review and editing, C.Y. and Y.Z.; Supervision, S.L.; Funding acquisition, C.Y. All authors have read and agreed to the published version of the manuscript.

**Funding:** This project is funded by the National Key Research and Development Program of China (No. 2020YFB0505802) and the Fundamental Research Funds for the Central Universities (2652017105).

**Data Availability Statement:** The GNSS data used in this paper are provided by CDDIS (https://cddis.nasa.gov/, accessed on 10 August 2021). The precise GNSS products are available from CODE (http://ftp.aiub.unibe.ch/CODE, accessed on 18 August 2021). The VMF1 troposphere mapping functions were obtained from VMF Data Server (http://doi.org/10.17616/R3RD2H, accessed on 20 August 2021). The ocean tide model was downloaded from the Ocean tide loading provider (http://holt.oso.chalmers.se/loading/index.html, accessed on 10 August 2021).

**Acknowledgments:** The authors thank the IGS for providing GNSS data and products.

**Conflicts of Interest:** The authors declare no conflict of interest.

## References


1. Hein, G.W. Status, perspectives and trends of satellite navigation. *Satell. Navig.* **2020**, *1*, 22. [CrossRef] [PubMed]
2. Dodson, A.H.; Moore, T.; Baker, F.D.; Swann, J.W. Hybrid GPS+GLONASS. *GPS Solut.* **1999**, *3*, 32–41. [CrossRef]
3. Stewart, M.P.; Tsakiri, M.; Wang, J.; Monico, J.F. The contribution of GLONASS measurements to regional and continental scale geodetic monitoring regimes. *Earth Planets Space* **2000**, *52*, 877–880. [CrossRef]
4. Bruyninx, C. Comparing GPS-only with GPS + GLONASS positioning in a regional permanent GNSS network. *GPS Solut.* **2007**, *11*, 97–106. [CrossRef]
5. Habrich, H. Evaluation of analysis options for GLONASS observations in regional GNSS networks. In *Geodetic Reference Frames*; Springer: Berlin/Heidelberg, Germany, 2009; pp. 121–129.
6. Cai, C.; Gao, Y. Precise point positioning using combined GPS and GLONASS observations. *Positioning* **2007**, *6*, 13–22. [CrossRef]
7. Alcay, S.; Inal, C.; Yigit, C.; Yetkin, M. Comparing GLONASS-only with GPS-only and hybrid positioning in various length of baselines. *Acta Geod. Geophys. Hung.* **2012**, *47*, 1–12. [CrossRef]
8. Nardo, A.; Huisman, L.; Teunissen, P.J.G. *GPS+GLONASS CORS Processing: The Asian-Pacific APREF Case//Earth on the Edge: Science for a Sustainable Planet*; Springer: Berlin/Heidelberg, Germany, 2014; pp. 239–246.
9. Zheng, Y.; Nie, G.; Fang, R.; Yin, Q.; Yi, W.; Liu, J. Investigation of GLONASS performance in differential positioning. *Earth Sci. Inform.* **2012**, *5*, 189–199. [CrossRef]
10. Cai, C.; Gao, Y. Modeling and assessment of combined GPS/GLONASS precise point positioning. *GPS Solut.* **2013**, *17*, 223–236. [CrossRef]
11. Yigit, C.O.; Gikas, V.; Alcay, S.; Ceylan, A. Performance evaluation of short to long term GPS, GLONASS and GPS/GLONASS post-processed PPP. *Surv. Rev.* **2014**, *46*, 155–166. [CrossRef]
12. Choy, S.; Zhang, S.; Lahaye, F.; Héroux, P. A comparison between GPS-only and combined GPS+GLONASS Precise Point Positioning. *J. Spat. Sci.* **2013**, *58*, 169–190. [CrossRef]
13. Mohammed, J.; Moore, T.; Hill, C.; Bingley, R.; Hansen, D. An assessment of static precise point positioning using GPS only, GLONASS only, and GPS plus GLONASS. *Measurement* **2016**, *88*, 121–130. [CrossRef]
14. Malik, J.S. Performance analysis of static precise point positioning using open-source GAMP. *Artif. Satell. J. Planet. Geod.* **2020**, *55*, 41–60. [CrossRef]
15. Hamed, M.; Abdallah, A.; Farah, A. Kinematic PPP using mixed GPS/GLONASS single-frequency observations. *Artif. Satell.* **2019**, *54*, 97–112. [CrossRef]
16. Deliktas, H.C. Investigation on the Contribution of GLONASS Observations to GPS Precise Point Positioning (PPP). Ph.D. Dissertation, The Ohio State University, Columbus, OH, USA, 2016.
17. Li, P.; Zhang, X. Integrating GPS and GLONASS to accelerate convergence and initialization times of precise point positioning. *GPS Solut.* **2014**, *18*, 461–471. [CrossRef]
18. Li, X.; Ge, M.; Dai, X.; Ren, X.; Fritsche, M.; Wickert, J.; Schuh, H. Accuracy and reliability of multi-GNSS real-time precise positioning: GPS, GLONASS, BeiDou, and Galileo. *J. Geod.* **2015**, *89*, 607–635. [CrossRef]

19. Abd Rabbou, M.; El-Rabbany, A. Performance analysis of precise point positioning using multi-constellation GNSS: GPS, GLONASS, Galileo and BeiDou. *Surv. Rev.* **2017**, *49*, 39–50. [CrossRef]
20. Pan, L.; Zhang, X.; Li, X.; Li, X.; Lu, C.; Liu, J.; Wang, Q. Satellite availability and point positioning accuracy evaluation on a global scale for integration of GPS, GLONASS, BeiDou and Galileo. *Adv. Space Res.* **2019**, *63*, 2696–2710. [CrossRef]
21. Estey, L.H.; Meertens, C.M. TEQC: The multi-purpose toolkit for GPS/GLONASS data. *GPS Solut.* **1999**, *3*, 42–49. [CrossRef]
22. Dach, R.; Hugentobler, U.; Fridez, P.; Meindl, M. *Bernese GPS Software Version 5.0*; Astronomical Institute, University of Bern: Bern, Switzerland, 2007.
23. Kouba, J. Implementation and testing of the gridded Vienna Mapping Function 1 (VMF1). *J. Geod.* **2008**, *82*, 193–205. [CrossRef]
24. re3data.org: VMF Data Server; Editing Status 2020-12-14; re3data.org-Registry of Research Data Repositories. Available online: https://www.re3data.org/repository/r3d100012025 (accessed on 26 July 2021).
25. Gu, S.; Zheng, F.; Gong, X.; Lou, Y.; Shi, C. *Fusing: A Distributed Software Platform for Real-Time High Precision Multi-GNSS Service*; IGS Workshop: Wuhan, China, 2018.
26. Lagler, K.; Schindelegger, M.; Böhm, J.; Krásná, H.; Nilsson, T. GPT2: Empirical slant delay model for radio space geodetic techniques. *Geophys. Res. Lett.* **2013**, *40*, 1069–1073. [CrossRef]
27. Dach, R.; Schmid, R.; Schmitz, M.; Thaller, D.; Schaer, S.; Lutz, S.; Steigenberger, P.; Wübbena, G.; Beutler, G. Improved antenna phase center models for GLONASS. *GPS Solut.* **2011**, *15*, 49–65. [CrossRef]
28. Liwosz, T. Effect of the GLONASS-specific receiver antenna phase center corrections on the results of European regional GNSS network. *Artif. Satell.* **2013**, *48*, 191. [CrossRef]
29. Wells, D.; Beck, N.; Kleusberg, A.; Krakiwsky, E.J.; Lachapelle, G.; Langley, R.B. *Guide to GPS Positioning*; Canadian GPS Assoc: Fredericton, NB, Canada, 1987.
30. Noll, C.E. The crustal dynamics data information system: A resource to support scientific analysis using space geodesy. *Adv. Space Res.* **2010**, *45*, 1421–1440. [CrossRef]
31. Villiger, R. (Ed.) *International GNSS Service Technical Report 2018 (IGS Annual Report)*; IGS Central Bureau and University of Bern; Bern Open Publishing: Bern, Switzerland, 2019.
32. Gong, X.; Gu, S.; Zheng, F.; Wu, Q.; Liu, S.; Lou, Y. Improving GPS and Galileo precise data processing based on calibration of signal distortion biases. *Measurement* **2021**, *174*, 108981. [CrossRef]

remote sensing

MDPI

Article

# Evaluation of Real-time Precise Point Positioning with Ambiguity Resolution Based on Multi-GNSS OSB Products from CNES

**Shi Du [1], Bao Shu [1,*], Wei Xie [1], Guanwen Huang [1], Yulong Ge [2] and Pan Li [1,3]**

1 College of Geology Engineering and Geomatics, Chang'an University, Xi'an 710054, China
2 School of Marine Science and Engineering, Nanjing Normal University, Nanjing 210023, China
3 Department of Geodesy, GeoForschungsZentrum (GFZ), Telegrafenberg, 14473 Potsdam, Germany
* Correspondence: baos613@chd.edu.cn; Tel.: +86-173-9276-8939

**Abstract:** Ambiguity resolution (AR) is a core technology that helps to speed up convergence time and increase positioning accuracy for precise point positioning (PPP), and the performance of PPP-AR is based on the quality of ambiguity resolution products. Real-time PPP-AR becomes a reality as users can now obtain publicly accessible real-time observable-specific signal bias (OSB) products from the Centre National d'Etudes Spatiales (CNES). Therefore, an analysis of the quality of OSB products and an evaluation of the performance of PPP-AR are required to promote the application of real-time positioning. For a total of 31 days between day of year (DOY) 121 and 151 in 2021, observation data were collected from 90 stations, and the OSB products were used to assess the experiments. As for the quality of the OSB products, the data availability (DA) of the GPS and Galileo satellites was greater than 97%, whereas that of BDS was less than 60%; the maximum fluctuation value (MAX) and standard deviation (STD) of the GPS, Galileo, and BDS satellites were 0.045 and 0.012; 0.081 and 0.028; and 0.292 and 0.085 cycles, respectively. In terms of ambiguity residuals using the OSB products, the wide-lane (WL) residual percentages within ±0.25 cycles for the GPS, Galileo, BDS-2, and BDS-3 systems were more than 92%, and the narrow-lane (NL) residual percentages within ±0.25 cycles for the four systems were 92%, 89%, 79%, and 60%, respectively. With regard to the performance of PPP-AR, the GPS+Galileo solution showed the best performance in the kinematic positioning mode, in which the mean root mean square (RMS) of positioning accuracy was 1.06, 1.27, and 2.85 cm for the east (E), north (N), and up (U) components, respectively, and the mean convergence time reached 9.6 min. In the static positioning mode, the mean convergence times of the GPS-only and GPS+Galileo solutions were 11.4 min and 8.0 min, respectively, and both of their mean RMS for positioning accuracy reached 0.79, 0.95, and 1.48 cm for the E, N, and U components, respectively. However, the addition of BDS did not further enhance the performance of multi-GNSS PPP-AR in either the kinematic or static positioning mode due to the poor quality of the real-time BDS products. More importantly, a prediction method was proposed to avoid re-convergence and to enhance the reliability of PPP-AR in the event of short-time missing real-time OSB products and to improve the positioning accuracy and the ambiguity fixed rate.

**Keywords:** real-time; precise point positioning; ambiguity resolution; OSB products

**Citation:** Du, S.; Shu, B.; Xie, W.; Huang, G.; Ge, Y.; Li, P. Evaluation of Real-time Precise Point Positioning with Ambiguity Resolution Based on Multi-GNSS OSB Products from CNES. *Remote Sens.* **2022**, *14*, 4970. https://doi.org/10.3390/rs14194970

Academic Editors: Chuang Shi, Xiaopeng Gong, Yidong Lou and Shengfeng Gu

Received: 2 September 2022
Accepted: 4 October 2022
Published: 6 October 2022

**Publisher's Note:** MDPI stays neutral with regard to jurisdictional claims in published maps and institutional affiliations.

**Copyright:** © 2022 by the authors. Licensee MDPI, Basel, Switzerland. This article is an open access article distributed under the terms and conditions of the Creative Commons Attribution (CC BY) license (https://creativecommons.org/licenses/by/4.0/).

## 1. Introduction

One method utilized by the global navigation satellite system (GNSS) to accomplish high-precision positioning is known as precise point positioning (PPP) [1], which has many advantages, including the need for only one receiver configuration as well as its flexible operation and wide-area application. To achieve centimeter-level positioning, PPP requires a convergence time of around 30 min, which makes it challenging to meet user demands in real time and severely restricts the marketing of PPP technology applications. Therefore, PPP ambiguity resolution (AR) technology has been suggested as a solution to the PPP convergence problem. The single-difference (SD) between satellites approach was first suggested by Ge et al. in 2008. The main idea was to estimate uncalibrated phase delay (UPD) products using the fractional portions of the float wide-lane (WL) and narrow-lane (NL) ambiguities. According to their experimental findings, more than 80% of the SD ambiguities from 450 stations over 14 days were fixed under the data test, and the positioning accuracy was increased by 30% in comparison to the float solution [2]. Meanwhile, the integer recovery clocks method and the decoupled clock model were proposed by Laurichese et al. [3,4] and Collins et al. [5,6], respectively, from the perspective of satellite clocks. Using the products provided by these methods, PPP-AR can also be implemented at the user end. Through theoretical derivation and a large amount of data analysis, Geng et al. and Shi et al. verified the equivalence of the three ambiguity resolution processes and the positioning performance [7,8]. The above and related studies enabled the implementation of PPP-AR, effectively shortening the convergence time and enhancing the positioning accuracy [9,10]. As a result, different forms of post-ambiguity resolution products have been released to users by several analysis centers. Chen et al. conducted an analysis of the performance of PPP-AR based on various ambiguity resolution products and demonstrated that these products could significantly increase positioning accuracy and reduce convergence times; however, the positioning performance varied, which suggests that the usefulness of ambiguity resolution products plays a role in PPP-AR performance [11].

The abovementioned research results indicate that PPP-AR technology can, to a certain extent, address the shortcomings of traditional PPP; however, these analyses were based on the post-processing mode. Various academics have conducted extensive studies on real-time PPP, including studies on the accuracy of real-time satellite orbit and clock offset products, the positioning performance of real-time PPP, real-time PPP for time transfer or tropospheric delay retrieval, and more [12–19]. Their findings have shown that real-time satellite orbit and clock offset products can be accurate enough to satisfy user needs, and real-time PPP positioning performance can achieve centimeter-level accuracy after convergence [13,16–19]. However, for the kinematic mode, the four-system positioning still requires longer than 15 min to converge to 0.1, 0.1, and 0.2 m for the east (E), north (N), and up (U) components, respectively [19]. Moreover, El-Mowafy presented a method that can guarantee real-time PPP with a 3D accuracy of less than a decimeter while real-time satellite orbit and clock offset products are unavailable [20]. With the demand for real-time AR products from users, CNES currently broadcasts real-time observable-specific signal bias (OSB) products to the world with open access, which allows for the user implementation of PPP-AR. However, there are a number of issues with real-time PPP-AR technology, including the inferior precision of real-time products compared to post-products, missing or outage real-time OSB corrections, etc. Thus, it is important to evaluate the quality of real-time OSB products and their performance in PPP-AR positioning in order to promote the application of real-time PPP.

In this contribution, the data quality of real-time multi-GNSS OSB products from CNES and the performance of the real-time PPP-AR were evaluated. In Section 2 of this paper, a dual-frequency mathematical model for PPP-AR based on OSB products is introduced. Then, in Section 3, the quality of the OSB products are analyzed using three metrics, the positioning accuracy and convergence time of the real-time PPP-AR are evaluated using various system combinations and positioning modes, and a method that can effectively avoid the interruption of short-term OSB products is validated. Finally, the conclusions are summarized.

## 2. Methodology

The dual-frequency ionosphere-free PPP model is first introduced in this section. Then, a method for recovering the integer features of ambiguities using OSB products is described. Finally, an AR method for an ionosphere-free combination is given.

## 2.1. Dual-frequency Ionosphere-Free PPP Model

The effects of first-order ionospheric delay can be removed using the dual-frequency ionosphere-free (IF) observation. The following are expressions for the code and carrier phase IF observations ($P^s_{r,IF}$ and $L^s_{r,IF}$, respectively) at a specific epoch:

$$\begin{cases} P^s_{r,IF} = \rho^s_r + t_r - t^s + T_r + (\alpha_{12}b_{r,1} + \beta_{12}b_{r,2}) - (\alpha_{12}b^s_1 + \beta_{12}b^s_2) + e^s_{r,IF} \\ L^s_{r,IF} = \rho^s_r + t_r - t^s + T_r + (\alpha_{12}B_{r,1} + \beta_{12}B_{r,2}) - (\alpha_{12}B^s_1 + \beta_{12}B^s_2) + \lambda_{IF}N^s_{r,IF} + \varepsilon^s_{r,IF} \end{cases} \tag{1}$$

$$\begin{cases} \lambda_{IF}N^s_{r,IF} = \alpha_{12}\lambda_1 N^s_{r,1} + \beta_{12}\lambda_2 N^s_{r,2} \\ \alpha_{12} = \frac{f_1^2}{f_1^2 - f_2^2}, \beta_{12} = -\frac{f_2^2}{f_1^2 - f_2^2} \end{cases} \tag{2}$$

where $s$ and $r$ denote the satellite and receiver, respectively; $\rho^s_r$ is the geometric distance between the satellite and the receiver (m); $t_r$ and $t^s$ are the receiver and satellite clock (m), respectively; $T_r$ is the slant tropospheric delay (m); $b_{r,i}$ and $b^s_i$ are the code hardware delay of the receiver and the satellite (m), respectively; $B_{r,i}$ and $B^s_i$ are the phase hardware delay of the receiver and the satellite (m), respectively; $f_i$ is the frequency; $\lambda_{IF}$ is the IF wavelength; $N^s_{r,IF}$ is the IF ambiguity (cycle); $e^s_{r,12}$ and $\varepsilon^s_{r,12}$ represent the sum of the measurement noises and the multipath effects for the code and carrier phase IF observations (m), respectively.

The satellite clock ($t^s$) in the PPP model is corrected using the clock offset products. Currently, the clock offset products are estimated via the IF combination, which causes the code IF hardware delay to be absorbed by the generated clock offset [21–23]. Therefore, the relationship between the clock offset products ($t^s_{IF}$) and the satellite clock ($t^s$) can be expressed as follows:

$$t^s_{IF} = t^s + (\alpha_{12}b^s_1 + \beta_{12}b^s_2) \tag{3}$$

By combining Equations (1) and (3), the linearized IF combined function model can be obtained as follows:

$$\begin{cases} p^s_{r,IF} = \mu^s_r \cdot x + t_{r,IF} - t^s_{IF} + T_r + e^s_{r,IF} \\ l^s_{r,IF} = \mu^s_r \cdot x + t_{r,IF} - t^s_{IF} + T_r + \lambda_{IF}\hat{N}^s_{r,IF} + \varepsilon^s_{r,IF} \end{cases} \tag{4}$$

where $p^s_{r,IF}$ and $l^s_{r,IF}$ denote the observed minus computed (OMC) IF observations of the code and carrier phase, respectively; $\mu^s_r$ is the unit vector from the receiver to the satellite, and $x$ is the coordinate of the estimated parameter. The receiver clock will absorb the code hardware delay of the receiver, which can be represented by the following expression:

$$t_{r,IF} = t_r + (\alpha_{12}b_{r,1} + \beta_{12}b_{r,2}) \tag{5}$$

Since the code and carrier phase observations in the PPP model share the same receiver clock parameter, where the parameter is based on the code observation, the code hardware delay will be added to the carrier phase observation [24,25]. Thus, the estimated ambiguity absorbs both the code and phase hardware delays, which can be written as follows:

$$\hat{N}^s_{r,IF} = N^s_{r,IF} + [\alpha_{12}(B_{r,1} - B^s_1 + b^s_1 - b_{r,1}) + \beta_{12}(B_{r,2} - B^s_2 + b^s_2 - b_{r,2})]/\lambda_{IF} \tag{6}$$

where $\hat{N}^s_{r,IF}$ and $N^s_{r,IF}$ represent the float and integer ambiguities, respectively. The integer feature of the IF ambiguity ($N^s_{r,IF}$) is destroyed, as can be seen from the equation above. Therefore, the integer feature for the IF ambiguity should first be recovered before PPP-AR.

## 2.2. Method for Recovering the Integer Feature of Ambiguity

The same function model between the server and the user is required for traditional AR products, which significantly limits the application of the AR products. With the development of AR technology, Laurichesse et al. proposed the use of undifferenced and uncombined OSB products, which comprise both code bias and phase bias [26]. Based on these OSB products, a variety of function models can be used by users to conduct single-frequency or multi-frequency PPP-AR, which helps to meet the current demand for multi-frequency, multi-system, and multi-model PPP-AR. By directly utilizing the OSB products in the code and phase observations, the satellite's code and phase hardware delays can be corrected. The following is an expression for the dual-frequency code bias ($\overline{b}_1^s$ and $\overline{b}_2^s$) and phase bias ($\overline{B}_1^s$ and $\overline{B}_2^s$) products [27]:

$$\begin{pmatrix} \overline{b}_1^s \\ \overline{b}_2^s \end{pmatrix} = \begin{pmatrix} \beta_{12}(b_1^s - b_2^s) \\ -\alpha_{12}(b_1^s - b_2^s) \end{pmatrix}, \begin{pmatrix} \overline{B}_1^s \\ \overline{B}_2^s \end{pmatrix} = \begin{pmatrix} \alpha_{12}b_1^s + \beta_{12}b_2^s - B_1^s \\ \alpha_{12}b_1^s + \beta_{12}b_2^s - B_2^s \end{pmatrix} \quad (7)$$

Since there is one code and phase bias for every code and phase observation, respectively, it is only necessary to add the OSB products directly to the raw observations when they are being employed. Additionally, the IF code and phase biases are formed, which can be expressed as follows:

$$\begin{cases} \overline{b}_{IF}^s = \alpha_{12}\overline{b}_1^s + \beta_{12}\overline{b}_2^s = 0 \\ \overline{B}_{IF}^s = \alpha_{12}\overline{B}_1^s + \beta_{12}\overline{B}_2^s = \alpha_{12}b_1^s + \beta_{12}b_2^s - \alpha_{12}B_1^s - \beta_{12}B_2^s \end{cases} \quad (8)$$

When $\overline{b}_{IF}^s$ and $\overline{B}_{IF}^s$ are combined with Equation (1), the new IF ambiguity can be stated as follows:

$$\hat{N}_{r,IF}^s = N_{r,IF}^s + [\alpha_{12}(B_{r,1} - b_{r,1}) + \beta_{12}(B_{r,2} - b_{r,2})]/\lambda_{IF} \quad (9)$$

Because of the negative impact of the code and phase hardware delays on the receiver, the IF ambiguity still does not have the integer feature. Usually, the SD between satellites can be used to eliminate the negative impact of the receiver, and the IF-SD ambiguity will recover the integer feature.

### 2.3. PPP-AR Process

The integer wide-lane (WL) and float narrow-lane (NL) ambiguities can be used to decompose the IF ambiguity in the IF combination, which translates as follows:

$$\hat{N}_{r,IF}^s = \left(\frac{f_2}{f_1 + f_2}\lambda_{WL}N_{r,WL}^s + \lambda_{NL}\hat{N}_{r,NL}^s\right)/\lambda_{IF} \quad (10)$$

where $N_{r,WL}^s$ is the integer WL ambiguity; $\hat{N}_{r,NL}^s$ is the float NL ambiguity; and $\lambda_{WL}$ and $\lambda_{NL}$ are the wavelengths of the WL and NL ambiguities, respectively, which can be formulated as follows:

$$\begin{cases} \lambda_{WL} = \frac{c}{f_1 - f_2} \\ \lambda_{NL} = \frac{c}{f_1 + f_2} \end{cases} \quad (11)$$

where $c$ denotes the speed of light.

The float NL ambiguity can be derived from Equation (10) when the float WL ambiguity is correctly fixed; the float WL ambiguity can be calculated through the Melbourne–Wübbena (MW) combination and fixed using the rounding method [2]. Then, the float WL and NL ambiguities can be summarized as follows:

$$\begin{cases} \hat{N}_{r,WL}^s = \left(\frac{L_{r,1}^s}{\lambda_1} - \frac{L_{r,2}^s}{\lambda_2} - \frac{\lambda_2 P_{r,1}^s + \lambda_1 P_{r,2}^s}{\lambda_{WL}(\lambda_2 + \lambda_1)}\right) \\ \hat{N}_{r,NL}^s = \frac{\lambda_{IF}\hat{N}_{r,\,IF}^s}{\lambda_{NL}} - \frac{\lambda_1 N_{r,\,WL}^s}{\lambda_2 - \lambda_1} \end{cases} \quad (12)$$

The WL ambiguity is affected by measurement noise and observation errors to a lesser extent due to its long wavelength, and it can be calculated with high accuracy using multi-epoch smoothing [28]. As a result, the rounding method can be used to fix it directly [2]. On the contrary, the float NL ambiguities from different satellites in the PPP model are highly correlated; hence, the LAMBDA method should be used for fixing. After inserting the fixed WL and NL ambiguities into (10), the fixed IF ambiguities are obtained. Then, a virtual observation is employed to constrain the filtering state, and the fixed solution is obtained.

It is important to note that when using the MW combination to obtain the float WL ambiguity, the antenna phase center correction of the receiver and satellite must be taken into account [29,30]. The formula is as follows:

$$\hat{N}^{s}_{r,WL} = \left( \frac{L^{s}_{r,1} + z^{s}_{r,1}}{\lambda_1} - \frac{L^{s}_{r,2} + z^{s}_{r,2}}{\lambda_2} - \frac{\lambda_2 \cdot (P^{s}_{r,1} + z^{s}_{r,1}) + \lambda_1 \cdot (P^{s}_{r,2} + z^{s}_{r,2})}{(\lambda_2 + \lambda_1) \cdot \lambda_{WL}} \right) \tag{13}$$

$$\begin{cases} z^{s}_{r,1} = z_{r,1} \cdot \sin\theta^{s}_{r} - z^{s}_{1} \\ z^{s}_{r,2} = z_{r,2} \cdot \sin\theta^{s}_{r} - z^{s}_{2} \end{cases} \tag{14}$$

where $z_{r,1}$ and $z_{r,2}$ are the vertical phase center offsets of the receiver antenna for the two frequencies, and similarly, $z^{s}_{1}$ and $z^{s}_{2}$ are those of the satellite antenna. $\theta^{s}_{r}$ denotes the elevation angle of the satellite $s$ with respect to the receiver $r$.

## 3. Real-Time PPP-AR Performance

In this section, we will first introduce the experimental data and processing strategies. Then, the quality of the real-time OSB products is analyzed. In addition, the performance of various PPP-AR combinations and modes based on the OSB products is assessed. Finally, a prediction method is proposed to effectively avoid the influence of short-term missing OSB products.

### 3.1. Data and Strategy

The post-store OSB products from day of year (DOY) 121 to 151 in 2021 were downloaded from http://www.ppp-wizard.net/products/REALTIME/ (accessed on 2 August 2022) to make it easier to analyze the real-time OSB products and PPP-AR.

The details of the dual-frequency and multi-GNSS real-time OSB products from CNES as of DOY 151 in 2021 are displayed in Table 1. It should be noted that the code bias was stable during the day [31]; therefore, the code bias was not analyzed.

**Table 1.** OSB product information provided by CNES.

| System | Frequency Number | Frequency Type | Code Bias | Phase Bias |
|---|---|---|---|---|
| GPS | 1 | L1 | C1C C1P C1W | L1C |
| | 2 | L2 | C2C C2S C2L C2X C2W | L2W |
| GLONASS | 1 | G1 | C1C C1P | |
| | 2 | G2 | C2C C2P | |
| Galileo | 1 | E1 | C1C | L1C |
| | 2 | E5a | C5Q | L5Q |
| BDS | 1 | B1I | C2I | L2I |
| | 2 | B3I | C6I | L6I |

**Note:** The sampling interval of post-store OSB products was 30 s. The "Frequency Number" only represents the index of the frequency for each system, which is convenient for the description below.

Figure 1 depicts the distribution of the 90 Multi-GNSS Experiment (MGEX) stations used to perform the PPP float solution and PPP-AR around the globe. These stations can receive GPS, Galileo, and BDS dual-frequency signals. The experiment time was from DOY 121 to 151 in 2021, which is a total of 31 days. To fully evaluate the real-time PPP-AR performance with different satellite systems, several combinations of static and kinematic positioning experiments were carried out. The combinations were as follows: 1) GPS-only; 2) GPS+Galileo; 3) GPS+Galileo+BDS. In the experimental analysis, incomplete observational data or data that did not pass quality checks (including detection, identification, and adaption (DIA) [32,33] and the Inter Quartile Range (IQR) method [34]) were excluded, where the excluded data were about 4% of the total data.

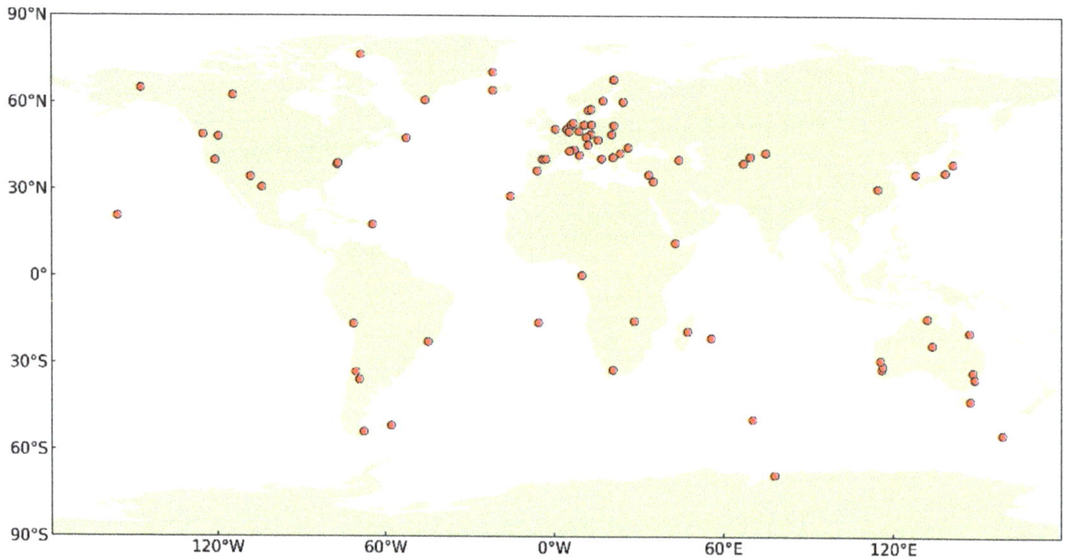


**Figure 1.** The 90 globally distributed MEGX stations used for the PPP-AR experiment.

All the PPP and PPP-AR experiments were performed with in-house software based on the secondary development of GAMP [35]. The phase windup was applied by the phase polarization effects. For the GPS L1/L2, Galileo E1/E5a, and BDS B1I/B3I frequencies, the igs14.atx file was utilized to correct the satellite phase center offset (PCO) and phase center variation (PCV); the receiver PCO and PCV of GPS were used for Galileo and BDS because those of Galileo and BDS were unavailable. It should be noted that the satellite orbit, clock offset, and OSB products from CNES were used in PPP-AR. The elevation and observable arc were used to define which ambiguity subsets should be employed, and the ratio test and bootstrapping success rate were used to validate whether the fixed subsets could be trusted. The detailed strategy for positioning is summarized in Table 2.

**Table 2.** Parameters for PPP configurations.

| Parameter | Configurations |
|---|---|
| Estimator | Extended Kalman filter |
| Observations model | Ionosphere-free combinations |
| Frequency | GPS: L1/L2 Galileo: E1/E5a BDS: B1I/B3I |
| Observation weighting | 0.3 m and 0.003 m for GPS/Galileo/BDS code and phase, respectively |
| Sampling rate and cutoff angle | 30 s and 7° |
| Orbits and clocks | Real-time products from CNES |
| Code and phase biases | Corrected with the OSB products from CNES |
| Station coordinate | Kinematic mode: white noise<br>Static mode: constant |
| Receiver clock | Estimated as white noise for each system |
| Tropospheric delay | A priori troposphere delay: Saastamoinen model [36]<br>Zenith wet tropospheric delay: estimated as a random walk |
| AR validation | Bootstrapping success rate threshold: 95%<br>Ratio test threshold: 2.0 |

### 3.2. *Quality Analysis of Phase Bias*

The two key metrics for analyzing the quality of real-time OSB products are data availability (DA) and stability, where DA can be defined as the ratio of available epochs to all epochs in a day. Figure 2 displays the DA results for the GPS, Galileo, and BDS satellites based on the mean DA of each satellite at two frequencies during a period of 31 days. As can be observed, all GPS and Galileo satellites had a DA of more than 90%, whereas the majority of BDS satellites had DAs of less than 60%. The major reason for the inferior DAs of BDS satellites can be inferred as the subpar quality of the real-time satellite orbit and the offset products for BDS [37,38], which hinder the generation of the phase bias products in real time.

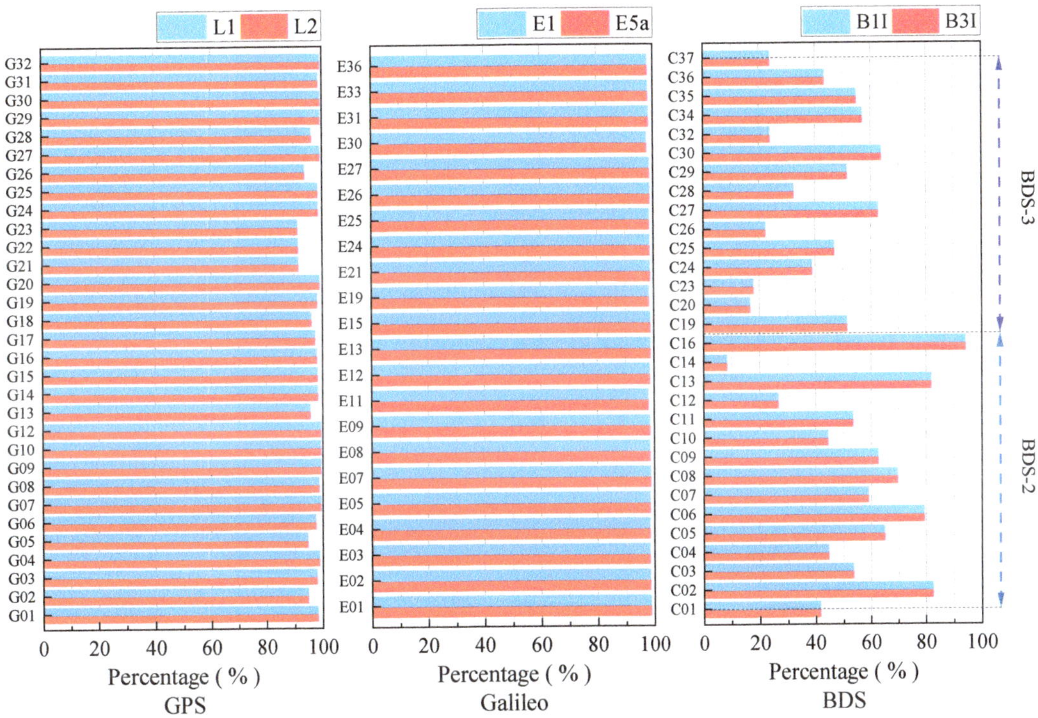


**Figure 2.** Mean DA of GPS (**left**), Galileo (**middle**), and BDS (**right**) satellites.

The mean DA of each satellite for each system is displayed in Table 3, which numerically reflects the DA of each system. Since frequency 1 and frequency 2 are two frequencies that are utilized to estimate the clock offset products of each system [39], the DAs of frequency 1 and frequency 2 are almostsame for each system. The overall DAs of the GPS and Galileo satellites were greater than 97%, while that of the BDS satellites was smaller than 50%. In conclusion, although real-time PPP-AR can be implemented using the OSB products, a consistent and reliable real-time PPP-AR service cannot be guaranteed due to its unstable DA, particularly in the case of BDS satellites. Therefore, a critical issue for real-time PPP-AR is how to avoid or weaken the influence of the absent phase bias products.

**Table 3.** Overall DA of GPS, Galileo, and BDS satellites.

| System | Frequency 1 | Frequency 2 |
|---|---|---|
| GPS | 97.6% | 97.6% |
| Galileo | 98.8% | 98.8% |
| BDS | 49.5% | 49.5% |

The maximum fluctuation value (MAX) and standard deviation (STD), where MAX is the absolute value of the difference between the maximum and minimum phase biases within one day, can be applied to evaluate the stability of real-time OSB products. Given that the results for the two frequencies were nearly identical, the 31-day data were counted to obtain the mean MAX and STD of frequency 1 for each satellite; the results are shown in Figure 3. With the exception of G01 and G21, all other GPS satellites had MAX and STD values that were better than 0.08 and 0.025 cycles, respectively. The MAX and STD of Galileo were only marginally inferior to those of GPS, with values at around 0.12 and 0.05 cycles, respectively. Apart from C07, C13, and C24, the MAX and STD of the remaining BSD MEO and IGSO satellites were approximately 0.15 and 0.05 cycles, respectively, whereas the MAX and STD of the GEO satellites were noticeably greater than those of the other satellites, exceeding 0.5 and 0.2 cycles, respectively. This was due to the fact that the quality of the satellite orbit and clock offset products for GEO satellites is poorer than that for MEO and IGSO satellites, leading to more satellite orbit and clock offset errors being absorbed by the phase bias. It is worth mentioning that the MAX and STD trends for each satellite were consistent, indicating that fewer outliers were included in the phase bias, which is beneficial to achieving a more robust PPP-AR performance.

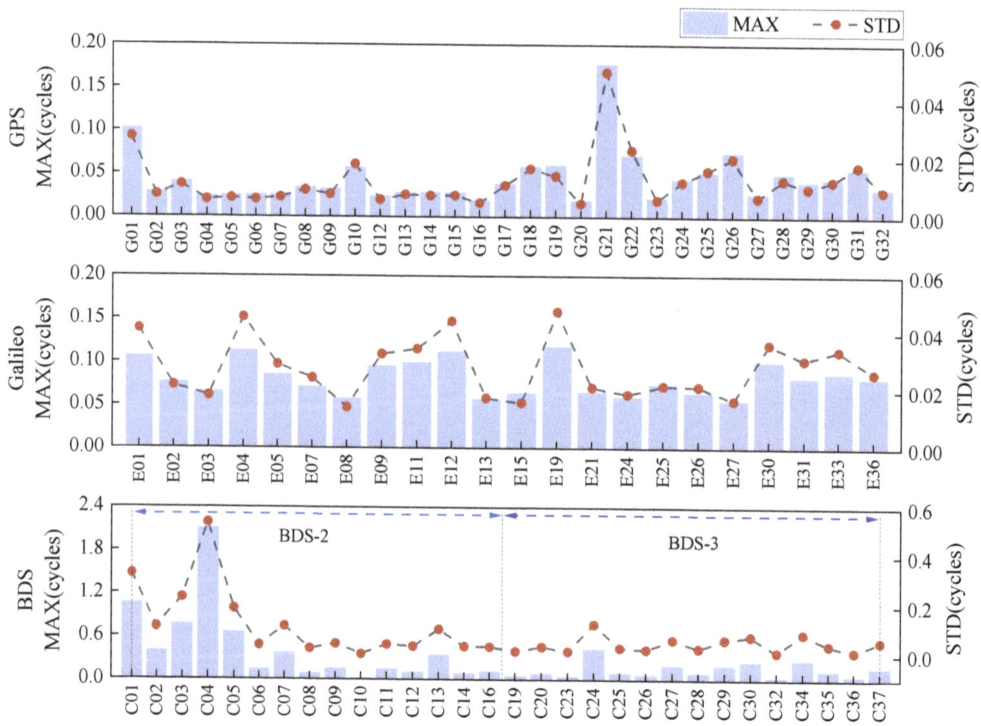


**Figure 3.** Mean MAX and STD of GPS (**top**), Galileo (**middle**), and BDS (**bottom**) satellites for frequency 1.

Similarly, the mean MAX and STD values of each system were calculated in order to depict the overall stability of each system numerically; the results are displayed in Table 4. The MAX and STD of the GPS, Galileo, and BDS systems were 0.045 and 0.012; 0.081 and 0.028; and 0.292 and 0.085 cycles, respectively, among which the BDS system was obviously inferior to the GPS and Galileo systems.

**Table 4.** Overall MAX and STD of GPS, Galileo, and BDS satellites.

| System (Units: Cycles) | Frequency 1 | | Frequency 2 | |
|---|---|---|---|---|
| | MAX | STD | MAX | STD |
| GPS | 0.045 | 0.012 | 0.045 | 0.012 |
| Galileo | 0.081 | 0.028 | 0.083 | 0.028 |
| BDS | 0.292 | 0.085 | 0.293 | 0.085 |

### 3.3. Distribution of Ambiguity Residuals Using OSB Products

One of the criterion indices for evaluating the quality of AR products is the distribution of ambiguity residuals [40,41]. The ambiguity should be close to an integer after applying the OSB product to conduct the SD between satellites; the rounding method can be utilized to determine the closest integers. The residuals denote the difference between the float values and the closest integers for the SD ambiguities, and these are specifically expressed as follows:

$$Residual = \Delta\hat{N}_r^s - rounding\left(\Delta\hat{N}_r^s\right) \tag{15}$$

where $\Delta\hat{N}_r^s$ denotes the SD float ambiguity. The converged float ambiguities were used in these statistics to reduce the negative effects of other errors. Furthermore, the BDS system was separated into BDS-2 and BDS-3 for statistical analysis in this subsection.

The WL and NL residual distributions for GPS, Galileo, BDS-2, and BDS-3 were plotted using the same 31-day data; they are illustrated in Figure 4, while the specific values are shown in Table 5. It can be seen from the figure that besides the NL residual distribution of the BDS-3, the WL and NL residual distributions for the four systems followed a normal distribution. Regarding the WL residuals for the four systems, the residual percentages within ±0.15 cycles were 83%, 84%, 86%, and 90%, respectively, and those within ±0.25 cycles were greater than 92%. As for the NL residuals, the residual percentages within ±0.25 for the GPS and Galileo systems were 92% and 89%, which were considerably better than those of 79% and 60% for BDS-2 and BDS-3, respectively. Inferior satellite orbit and clock offset products, model residuals, and other factors can be blamed for the poor NL residuals of BDS-2 and BDS-3. It is evident that the ambiguity residuals of GPS showed the best performance, while those of Galileo were slightly worse than those of GPS; those of BDS-2 and BDS-3 were the worst, which is consistent with the quality analysis of the OSB products. Considering the DA, MAX, STD, and ambiguity residuals, it can be concluded that the positioning accuracy and convergence time will be impacted by BDS when performing multi-GNSS PPP-AR based on the OSB products.

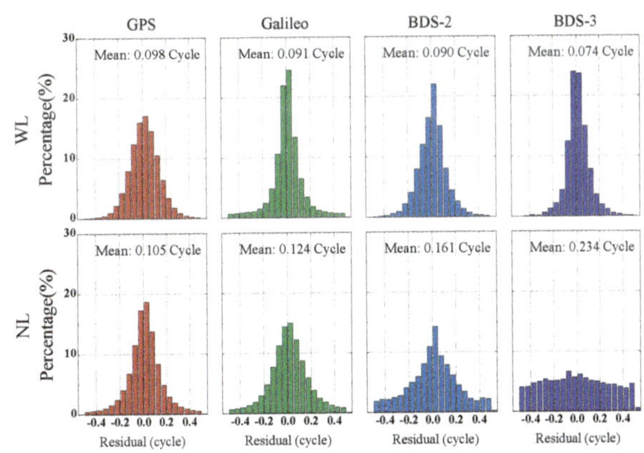


**Figure 4.** WL (**top row**) and NL (**bottom row**) residual distributions for GPS (**red left column**), Galileo (**green left column**), BDS-2 (**light blue right column**), and BDS-3 (**navy blue right column**).

**Table 5.** WL and NL residual statistic results for the four systems.

| System | WL | | NL | |
|---|---|---|---|---|
| | With ± 0.15 | With ± 0.25 | With ± 0.15 | With ± 0.25 |
| GPS | 82.8% | 95.8% | 80.4% | 92.1% |
| Galileo | 84.3% | 92.3% | 74.3% | 88.7% |
| BDS-2 | 85.7% | 95.9% | 61.5% | 79.0% |
| BDS-3 | 89.3% | 97.2% | 39.8% | 60.1% |

### 3.4. Performance Analysis of PPP-AR

The convergence of real-time positioning is defined as having positioning errors in the east (E), north (N), and up (U) components smaller than 10 cm for 10 consecutive epochs. After convergence, the root mean square (RMS) of the ENU components represents the positioning accuracy, where the reference coordinates for each station are obtained from the IGS weekly solutions.

The average kinematic positioning accuracy results for the three combinations at 14 typical stations during the 31-day period are shown in Figure 5, where the float and fixed solutions are represented in the figure by the white diagonal and solid lines, respectively. The positioning accuracy of PPP-AR, especially for GPS-only solutions, is greatly increased when compared to that of float solutions, as can be seen in the figure. The mean positioning accuracy results of the kinematic and static PPP solutions for the three combinations were calculated using data from the 90 stations over 31 days to further reflect the improvement of PPP-AR in positioning accuracy; the results are given in Table 6.

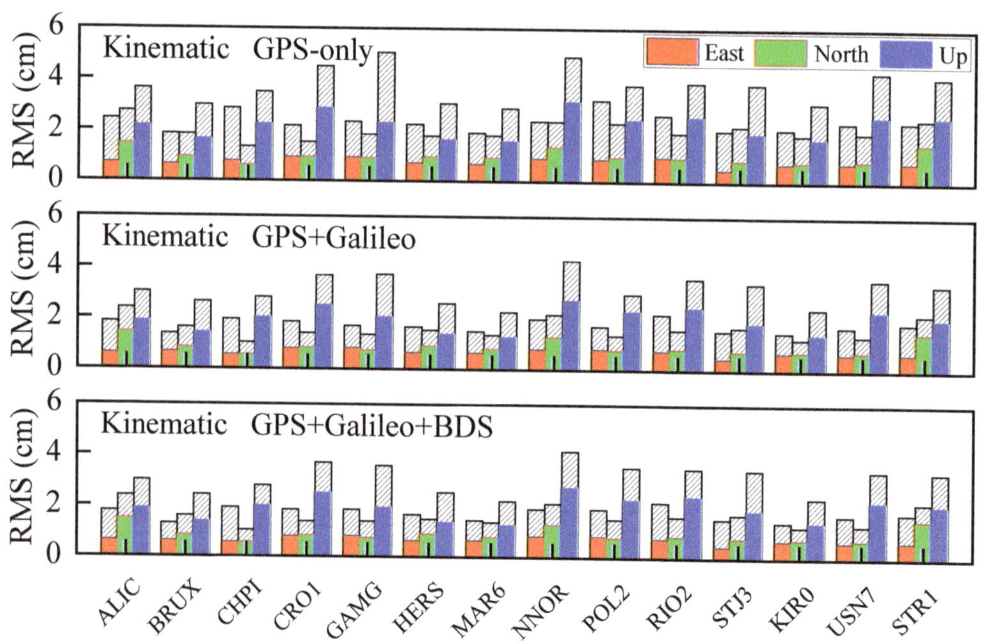


**Figure 5.** Statistical diagram of mean 31-day kinematic results from 14 stations for the GPS-only (**top**), GPS+Galileo (**middle**), and GPS+Galileo+BDS (**bottom**) solutions (white diagonal line—float solution; solid line—fixed solution).

**Table 6.** Statistical results for the kinematic and static position accuracy of the three combinations.

| Mode | System | Float (cm) | | | Fixed (cm) | | | Fixed Rate |
|---|---|---|---|---|---|---|---|---|
| | | E | N | U | E | N | U | |
| Kinematic | G | 2.44 | 2.03 | 4.13 | 1.15 (53%) | 1.38 (32%) | 3.14 (24%) | 97.8% |
| | GE | 1.90 | 1.64 | 3.46 | 1.06 (44%) | 1.27 (23%) | 2.85 (17%) | 98.5% |
| | GEC | 1.89 | 1.64 | 3.43 | 1.06 (44%) | 1.27 (22%) | 2.85 (17%) | 98.3% |
| Static | G | 1.42 | 1.14 | 1.73 | 0.79 (44%) | 0.95 (16%) | 1.48 (15%) | 98.8% |
| | GE | 1.18 | 1.09 | 1.61 | 0.78 (34%) | 0.94 (14%) | 1.42 (12%) | 98.7% |
| | GEC | 1.18 | 1.09 | 1.59 | 0.77 (34%) | 0.94 (14%) | 1.41 (11%) | 98.3% |

When comparing the float solution in the GPS-only, the kinematic positioning accuracy of PPP-AR improved by 53%, 32%, and 24% (from 2.44, 2.03, and 4.13 cm to 1.15, 1.38, and 3.14 cm) for the E, N, and U components, respectively; on the other hand, it improved by 44%, 16%, and 15% (from 1.42, 1.14, and 1.73 cm to 0.79, 0.95, and 1.48 cm) for the E, N, and U components, respectively, in the static mode. With respect to the GPS+Galileo solution, the kinematic and static fixed solutions for the E, N, and U components improved from 1.90 to 1.06, 1.64 to 1.27, and 3.46 to 2.85 cm, and from 1.18 to 0.78, 1.09 to 0.94, and 1.61 to 1.42 cm, respectively. As demonstrated above, PPP-AR enhanced the positioning accuracy, most notably for the E component, while the GPS+Galileo solution improved both the float and fixed solutions to a certain extent in comparison to the GPS-only solution, particularly in the kinematic mode. However, the improvement in positioning accuracy was limited after the BDS satellites were involved in the PPP solutions; this can be attributed to the inferior quality of the DA of the BDS OSB products, which resulted in the participation of only a few BDS satellites in positioning. Meanwhile, the real-time satellite orbit and clock offset products for the BDS satellites were inaccurate, and the positioning accuracy could not be effectively improved under equal weight processing for all systems in this study. To summarize, in kinematic mode, the positioning accuracy of the float solutions after convergence reached about 2, 2, and 4 cm for the E, N, and U components, respectively, while the fixed solutions further improved the positioning accuracy to 1, 1, and 3 cm for the E, N, and U components, respectively. In static mode, the fixed solution improved the float solution from 1~2 cm to 7~9 mm for the E and N components, while the U component was greater than 1.5 cm. The mean ambiguity fixed rates for the kinematic and static modes of the three combinations were more than 97%.

Using the same data from 90 stations over 31 days, the kinematic and static convergence time results for the three combinations were counted using a frequency histogram, where the white diagonal and solid lines represent the convergence times for the float and fixed solutions, respectively. Each panel in Figure 6 represents the convergence time of the float solutions after promotion to fixed solutions; they are divided into the mean convergence time, the 10 min convergence ratio, the 20 min convergence ratio, and the 30 min convergence ratio. Since 10 epochs were used for MW smoothing in this study, the statistical analysis of convergence time was started after 5 min.

The mean convergence time of the kinematic and static float solutions for the GPS-only, GPS+Galileo, and GPS+Galileo+BDS solutions were 28.8, 19.7, and 20.4 min and 18.9, 14.7, and 15.0 min, respectively, while the results of the fixed solutions were 16.8, 9.6, and 9.89 min and 11.4, 8.0, and 8.1 min, respectively. The figure shows that the portion of each convergence period for the GPS+Galileo solution was higher than that of the GPS-only solution, especially at 10 min. This is because the satellite space geometry configuration is improved when more satellites are involved in PPP-AR. However, the convergence time of the GPS+Galileo and GPS+Galileo+BDS solutions was almost the same because the BDS real-time products were of poor accuracy and were frequently missing. To summarize, the convergence time of the fixed solution was much faster than that of the float solution for each convergence period, which demonstrates the effectiveness of PPP-AR in increasing the speed of convergence; in addition, the multi-GNSS PPP-AR further hastened the convergence time.

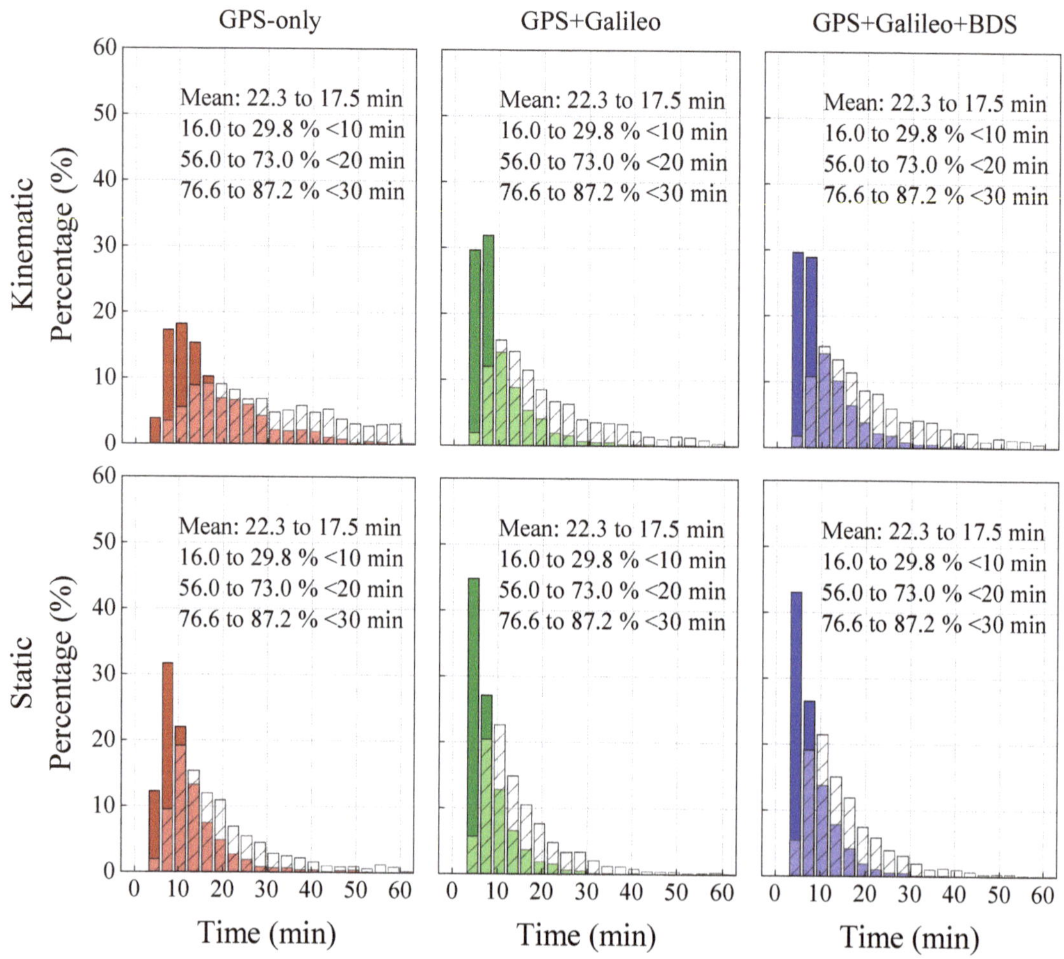


**Figure 6.** Mean convergence time of the kinematic (**top row**) and static (**bottom row**) solutions for the GPS-only (**red left column**), GPS+Galileo (**green middle column**), and GPS+Galileo+BDS (**blue right column**) systems (white diagonal line—float solution; solid line—fixed solution).

### *3.5. Dealing with the Missing Phase Bias*

It should be pointed out that during the statistical analysis of positioning accuracy and convergence time, we found a problem where all stations re-converged in some periods—as shown in Figure 7, where the float and fixed solutions are represented in black and red, respectively—which greatly reduced the reliability of real-time PPP-AR services.

After the data analysis, it was found that this was mainly due to the missing real-time OSB products. Therefore, a polynomial fitting method was proposed to compensate for the temporary absence of the OSB products, which utilized the previous data in order to guarantee that the positioning result was reliable. The polynomial fitting method can be applied to fit an $n$-order polynomial based on $k$ known values, allowing for the value at the next epoch to be predicted, which can be expressed as follows:

$$\varphi_i = a_0 + a_1(t_i - t_0) + a_2(t_i - t_0)^2 + \ldots + a_n(t_i - t_0)^n, \ (i = 1,2,3,\ldots,k; k > n+1) \quad (16)$$

where $n$ represents the fitting order (second-order fitting was adopted in this study); $a_i$ denotes the $i$-order fitting coefficient; $t_0$ and $t_i$ represent the time of fitting and interpolated values, respectively; and $\varphi_i$ represents the interpolated values.

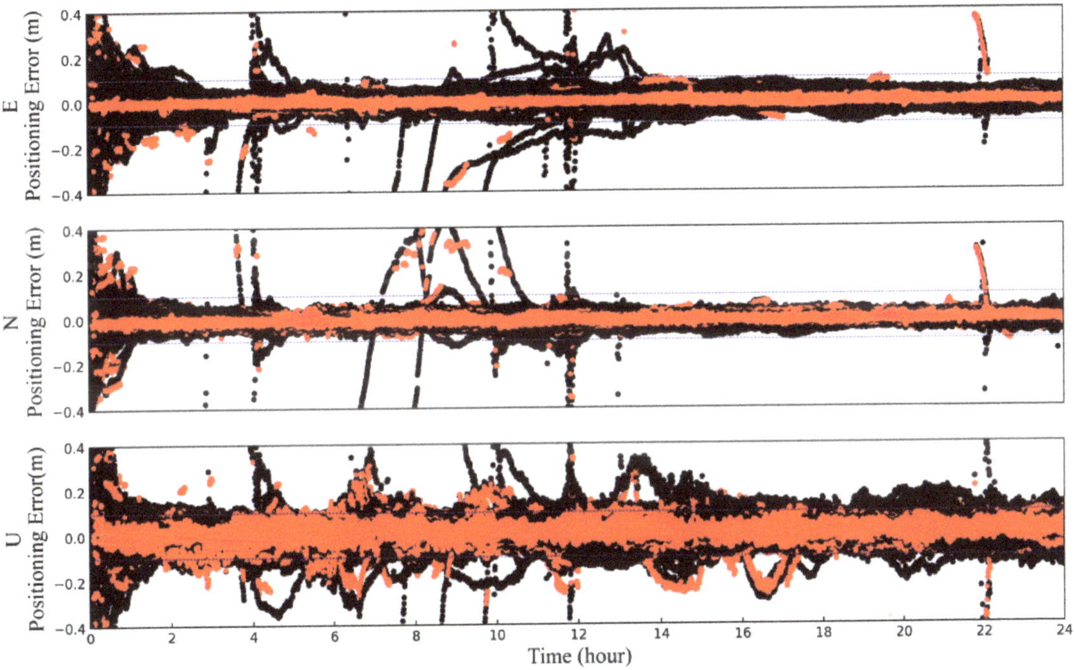


**Figure 7.** The kinematic positioning errors at the same 14 stations in Figure 5 for the GPS-only system from DOY 121 to 151 in 2021 (black—float solution; red—fixed solution).

By substituting the $k$ values into Equation (16) to perform polynomial fitting and using the least square method to obtain the coefficient of the polynomial, the polynomial can be employed to predict the missing value. It should be noted that the principle of the OSB and UPD is the same, and they can be converted into each other [42,43]. The missing phase bias products were only predicted within 15 min in this study because it has been demonstrated that the NL UPD products are stable within 15 min [2].

For instance, on DOY 138 in 2021, the phase bias products of all satellites were missing from 4:00:30 to 4:01:30 and from 11:43:00 to 11:44:30; the positioning errors at ALIC are plotted in Figure 8, where the upper figure represents the original results, and the lower figure represents the results using the proposed method. In comparison to the original positioning results, the compensated positioning results showed restored reliability while simultaneously avoiding re-convergence. Of course, the positioning accuracy and ambiguity fixed rate were also improved. It should be mentioned that the performance analysis in Section 3.4 was based on the method in Section 3.5.

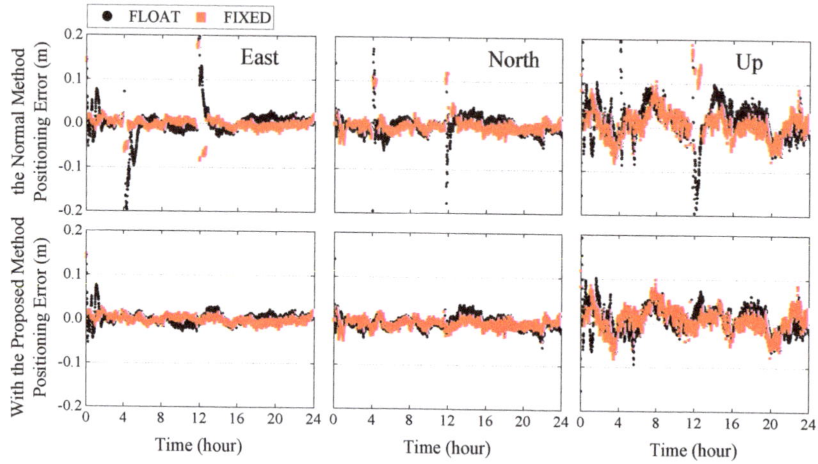


**Figure 8.** Kinematic positioning errors at ALIC for GPS-only on DOY 138 in 2021.

## 4. Discussion

In this study, it can be seen that the BDS system still plays an auxiliary role in multi-GNSS PPP, although the visible BDS satellite number is very considerable. The main factors affecting the PPP performance are the accuracy and stability of precise orbit, clock, and OSB products. In fact, the quality of the OSB products depends on the upstream orbit and clock products. The average influence of GFZ real-time orbit error on NL UPD was from 0.06 to 0.78 cycles for four different GNSS systems [44]. In terms of CNES real-time orbit products, the 3D orbit RMS error is typically 5, 10, 18, 18, and 36 cm for GPS, GLONASS, Galileo, BDS MEO, and IGSO satellites, respectively [45]; the BDS orbit error was obviously higher than that of GPS. As a result, the effective solution for improving the BDS PPP ambiguity fixed rate is to reduce the effects of orbital errors. Estimating OSB products that can compensate for the BDS real-time orbit error may be a feasible solution in our future work.

Moreover, the stochastic model setting is also a vital point for multi-GNSS positioning as the errors of precise products vary in different GNSS systems or in different satellites ofthe same system. The errors of the CNES real-time orbit, clock, and OSB products were unknown, and equal weight was thus set for all satellite systems in this study. Therefore, due to the low quality of the BDS real-time products currently provided by CNES, the positioning accuracy became even worse after jointing the BDS into GPS PPP. Consequently, the integrity monitoring of satellite precise products is very important for a reliable real-time PPP service [46,47]. In addition, the post-store real-time precise products from CNES were used in this study, which prevented the possible communication delay in real-time applications. The extrapolated orbit and clock offset should be used when a communication delay occurs, which would allow the performance of PPP to drop rapidly. However, the extrapolation error will be different for different GNSS systems and different satellites [48]. This should also be considered in the stochastic model setting for multi-GNSS real-time PPP applications.

## 5. Conclusions

As a key product for real-time PPP-AR, real-time OSB products are currently broadcasted by CNES. Based on the dual-frequency ionospheric-free observation model, the integer feature of ambiguity was recovered by using the OSB products and PPP-AR could be performed. The quality of the OSB products was analyzed, and the performance of the dual-frequency PPP-AR was evaluated. The experimental results showed that:

(1) The mean DA of the GPS and Galileo satellites was above 97% according to the 31-day statistical analysis of the OSB products, whereas that of BDS was less than 60%. The

mean MAX and STD of the GPS, Galileo, and BDS satellites were 0.045 and 0.012; 0.081 and 0.028; and 0.292 and 0.085 cycles, respectively. By analyzing the residual distribution of the OSB products, the WL residual percentages within ±0.25 cycles for the GPS, Galileo, BDS-2, and BDS-3 satellites were found to be more than 92%, while the NL residual percentages within ±0.25 cycles for the four systems were 92%, 89%, 79%, and 60%, respectively. Considering the DA, MAX, STD, and ambiguity residuals, we concluded that the reliability of PPP-AR, including the BDS system, may not be guaranteed using the OSB products from CNES, especially for BDS-3. In fact, the experiment described in this paper also confirmed that the advancement of positioning accuracy was constrained after BDS satellites had been incorporated into multi-GNSS PPP-AR.

(2) Due to the accuracy restrictions of real-time products, real-time PPP-AR cannot greatly increase positioning accuracy; however, it can dramatically speed up convergence time. In the kinematic mode, the GPS+Galileo fixed solution reduced the mean convergence time from 19.7 to 9.6 min and increased the convergence ratio within 10 min from 22.2% to 70.2% when compared to the float solution. After convergence, the positioning accuracy of the GPS+Galileo float solution reached about 2, 2, and 4 cm for the E, N, and U components, respectively, while that of the GPS+Galileo fixed solution was further enhanced to 1, 1, and 3 cm for the E, N, and U components, respectively. In the static mode, the overall fixed solution improved the float solution from 1~2 cm to 7~9 mm for the E and N components, and the U component was better than 1.5 cm. The mean convergence times of the GPS-only and GPS+Galileo solutions were 11.4 min and 8.0 min, respectively. The mean ambiguity fixed rates for the kinematic and static modes for the three combinations were more than 97%. In conclusion, the GPS+Galileo solution showed the best performance in the kinematic mode when using the OSB products from CNES, while the GPS-only solution can achieve almost the same performance as the GPS+Galileo solution in the static mode.

(3) A prediction method was proposed to compensate for the short-term missing OSB products, and our experiment showed that the positioning results were more reliable and avoided re-convergence.

**Author Contributions:** Conceptualization, S.D., B.S. and G.H.; methodology, S.D. and B.S.; validation, S.D. and W.X.; formal analysis, S.D. and W.X.; writing—original draft preparation, S.D.; writing—review and editing, S.D., B.S., W.X., G.H., Y.G. and P.L.; visualization, S.D., B.S. and W.X.; project administration, G.H.; funding acquisition, G.H. All authors have read and agreed to the published version of the manuscript.

**Funding:** This research was funded by the National Natural Science Foundation of China (41941019; 42004024), the National Key R&D Program of China (2021YFB2600603; 2021YFC3000503), and the Key R&D Program of Shaanxi Province (2022ZDLSF07-12).

**Data Availability Statement:** The CDDIS website (https://cddis.nasa.gov/archive/gnss, accessed on 2 August 2022) was used to obtain the MGEX data, which include the stations, DCB, and ATX. The precise products obtained from CNES (http://www.ppp-wizard.net/products/REALTIME/, accessed on 2 August 2022) include the satellite orbit, clock offset, and OSB.

**Acknowledgments:** The authors gratefully acknowledge IGS MGEX for providing the GNSS data. We also acknowledge the CNES for providing real-time precise products.

**Conflicts of Interest:** The authors declare no conflict of interest.

## References

1. Zumberge, J.F.; Heflin, M.B.; Jefferson, D.C.; Watkins, M.M.; Webb, F.H. Precise Point Positioning for the Efficient and Robust Analysis of GPS Data from Large Networks. *J. Geophys. Res. Soild Earth* **1997**, *102*, 5005–5017. [CrossRef]
2. Ge, M.; Gendt, G.; Rothacher, M.; Shi, C.; Liu, J. Resolution of GPS carrier-phase ambiguities in Precise Point Positioning (PPP) with daily observations. *J. Geod.* **2008**, *82*, 389–399. [CrossRef]
3. Mercier, F.; Laurichesse, D. Zero-difference ambiguity blocking, Properties of satellite/receiver widelane biases. In Proceedings of the ENC-GNSS 2008, Toulouse, France, 23–25 April 2008.
4. Laurichesse, D.; Mercier, F.; Berthias, J.-P.; Broca, P.; Cerri, L. Integer Ambiguity Resolution on Undifferenced GPS Phase Measurements and Its Application to PPP and Satellite Precise Orbit Determination. *Navigation* **2009**, *56*, 135–149. [CrossRef]

5. Collins, P.; Lahaye, F.; Héroux, P.; Bisnath, S. Precise Point Positioning with Ambiguity Resolution using the Decoupled Clock Model. In Proceedings of the 21st International Technical Meeting of the Satellite Division of the Institute of Navigation (ION GNSS 2008), Savannah, GA, USA, 16–19 September 2008; pp. 1315–1322.
6. Collins, P.; Bisnath, S.; Lahaye, F.; Héroux, P. Undifferenced GPS Ambiguity Resolution Using the Decoupled Clock Model and Ambiguity Datum Fixing. *Navigation* **2010**, *57*, 123–135. [CrossRef]
7. Geng, J.; Meng, X.; Dodson, A.H.; Teferle, F.N. Integer ambiguity resolution in precise point positioning: Method comparison. *J. Geod.* **2010**, *84*, 569–581. [CrossRef]
8. Shi, J.; Gao, Y. A comparison of three PPP integer ambiguity resolution methods. *GPS Solut.* **2014**, *18*, 519–528. [CrossRef]
9. Wang, J.; Huang, G.; Zhang, Q.; Gao, Y.; Gao, Y.; Luo, Y. GPS/BDS-2/Galileo Precise Point Positioning Ambiguity Resolution Based on the Uncombined Model. *Remote Sens.* **2020**, *12*, 1853. [CrossRef]
10. Zhao, Q.; Guo, J.; Liu, S.; Tao, J.; Hu, Z.; Chen, G. A variant of raw observation approach for BDS/GNSS precise point positioning with fast integer ambiguity resolution. *Satell. Navig.* **2021**, *2*, 29. [CrossRef]
11. Chen, C.; Xiao, G.; Chang, G.; Xu, T.; Yang, L. Assessment of GPS/Galileo/BDS Precise Point Positioning with Ambiguity Resolution Using Products from Different Analysis Centers. *Remote Sens.* **2021**, *13*, 3266. [CrossRef]
12. Gao, Y.; Chen, K. Performance Analysis of Precise Point Positioning Using Rea-Time Orbit and Clock Products. *J. Glob. Position. Syst.* **2004**, *3*, 95–100. [CrossRef]
13. Capilla, R.M.; Berné-Valero, J.L.; Hermosilla-Rodrigo, A. Impact of multi-constellation products and ambiguity resolution in Precise Point Positioning for real-time measurements. *Measurement* **2017**, *100*, 183–193. [CrossRef]
14. Wang, Z.; Li, Z.; Wang, L.; Wang, X.; Yuan, H. Assessment of Multiple GNSS Real-Time SSR Products from Different Analysis Centers. *ISPRS Int. J. Geo-Inf.* **2018**, *7*, 85. [CrossRef]
15. Alkan, R.M.; Erol, S.; İlçi, V.; Ozulu, İ.M. Comparative analysis of real-time kinematic and PPP techniques in dynamic environment. *Measurement* **2020**, *163*, 107995. [CrossRef]
16. Zhao, X.; Ge, Y.; Ke, F.; Liu, C.; Li, F. Investigation of real-time kinematic multi-GNSS precise point positioning with the CNES products. *Measurement* **2020**, *166*, 108231. [CrossRef]
17. Bulbul, S.; Bilgen, B.; Inal, C. The performance assessment of Precise Point Positioning (PPP) under various observation conditions. *Measurement* **2021**, *171*, 108780. [CrossRef]
18. Ge, Y.; Chen, S.; Wu, T.; Fan, C.; Qin, W.; Zhou, F.; Yang, X. An analysis of BDS-3 real-time PPP: Time transfer, positioning, and tropospheric delay retrieval. *Measurement* **2021**, *172*, 108871. [CrossRef]
19. Li, B.; Ge, H.; Bu, Y.; Zheng, Y.; Yuan, L. Comprehensive assessment of real-time precise products from IGS analysis centers. *Satell. Navig.* **2022**, *3*, 12. [CrossRef]
20. El-Mowafy, A.; Deo, M.; Kubo, N. Maintaining real-time precise point positioning during outages of orbit and clock corrections. *GPS Solut.* **2017**, *21*, 937–947. [CrossRef]
21. Fu, W.; Yang, Y.; Zhang, Q.; Huang, G. Real-time estimation of BDS/GPS high-rate satellite clock offsets using sequential least squares. *Adv. Space Res.* **2018**, *62*, 477–487. [CrossRef]
22. Xie, W.; Huang, G.; Fu, W.; Li, P.; Cui, B. An efficient clock offset datum switching compensation method for BDS real-time satellite clock offset estimation. *Adv. Space Res.* **2021**, *68*, 1802–1813. [CrossRef]
23. Fu, W.; Wang, L.; Chen, R.; Han, Y.; Zhou, H.; Li, T. Combined BDS-2/BDS-3 real-time satellite clock estimation with the overlapping B1I/B3I signals. *Adv. Space Res.* **2021**, *68*, 4470–4483. [CrossRef]
24. Defraigne, P.; Bruyninx, C. On the link between GPS pseudorange noise and day-boundary discontinuities in geodetic time transfer solutions. *GPS Solut.* **2007**, *11*, 239–249. [CrossRef]
25. Fu, W.; Huang, G.; Zhang, Q.; Gu, S.; Ge, M.; Schuh, H. Multi-GNSS real-time clock estimation using sequential least square adjustment with online quality control. *J. Geod.* **2019**, *93*, 963–976. [CrossRef]
26. Laurichesse, D.; Langley, R. Handling the biases for improved triple frequency PPP convergence. In Proceedings of the GPS World 2015, Cleveland, OH, USA, 26 April 2015.
27. Liu, T.; Chen, H.; Chen, Q.; Jiang, W.; Laurichesse, D.; An, X.; Geng, T. Characteristics of phase bias from CNES and its application in multi-frequency and multi-GNSS precise point positioning with ambiguity resolution. *GPS Solut.* **2021**, *25*, 58. [CrossRef]
28. Shu, B.; Liu, H.; Wang, L.; Huang, G.; Zhang, Q.; Yang, Z. Performance improvement of real-time PPP ambiguity resolution using a regional integer clock. *Adv. Space Res.* **2021**, *67*, 1623–1637. [CrossRef]
29. Geng, J.; Guo, J.; Wang, C.; Zhang, Q. Satellite antenna phase center errors: Magnified threat to multi-frequency PPP ambiguity resolution. *J. Geod.* **2021**, *95*, 72. [CrossRef]
30. Geng, J.; Yang, S.; Guo, J. Assessing IGS GPS/Galileo/BDS-2/BDS-3 phase bias products with PRIDE PPP-AR. *Satell. Navig.* **2021**, *2*, 17. [CrossRef]
31. Shu, B.; Liu, H.; Xu, L.; Gong, X.; Qian, C.; Zhang, M.; Zhang, R. Analysis of satellite-induced factors affecting the accuracy of the BDS satellite differential code bias. *GPS Solut.* **2017**, *21*, 905–916. [CrossRef]
32. Teunissen, P.J.G. Distributional Theory for the DIA Method. *J. Geod.* **2018**, *92*, 59–80. [CrossRef]
33. Yang, L.; Shen, Y.; Li, B.; Rizos, C. Simplified Algebraic Estimation for the Quality Control of DIA Estimator. *J. Geod.* **2021**, *95*, 14. [CrossRef]
34. Xie, W.; Huang, G.; Fu, W.; Shu, B.; Cui, B.; Li, M.; Yue, F. A Quality Control Method Based on Improved IQR for Estimating Multi-GNSS Real-Time Satellite Clock Offset. *Measurement* **2022**, *201*, 111695. [CrossRef]

35. Zhou, F.; Dong, D.; Li, W.; Jiang, X.; Wickert, J.; Schuh, H. GAMP: An open-source software of multi-GNSS precise point positioning using undiferenced and uncombined observations. *GPS Solut.* **2018**, *22*, 33. [CrossRef]
36. Saastamoinen, J. Contributions to the Theory of Atmospheric Refraction. *J. Geod.* **1972**, *105*, 279–298. [CrossRef]
37. Zhang, L.; Yang, H.; Gao, Y.; Yao, Y.; Xu, C. Evaluation and analysis of real-time precise orbits and clocks products from different IGS analysis centers. *Adv. Space Res.* **2018**, *61*, 2942–2954. [CrossRef]
38. Ouyang, C.; Shi, J.; Huang, Y.; Guo, J.; Xu, C. Evaluation of BDS-2 real-time orbit and clock corrections from four IGS analysis centers. *Measurement* **2021**, *168*, 108441. [CrossRef]
39. Li, X.; Liu, G.; Li, X.; Zhou, F.; Feng, G.; Yuan, Y.; Zhang, K. Galileo PPP rapid ambiguity resolution with fve-frequency observations. *GPS Solut.* **2020**, *24*, 24. [CrossRef]
40. Wang, J.; Huang, G.; Yang, Y.; Zhang, Q.; Gao, Y.; Xiao, G. FCB estimation with three different PPP models: Equivalence analysis and experiment tests. *GPS Solut.* **2019**, *23*, 93. [CrossRef]
41. Wang, J.; Zhang, Q.; Huang, G. Estimation of fractional cycle bias for GPS/BDS-2/Galileo based on international GNSS monitoring and assessment system observations using the uncombined PPP model. *Satell. Navig.* **2021**, *2*, 9. [CrossRef]
42. Laurichesse, D. Phase Biases Estimation for Undifferenced Ambiguity Resolution. In Proceedings of the PPP-RTK & Open Standards Symposium, Toulouse, France, 12–13 March 2012.
43. Banville, S.; Geng, J.; Loyer, S.; Schaer, S.; Springer, T.; Strasser, S. On the Interoperability of IGS Products for Precise Point Positioning with Ambiguity Resolution. *J. Geod.* **2020**, *94*, 10. [CrossRef]
44. Li, Y.; Gao, Y.; Li, B. An Impact Analysis of Arc Length on Orbit Prediction and Clock Estimation for PPP Ambiguity Resolution. *GPS Solut.* **2015**, *19*, 201–213. [CrossRef]
45. Kazmierski, K.; Sośnica, K.; Hadas, T. Quality Assessment of Multi-GNSS Orbits and Clocks for Real-Time Precise Point Positioning. *GPS Solut.* **2018**, *22*, 11. [CrossRef]
46. Fujita, S.; Sato, Y.; Miya, M.; Ota, K.; Hirokawa, R.; Takiguchi, J. Design of Integrity Function on Centimeter Level Augmentation Service (CLAS) in Japanese Quasi-Zenith Satellite System. In Proceedings of the 29th International Technical Meeting of the Satellite Division of the Institute of Navigation (ION GNSS+ 2016), Portland, OR, USA, 12–16 September 2016; pp. 3258–3263.
47. Weinbach, U.; Brandl, M.; Chen, X.; Landau, H.; Pastor, F.; Reussner, N.; Rodriguez-Solano, C. Integrity of the Trimble CenterPoint RTX Correction Service. In Proceedings of the 31st International Technical Meeting of the Satellite Division of the Institute of Navigation (ION GNSS+ 2018), Miami, Florida, 24–28 September 2018; pp. 1902–1909.
48. Shu, B.; Liu, H.; Feng, Y.; Xu, L.; Qian, C.; Yang, Z. Analysis of Factors Affecting Asynchronous RTK Positioning with GNSS Signals. *Remote Sens.* **2019**, *11*, 1256. [CrossRef]

*remote sensing*

*Article*

# Analysis of Regional Satellite Clock Bias Characteristics Based on BeiDou System

**Wenxuan Liu [1], Hu Wang [1,*], Hongyang Ma [2], Yingyan Cheng [1], Pengyuan Li [3], Bo Li [4] and Yingying Ren [3]**

1 Chinese Academy of Surveying & Mapping, Beijing 100036, China
2 School of Geomatics Science and Technology, Nanjing Tech University, Nanjing 210037, China
3 College of Surveying and Geo-Informatics, Tongji University, Shanghai 200092, China
4 School of Geomatics, Liaoning Technical University, Fuxin 123000, China
* Correspondence: wanghu@casm.ac.cn; Tel.: +86-15652077998

**Abstract:** With the continuous development of the Global Navigation Satellite System (GNSS), the calculation theory and strategy of the global Satellite Clock Bias (SCB) tends to be mature. However, in some eventualities with restricted conditions, the calculation and application of the global SCB are limited; hence, the application of regional SCB is derived. This paper focuses on the quality of regional SCB products in different regions, calculates three groups of regional SCB products, and analyzes their properties and application effects. We expand the double-differenced assessment method for SCB and extend satellite clock accuracy assessment to regional satellite clock products. Additionally, the Regional Effect Bias (REB) is introduced to analyze the influence of the relative position of satellite geometry on the SCB products due to the regional effects. The conclusions are as follows: (1) In low-latitude regions, SCB products have a high degree of completeness and a large number of satellite observations, which is conducive to expanding the positioning application range of regional SCB; (2) the low-latitude regions SCB will be affected by ionospheric activity, and the accuracy will be slightly lower than that of satellite clocks deviation in mid-latitudes; (3) in this paper, the REB in this area is in the level of $10^{-7}$. The experiment displays the result that the values of REB in low-latitude areas are larger, leading to fluctuated Precise Point Position (PPP) results. However, there are fewer stations in the mid-latitude regions, which will also affect the accuracy of PPP; (4) the accuracy of the positioning results of the regional satellite clock deviation in the Chinese region is higher than that of the global clock.

**Keywords:** precise point positioning; clock estimation; station network

**Citation:** Liu, W.; Wang, H.; Ma, H.; Cheng, Y.; Li, P.; Li, B.; Ren, Y. Analysis of Regional Satellite Clock Bias Characteristics Based on BeiDou System. *Remote Sens.* **2022**, *14*, 6047. https://doi.org/10.3390/rs14236047

Academic Editors: Shengfeng Gu, Xiaopeng Gong and Chuang Shi

Received: 26 October 2022
Accepted: 25 November 2022
Published: 29 November 2022

**Publisher's Note:** MDPI stays neutral with regard to jurisdictional claims in published maps and institutional affiliations.

**Copyright:** © 2022 by the authors. Licensee MDPI, Basel, Switzerland. This article is an open access article distributed under the terms and conditions of the Creative Commons Attribution (CC BY) license (https://creativecommons.org/licenses/by/4.0/).

## 1. Introduction

After the great success of the Global Positing System (GPS) and Russian Global'naya Navigatsionnaya Sputnikova Sistema (GLONASS), Europe and China have established the Galileo Satellite Navigation System (Galileo) and BeiDou Navigation Satellite System (BDS) separately. In addition to the above four Global Navigation Satellite Systems (GNSS), India and Japan have also established their regional satellite systems, namely, the Indian Regional Navigation Satellite System (IRNSS) and Quasi-Zenith Satellite System (QZSS). The construction of China's BeiDou satellite navigation system is based on a three-step strategy. The BeiDou-1 system (BDS-1) was completed at the end of 2000 to provide services to the east Asia region; the BeiDou-2 system (BDS-2) was completed at the end of 2012 to provide services to the Asia-Pacific region, and the BeiDou-3 system (BDS-3) was completed in 2020 to provide services to the globe.

The compensation of Satellite Clock Biases (SCB) is an important part of GNSS high-precision data processing. Currently, several Analysis Centers (ACs) around the world provide a variety of GNSS clock bias products, e.g., post-processing SCB products, which play a key role in Positioning, Navigation, and Timing (PNT) [1]. The application of post-processing PPP in scientific research includes geodesy, atmospheric monitoring [2], and plate drift monitoring [3]. In recent years, with the improvement of ultra-rapid satellite orbit accuracy, orbit determination accuracy has met the need for real-time positioning [4]. With the expanded real-time application scenario, the real-time precise point positioning (RT-PPP) technology based on the real-time clock [5–8] has been applied in vehicle navigation [9], autonomous driving, aerial triangulation [10], time transfer [11], deformation monitoring [12], and other fields. Therefore, real-time SCB has become an important research area.

The International GNSS Service (IGS) usually uses fixed orbits and known station coordinates to calculate the post-processing SCB, and the posterior SCB accuracy can reach 0.1 ns [13]. Some ACs use Orbit Determination and Time Synchronization (ODTS) method, for which the orbit product and the SCB product are calculated simultaneously [14]. However, the resulting post-processing SCB product contains the assimilated orbit error [15]. Common methods of studying real-time SCB include the undifferenced method (UD), epoch-differenced method (ED), undifferenced range, and epoch-differenced phase mixed-difference model (MD). The mixed-difference model can eliminate a large number of ambiguity parameters, and the dimension of the obtained equation matrix is small, which can meet the calculation efficiency requirement of real-time estimation [16]. However, the MD method requires additional clock deviation reference. The UD method is to estimate the satellite clock bias, receiver clock bias, atmospheric delay error, and ambiguity parameters simultaneously, and the accuracy of the SCB obtained in this way has higher accuracy, but the ambiguity parameters need to be computed for each epoch and the data calculating efficiency is inefficient [17]. Li implemented a real-time precision clock correction estimation algorithm based on undifferenced carrier phase observations and introduced clock error reference while improving the calculation speed of SCB [18].

To provide RT-PPP services, it is necessary to estimate the precise satellite clocks and orbits quickly and publish them to users [19–21]. Yan et al. applied a regional Continuous Operational Reference System (CORS) service method for SCB estimation [22]. The method is based on ground-tracking station networks covering a region. The system control center receives the real-time observation data from each station and calculates the SCB products precisely. In regional CORS networks using low-cost receivers, PPP using uncombined GNSS observations with ionospheric delay parameters obtained from CORS has good results [8]. However, with the system expansion of the CORS network, the parameters, including phase ambiguity, receiver clock bias, and station Zenith Tropospheric Delay (ZTD), additionally increase. This will reduce computational efficiency. Pan et al. used the broadcast ephemeris to estimate the SCB together with the orbit through the regional CORS station, to realize the regional PPP completely autonomously [23].

In summary, the current research on regional clocks focuses on calculating regional SCB based on regionally distributed CORS to expand the calculation and application scenarios of PPP under regional conditions. However, there are few studies on the characteristics of regional SCB in different regions. In this paper, we fix the orbit products and calculate different regional SCB products by different regional stations in China. The real-time SCB between the global and regional networks in China is computed by the mixed-differenced method (MD). Firstly, this paper counts the epoch which can calculate the SCB of each satellite in different regions within a day to determine whether the data of the SCB of each region can approach the application requirements of PPP. Secondly, the calculated regional SCB and global SCB products are used to evaluate the accuracy of each regional satellite clock. Then, the quadratic fitting coefficient of global SCB is used to study the influence of station region distribution on satellite clock products. Finally, this paper selects stations in different regions and uses PPP to verify the SCB products' quality in each region.

This paper studies the calculation of real-time BeiDou satellite clock deviation in regional networks and analyzes the characteristics of SCB products in the regional network. It expands the application of the BeiDou system and compensates for the previous deficiency of only using the global station net to estimate the SCB products. We study the regional SCB products and their PPP performance. This article is organized as follows. The Section 1 is the introduction, which briefly describes the research status of SCB and the research content of this paper. In Section 2, we briefly introduce the mixed-difference method of SCB calculation and the evaluation strategy of the regional SCB. In Section 3, we calculate the completeness of three different regional SCB products and evaluate their accuracy. In Section 4, we conduct PPP experiments with regional SCB and draw experimental results, and the relevant discussions and conclusions are presented in Sections 5 and 6 separately.

## 2. Data Process

### 2.1. SCB Calculation

The regional SCB in our paper is estimated by mixing undifferenced range and epoch-differential phase, which is usually used for global rapid SCB calculation of GPS satellites [8]. The epoch-difference method is used to remove the ambiguity parameters, and the clock deviation can be corrected by the initial clock deviation estimated by the range. The combination of the two steps can improve the efficiency of the solution on the premise of maintaining accuracy.

The definition equation of the MD model is:

$$\delta t(i) = \delta t(i_0) + \sum_{j=i_0+1}^{i} \Delta\delta_t(j) \tag{1}$$

The clock changes and therefore the accumulated clock corrections $\delta t(i)$ at the epoch number $i$ can be estimated rather precisely; they are biased by the initial clock offset $\delta t(i_0)$ at the starting epoch $i_0$; $\Delta\delta t(j)$ is the cumulative value of the clock bias; the $\Delta$ is the difference operator two adjacent epochs; and the epoch-difference clock $\Delta\delta t(i)$ can be expressed as:

$$\Delta\delta t(i) = \delta t(i) - \delta t(i-1) \tag{2}$$

In order to make the formula concise, we remove the true value of the satellite-receiver distance as the observation value from both sides of the equations of Equations (3) and (4). The observation equations of undifferenced range and epoch-difference phase are:

$$v_{Pc}(i) = \delta_{t_r}(i) - \delta_{t_s}(i) + m(i)\delta T(i) + \varepsilon_{Pc}(i) \tag{3}$$

$$v_{\Delta Lc}(i) = \Delta\delta_{t_r}(i) - \Delta\delta_{t_s}(i) + \Delta m(i)\delta T(i) + \Delta\varepsilon_{\Delta Lc}(i) \tag{4}$$

Since the phase observation value in Equation (4) adopts the epoch-difference phase, the phase observation value and other parameters related to the epoch-difference are denoted by $\Delta$. Where $v_{Pc}$, $\varepsilon_{Pc}$ are the range observation and the residual of the range observation, respectively; $v_{\Delta Lc}$, $\Delta\varepsilon_{Lc}$ are, respectively, the observation of phase and the phase residual of the observation; $\delta T$ and $m$ represent Zenith Total Delay (ZTD) and its mapping function, respectively; $\delta_{t_r}$ and $\delta_{t_s}$ are receiver and satellite clock biases.

Substituting the receiver clock bias $\delta_{t_r}(i)$ and satellite clock bias $\delta_{t_s}(i)$ in Equation (3) by Equation (1), accumulated from epoch $i_{r0}$ and $i_{s0}$, we have:

$$\begin{aligned} v_{Pc}(i) = {} & \Delta\delta_{t_r}(i) - \Delta\delta_{t_s}(i) + m(i)\delta T(i) + \delta_{t_r}(i_{r0}) - \delta_{t_s}(i_{s0}) + \varepsilon_{Pc}(i) \\ & + \sum_{j=i_{r0}+1}^{i-1} \Delta\delta t_r(j) - \sum_{j=i_{s0}+1}^{i-1} \Delta\delta t_s(j) \end{aligned} \tag{5}$$

The last two terms of the above equation can be replaced by the clock deviation estimated at the previous epochs; the sum of the last three terms of the distance observation equation is expressed as $\bar{\varepsilon}_{p_c}(i)$. The Equation (5) can be expressed as:

$$v_{Pc}(i) = \Delta\delta_{t_r}(i) - \Delta\delta_{t_s}(i) + m(i)\delta T(i) + \delta_{t_r}(i_{r0}) - \delta_{t_s}(i_{s0}) + \bar{\varepsilon}_{p_c}(i) \tag{6}$$

In this paper, Equations (4) and (6) are used to estimate the initial clock deviation of each station and satellite.

The epoch-difference method can be used to accurately estimate the clock bias variation and the ZTD [15,24,25]. Therefore, when the mixed algorithm computes the clock bias, Equation (4) is first used to estimate the clock biases and ZTDs of each epoch, and then these estimates will be used to correct the range observations so that only the initial clock deviations remain in the range observations. The corresponding observation equations can be obtained from Equation (5) by putting the clock offset parameters at epoch $i$ into the accumulated clocks as follows:

$$v_{Pc}(i) = \delta_{t_r}(i_{r0}) - \delta_{t_s}(i_{s0}) + m(i)\delta T(i) + \varepsilon_{Pc}(i) + \sum_{j=i_{r0}+1}^{i} \Delta\delta t_r(j) - \sum_{j=i_{s0}+1}^{i} \Delta\delta t_s(j) \quad (7)$$

When calculating the initial value, there are no accumulated terms in Equation (7), so we remove the last two terms. The third and fourth terms of Equation (7) are represented by $\widetilde{\varepsilon}_{Pc}\ (i)$; then, Equation (7) can be expressed as:

$$v_{Pc}(i) = \delta t_r(i_{r0}) - \delta t_s(i_{s0}) + \widetilde{\varepsilon}_{Pc}\ (i) \quad (8)$$

Through the observation equations established by Equations (4) and (8), the least squares method is used to estimate the SCB of the observed data. The differenced clock and ZTD parameters are estimated using the epoch-differenced phases with Equation (4), and the initial clock biases are estimated using the undifferenced ranges with Equation (8).

### 2.2. SCB Evaluation

Due to the curvature of the earth and the restrictions of regional location, it is difficult for the regional network stations to completely receive the observation signals of the global satellites as the global network. When the satellite's trajectory vanishes from the network's sky, that can result in the interruption and also the absence of the estimation of the satellite clock error product during this period. Therefore, the estimation of most satellite clock products is incomplete, and only some periods' SCB products exist. Therefore, this paper first evaluates whether the completeness of the BeiDou satellite clock product will meet the requirements of PPP.

After the evaluation of the completeness, we compare the product accuracy of the BeiDou SCB. When evaluating the accuracy of the clock product, it is necessary to eliminate the time-scale differences; that is, the deviation of the system from the reference clock product [26,27]. The time scale difference comes from the base clock selected by the system. The SCB product model can be expressed as [15]:

$$C_a^S = O_a + O_a^S + T^S + R_a^S + \varepsilon_a^S \quad (9)$$

In Equation (9), the superscript S and the subscript a represent the satellite and the analysis center, respectively; $C_a^S$ is the clock product; $O_a$ is the time scale differences introduced by the reference clock when calculating the clock bias. $O_a^S$ is the initial clock bias; it is involved in the across-time clock correction between epochs. To restore the satellite clock correction, we must introduce an initial clock. With the limited accuracy of the initial clock, there is a systematic bias between the introduced and theoretical initial clock. The systematic bias is called the initial clock bias $O_a^S$. $T^S$ is the phase estimation clock correction [15]; $R_a^S$ is the effect caused by the assimilated orbital error [28]. Due to the correlation of the orbit and clock offset, most of radial orbital errors can be absorbed by the clock offset. These radial orbit errors are denoted as $R_a^S$ in the clock product; $\varepsilon_a^S$ represents the noise.

In this paper, we use the Single Satellite method (SSM); that is, a satellite is selected to construct inter-satellite differences (SD) to eliminate the time scale differences [29]. $S_0$ is selected as the reference star for inter-satellite difference:

$$C_{a_1}^{\Delta S} = C_{a_1}^{S} - C_{a_1}^{S_0} \quad \begin{aligned} &= \left(O_{a_1} + O_{a_1}^{S} + T^{S} + R_{a_1}^{S} + \varepsilon_{a_1}^{S}\right) - \left(O_{a_1} + O_{a_1}^{S_0} + T^{S_0} + R_{a_1}^{S_0} + \varepsilon_{a_1}^{S_0}\right) \\ &= \left(O_{a_1}^{S} - O_{a_1}^{S_0}\right) + \left(T^{S} - T^{S_0}\right) + \left(R_{a_1}^{S} - R_{a_1}^{S_0}\right) + \left(\varepsilon_{a_1}^{S} - \varepsilon_{a_1}^{S_0}\right) = O_{a_1}^{\Delta S} + T^{\Delta S} + R_{a_1}^{\Delta S} + \varepsilon_{a_1}^{\Delta S} \end{aligned} \tag{10}$$

In Equation (10), $\Delta$ is the difference operator, $S$ is the satellite to be evaluated, $S_0$ is the reference satellite, $\Delta S$ represents the difference between satellites and reference satellite, and $O_a$ is the time scale differences. After time scale differences are eliminated, the product-differenced (PD) between different analysis centers is:

$$C_{\Delta a}^{\Delta S} = C_{a_2}^{\Delta S} - C_{a_1}^{\Delta S} \quad \begin{aligned} &= \left(O_{a_2}^{\Delta S} + T^{\Delta S} + R_{a_2}^{\Delta S} + \varepsilon_{a_2}^{\Delta S}\right) - \left(O_{a_1}^{\Delta S} + T^{\Delta S} + R_{a_1}^{\Delta S} + \varepsilon_{a_1}^{\Delta S}\right) \\ &= \left(O_{a_2}^{\Delta S} - O_{a_1}^{\Delta S}\right) + \left(R_{a_2}^{\Delta S} - R_{a_1}^{\Delta S}\right) + \left(\varepsilon_{a_2}^{\Delta S} - \varepsilon_{a_1}^{\Delta S}\right) = O_{\Delta a}^{\Delta S} + R_{\Delta a}^{\Delta S} + \varepsilon_{\Delta a}^{\Delta S} \end{aligned} \tag{11}$$

In Equation (10), $\Delta a$ represents the difference between products of different analysis centers, $(\cdot)_{\Delta a}^{\Delta S}$ is the second difference operator. The initial clock deviation $O_a^S$ is a constant offset in the continuous arc, which only affects the convergence time of PPP [22]. The assimilated orbit error $R_a^S$ is periodic and the effect can be eliminated by combining with the corresponding orbit products in PPP processing. It can be seen from the above formulas that the SCB product can be decomposed into the constant item, initial deviation, assimilated orbital period item, and noise in the form after double difference.

All SCB products in this paper are calculated by GFZ orbit; therefore, the influence of assimilation orbit error acting on SCB is the same, and the second difference $R_{\Delta a}^{\Delta S}$ is 0. The initial clock offset is a constant, and so is its second difference $O_{\Delta a}^{\Delta S}$. Therefore, the Standard Deviation (STD) can be used to measure the noise level and evaluate the accuracy of BeiDou satellite clock in this paper.

After accurately estimating BeiDou clock products, the phase, frequency, clock drift, and noise parameters of the BeiDou satellite clock can be obtained by quadratic term fitting of the global SCB, as shown in Equation (12):

$$\Delta C_i = a_0 + a_1(t_i - t_0) + a_2(t_i - t_0)^2 + \varepsilon_i \,, i = 1, 2, 3 \ldots, n \tag{12}$$

where $\Delta C_i$ is the SCB data in the epoch $i$; $a_0$, $a_1$, and $a_2$ are the clock offset, frequency offset, and frequency drift separately; $\varepsilon_i$ is the noise fitting residual; $n$ is the number of epochs of the clock difference of the fitting period. Combined with Equation (12) and the least squares method to process the global clock deviation sequence of the fitting period, the clock error, frequency offset, and frequency drift parameters of the satellite clock in this period can be obtained. Substitute these parameters of the corresponding fitting period to $\Delta TC_i = a_0 + a_1(t_i - t_0) + a_2(t_i - t_0)^2$ to obtain the theoretical clock deviation $\Delta TC_i$ at each epoch; by comparing the SCB products in different regions with the theoretical value of the SCB, the influence of the regional station network on the SCB calculation can be studied.

## 3. Product Analysis

The original observation data of 120 MGEX stations that can receive BeiDou observation data from all around the world are selected in this paper to calculate the global SCB. The specific SCB estimation strategy is shown in Table 1.

In this paper, we calculate global SCB in the strategy shown in Table 1. To verify the clock product, we compare the consistency between our global product and the global products of CODE, WHU, and GFZ. The consistency between products is shown in Figure 1. It can be seen from this figure that the consistency level of the SCB products calculated in this paper differs from other analysis centers at about 0.1 ns, which can verify the correctness of the programs and algorithms in this paper.

**Table 1.** Processing strategy and parameter model of SCB.

| Items | Parameters | | Models for SCB Determination |
|---|---|---|---|
| Observation information | Observation | | UD ionosphere-free range and ED ionosphere-free carrier-phase observation |
| | Prior information | | P1: 1.0 m; L1: 0.01 cycle |
| | Elevation mask | | 7° |
| | Observation weight | | $\mathrm{p}=1,\ \mathrm{elev}\ >\ 30°,\ \mathrm{p}=2\sin(\mathrm{elev}),\ \mathrm{elev}\leq 30°$ |
| Correction | Phase rotation correction | | Model Correct [30] |
| | PCO | Satellite | GPS PCO: IGS08.atx [31]; BeiDou GEO PCO:IGS08.atx; BeiDou IGSO/MEO PCO:ESA Mode [32] |
| | | Receiver | GPS, BeiDou PCO: IGS08.atx |
| | PCV | Satellite | GPS PCV: IGS08.atx; BeiDou PCV: corrected |
| | | Receiver | GPS, BeiDou PCV:IGS08.atx |
| | Tides | | Ocean tides; solid earth tides; solid earth pole tides: IERS conventions 2010 |
| Parameters estimation | Relativistic effects | | IERS conventions 2003 |
| | Reference clock | | One satellite clock |
| | Satellite orbit | | Fixed |
| | Station coordinate | | Fixed |
| | Tropospheric delay | | Saastamoinen model + GMF mapping function random-walk process for each epoch |
| | Satellite/receiver clock | | Estimated as white noise |
| | Ambiguity | | Estimated in un-differenced, eliminated by phase epoch-differenced/estimated if exit cycle slips |

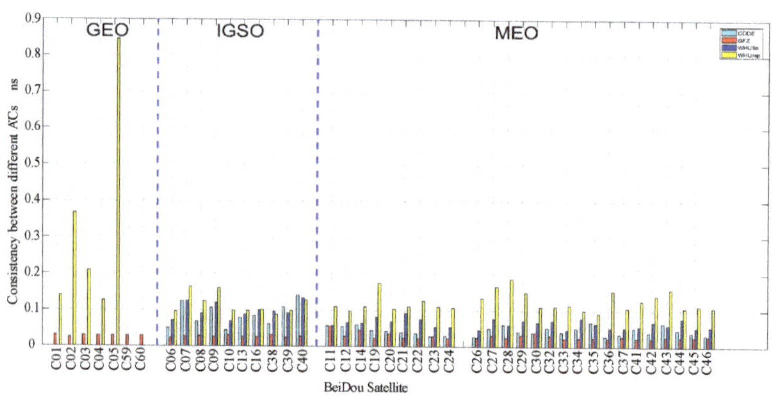


**Figure 1.** The consistency verification.

Three research regions are selected for calculation and SCB, namely, the China regional network, the North China regional network, and the South China regional network. Among 210 national stations distributed nationwide, 50 stations are selected by region to calculate the regional SCB of the BeiDou satellite in each region. The distribution of national stations in the three selected regions is shown in Figures 2–4.

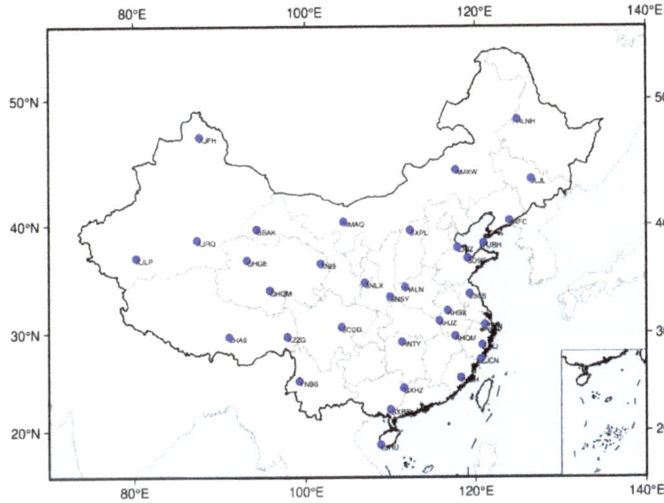


**Figure 2.** Station distribution in China.

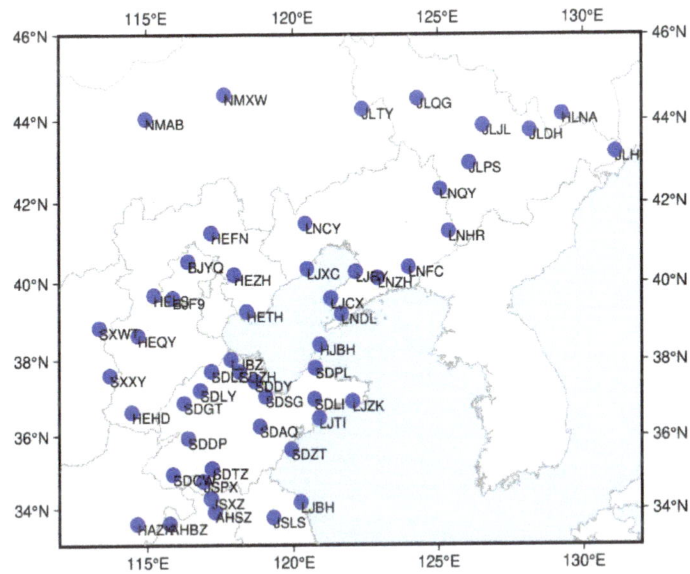


**Figure 3.** Station distribution in North China.

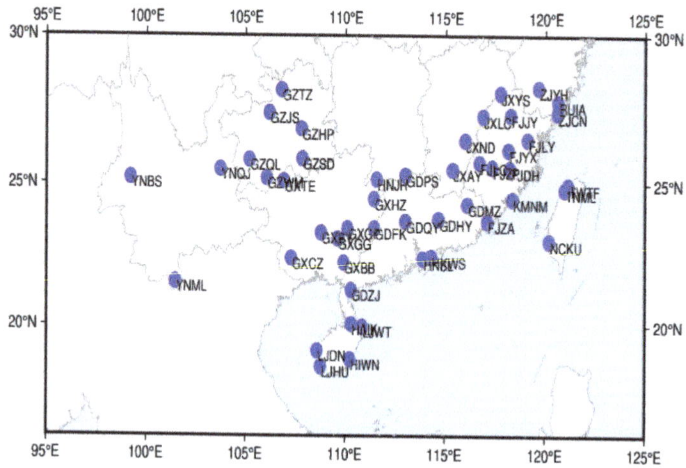


**Figure 4.** Station distribution in South China.

### *3.1. Completeness of Regional Clock Products*

Global satellites can be tracked by globally networked stations as they are rotating around the earth, but our application scenario is limited to regionally networked stations. When a satellite is far from the study area, it cannot be tracked by any ground station in that area due to the curvature of the Earth and elevation angle. This phenomenon will result in the satellite clock bias sequence not being able to be obtained completely when only using the observation value of the regional station to calculate the satellite clock bias, causing the satellite clock bias product incompleteness. The most intuitive impact of this incompleteness on the user is the reduction of the number of observations. Too few observations may let users cannot make PPP processing.

Therefore, it is necessary to analyze the completeness of the satellite clock product to study whether it is close to the requirements of PPP. The Geosynchronous Earth Orbit (GEO) and Inclined Geosynchronous Orbit (IGSO) of the BeiDou system's period is 23h56min and the Middle Earth Orbit (MEO)'s period is 12h56min; hence, the satellite-receiver geometry is almost the same every day. The discontinuous situation of regional BeiDou SCB is statistically analyzed by using the SCB product on the day of year 214 in 2021. The results are shown in Tables 2–4. The coverage period diagram is drawn and the results are shown in Figures 5–7 (blue for GEO, yellow for MEO, and red for IGSO).

**Table 2.** Daily coverage of regional SCB data in China.

| GEO | | IGSO | | MEO | | | | | |
|---|---|---|---|---|---|---|---|---|---|
| C01 | 0.00% | C06, C39 | 0.00% | C11 | 51.53% | C25 | 38.19% | C35 | 32.01% |
| C02 | 100.00% | C07 | 95.80% | C12 | 50.21% | C26 | 51.08% | C36 | 26.49% |
| C03 | 100.00% | C08 | 81.94% | C14 | 41.94% | C27 | 27.81% | C37 | 25.38% |
| C04 | 100.00% | C09 | 87.67% | C19 | 39.93% | C28 | 39.83% | C41 | 28.09% |
| C05 | 100.00% | C10 | 94.03% | C20 | 44.06% | C29 | 29.31% | C42 | 33.02% |
| C59 | 100.00% | C13 | 84.55% | C21 | 38.26% | C30 | 27.60% | C43 | 39.90% |
| C60 | 100.00% | C16 | 89.65% | C22 | 36.42% | C32 | 32.05% | C44 | 36.56% |
| | | C38 | 77.74% | C23 | 38.47% | C33 | 29.48% | C45 | 36.46% |
| | | C40 | 76.49% | C24 | 46.63% | C34 | 40.17% | C46 | 17.43% |

**Table 3.** Daily coverage of regional SCB data in North China.

| GEO | | IGSO | | MEO | | | |
|---|---|---|---|---|---|---|---|
| C01 | 0.00% | C06 | 0.00% | C11 | 41.01% | C24 | 31.60% |
| C02 | 100.00% | C07 | 78.30% | C12 | 40.17% | C25 | 33.78% |
| C03 | 100.00% | C08 | 73.26% | C14 | 34.24% | C26 | 39.48% |
| C04 | 100.00% | C09 | 73.89% | C19 | 28.44% | C27 | 21.01% |
| C05 | 41.74% | C10 | 76.15% | C20 | 32.29% | C28 | 27.92% |
| C59 | 0.00% | C13 | 73.09% | C21 | 33.96% | C29 | 27.71% |
| C60 | 0.00% | C16 | 76.08% | C22 | 32.53% | C30 | 22.67% |
| | | C38–C40 | 0.00% | C23 | 32.88% | C32–C37 | 0.00% |
| | | | | | | C41–C46 | 0.00% |

**Table 4.** Comparison of daily SCB coverage between china and the North China region.

| MEO | | | | IGSO | |
|---|---|---|---|---|---|
| C11 | 10.52% | C24 | 15.03% | C07 | 17.50% |
| C12 | 10.03% | C25 | 4.41% | C08 | 8.68% |
| C14 | 7.71% | C26 | 11.60% | C09 | 13.78% |
| C19 | 11.49% | C27 | 6.81% | C10 | 17.88% |
| C20 | 11.77% | C28 | 11.91% | C13 | 11.46% |
| C21 | 4.31% | C29 | 1.60% | C16 | 13.58% |
| C22 | 3.89% | C30 | 4.93% | | |
| C23 | 5.59% | | | | |

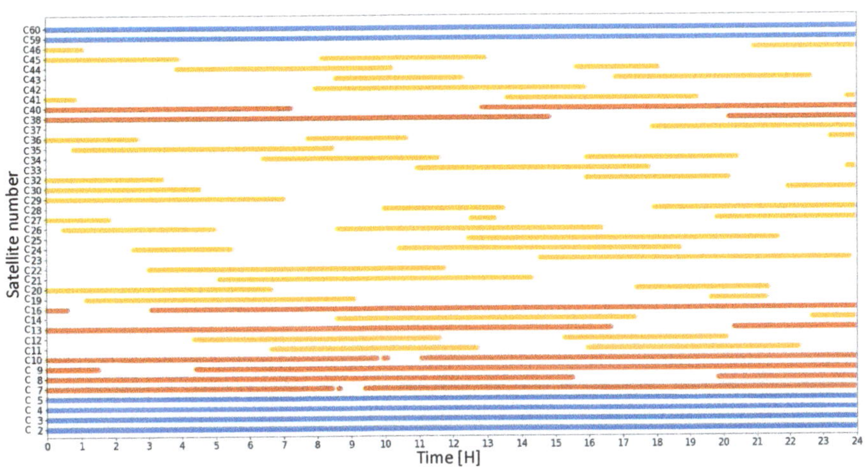


**Figure 5.** China regional SCB data coverage period.

As can be seen from Figure 5 and Table 2, most of the SCB products of BDS satellites have been computed in China. The relative position of the GEO satellites to the ground station hardly changes, so the daily SCB is completely calculated. There are some defects in the IGSO satellite clock deviation; according to the statistical data in Table 2, the defects are about 5–25%. All MEO satellites can be observed in China and about 17–51% of clock bias in the epoch can be calculated every day. Most of satellites signals can be received about 40% epochs of a day. BDS-2 satellites have less vacancy than BDS-3 satellites. This is because the BDS-2 satellite positioning system was originally designed to serve the Asia-Pacific region and the trajectory of its sub-satellite points covered mostly in China, with relatively complete station observations.

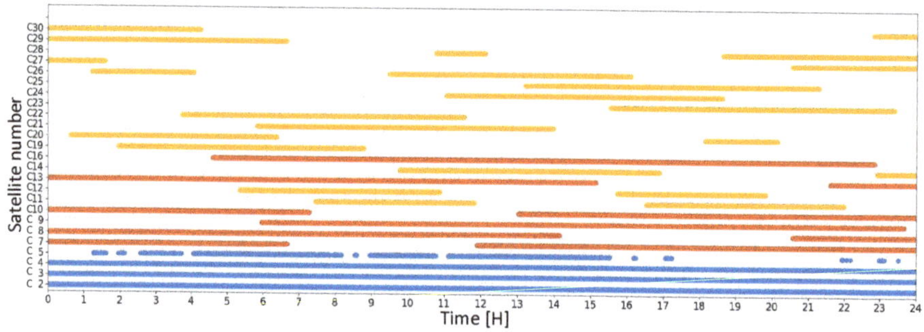


**Figure 6.** North China regional SCB data coverage period.

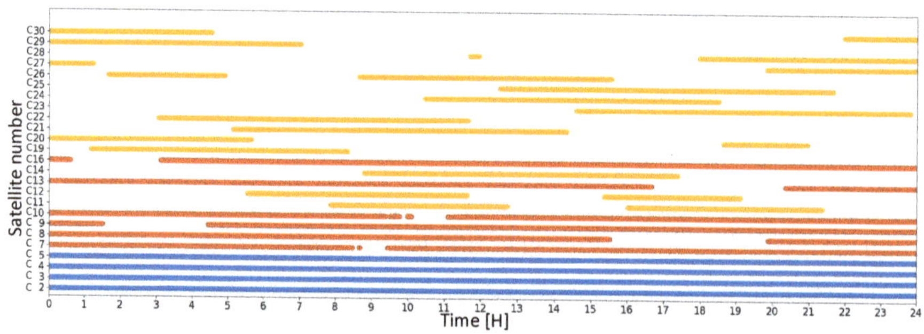


**Figure 7.** South China regional SCB data coverage period.

Figure 4 shows that the stations in the North China region are distributed in the mid-latitude area north of 34 °N. However, the GEO satellites are above the equator, causing the poor signal intensity of some GEO satellites with low altitudes, resulting in the discontinuous observation of C05 in Figure 6. Due to the western position of the projection point of the satellite on the ground of C59 and C60, the observation stations are concentrated in the east of China, which also leads to the lack of observation of the C59 and C60 satellites. Some IGSO satellites (C38–C40) and MEO satellites (C32–C37, C41–C46) of BDS-3 provide services to the Western Hemisphere. These satellites cannot be observed by stations in North China, so the SCB data is missing. It can be seen from Table 3 that about 21–41% of the observed satellites can be used for SCB calculation every day. Table 4 shows that the SCB calculated every day in this region is smaller than that in China. The missing epoch of MEO satellites accounts for about 2–15% of the total epoch per day, and the reason for the lack is mainly due to the smaller distribution area of survey stations. The missing epoch of IGSO satellites accounts for about 9–18% of the total epoch per day; the main reason for the lacking is the higher latitude.

It can be seen from Figure 4 that the stations in South China are distributed in low-latitude areas south of 28 °N and the GEO satellite is above the equator, with good observation conditions. The longitude range of the station is similar to that of the North China region in Figure 3, and the solution of C59 and C60 is also lacking. The lack of calculation data for IGSO and MEO satellites is similar to that of the North China region, indicating that whether the calculation of clock products of IGSO and MEO satellites in China is null is closely related to longitude. For the IGSO and MEO satellites whose solution results are not null, it can be seen from Tables 5 and 6 that the daily coverage rate of clock products in South China is higher than that in North China. The daily coverage rate and period of clock products of some stations in South China are the same as the clock product of the China region, which can also be seen in Figure 7.

**Table 5.** Daily coverage of regional SCB data in South China.

| GEO | | IGSO | | MEO | | | |
|---|---|---|---|---|---|---|---|
| C01 | 0.00% | C06 | 0.00% | C11 | 42.71% | C24 | 33.61% |
| C02 | 100.00% | C07 | 95.73% | C12 | 41.01% | C25 | 38.19% |
| C03 | 100.00% | C08 | 81.94% | C14 | 35.80% | C26 | 42.50% |
| C04 | 100.00% | C09 | 87.67% | C19 | 29.83% | C27 | 22.60% |
| C05 | 100.00% | C10 | 93.75% | C20 | 32.64% | C28 | 26.28% |
| C59 | 0.00% | C13 | 84.55% | C21 | 38.26% | C29 | 29.31% |
| C60 | 0.00% | C16 | 89.65% | C22 | 35.94% | C30 | 27.50% |
| | | C38–C40 | 0.00% | C23 | 38.47% | C32–C37 | 0.00% |
| | | | | | | C41–C46 | 0.00% |

**Table 6.** Comparison of daily coverage of SCB data in each region.

| China Clock Minus South China Clock | | | | | | North China Clock Minus South China Clock | | | | | |
|---|---|---|---|---|---|---|---|---|---|---|---|
| MEO | | | | IGSO | | MEO | | | | IGSO | |
| C11 | 8.82% | C25 | 0.00% | C06 | 0.00% | C11 | −1.70% | C25 | −4.41% | C06 | 0.00% |
| C12 | 9.20% | C26 | 8.58% | C07 | 0.07% | C12 | −0.83% | C26 | −3.02% | C07 | −17.43% |
| C14 | 6.15% | C27 | 5.21% | C08 | 0.00% | C14 | −1.56% | C27 | −1.60% | C08 | −8.68% |
| C19 | 10.10% | C28 | 13.54% | C09 | 0.00% | C19 | −1.39% | C28 | 1.63% | C09 | −13.79% |
| C20 | 11.42% | C29 | 0.00% | C10 | 0.28% | C20 | −0.35% | C29 | −1.60% | C10 | −17.60% |
| C21 | 0.00% | C30 | 0.10% | C13 | 0.00% | C21 | −4.31% | C30 | −4.83% | C13 | −11.46% |
| C22 | 0.49% | C32–C37 | 32.05% | C16 | 0.00% | C22 | −3.40% | | | C16 | −13.58% |
| C23 | 0.00% | C41–C46 | 29.48% | | | C23 | −5.59% | | | | |
| C24 | 13.02% | | | | | C24 | −2.01% | | | | |

In this section, we make statistics on the number of satellites that can be calculated for SCB products in each region (hereafter referred to as the number of observable satellites); the results are shown in Figure 8. As can be seen from Figure 8, in the North China region (middle-latitude region), the number of observable satellites is less than the South China region (low-latitude region). This confirms the conclusion in Table 6 that the stations in the low-latitude region contribute more to the calculation of SCB, and more visible satellites can be observed. More than 10 satellites in the three regions can be calculated to reach the requirements of precise point positioning.

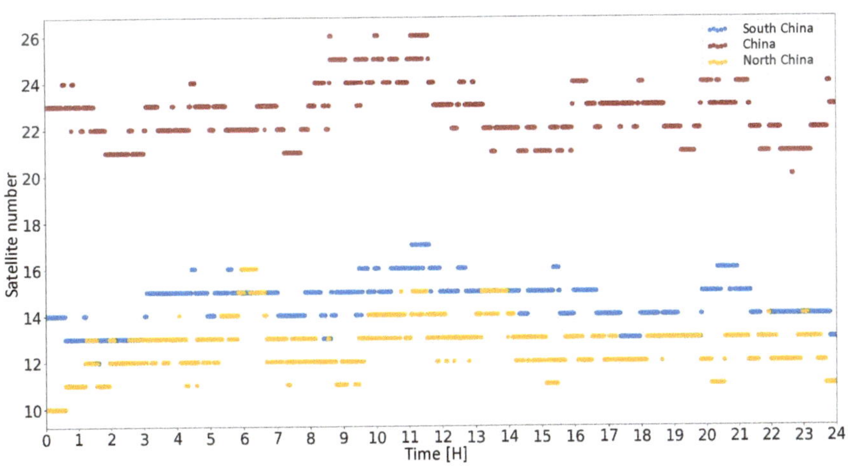


**Figure 8.** Regional SCB data's solvable satellites.

### 3.2. *Accuracy of Regional Clock Products*

In this paper, the global and regional SCB are calculated by the fixed orbit. The study time is 214–243 days in 2021. The global SCB product is used as the reference, and the standard deviation (STD) of the double-difference SCB of each satellite within the study time is calculated to evaluate the accuracy level of satellite clocks in each region. In the calculation of the double difference, C14 is used as the reference satellite.

It can be seen from Figure 9 that the accuracy level of SCB products is affected by the satellite orbit; the order from high to low is MEO, IGSO, and GEO. The accuracy level of SCB products is also affected by the station region. The accuracy of China regional SCB products in each region is less than 0.7 ns for the GEO satellite, 0.3 ns for the IGSO satellite, and 0.2 ns for the MEO satellite. The accuracy of SCB products in the North China region is lower than that in China. The STD of the GEO satellite is within 0.7 ns, the IGSO satellite is within 0.3 ns, and the MEO satellite is within 0.2 ns. In the regional SCB, the accuracy of clock products in South China is lower than that in the North China region. The STD of GEO satellite clock products calculated in the South China region is within 0.7 ns and that of IGSO and MEO satellite clock products is within 0.3 ns. The stations in South China have lower latitudes and are greatly affected by the ionosphere. Although the ionospheric elimination algorithm is used in this experiment, the active ionospheric variation will still affect the float ambiguity resolution, cycle slip detection, and the quality of ionospheric elimination, resulting in the instability of SCB observation data in the South China region, and the accuracy is worse than that in the North China region.

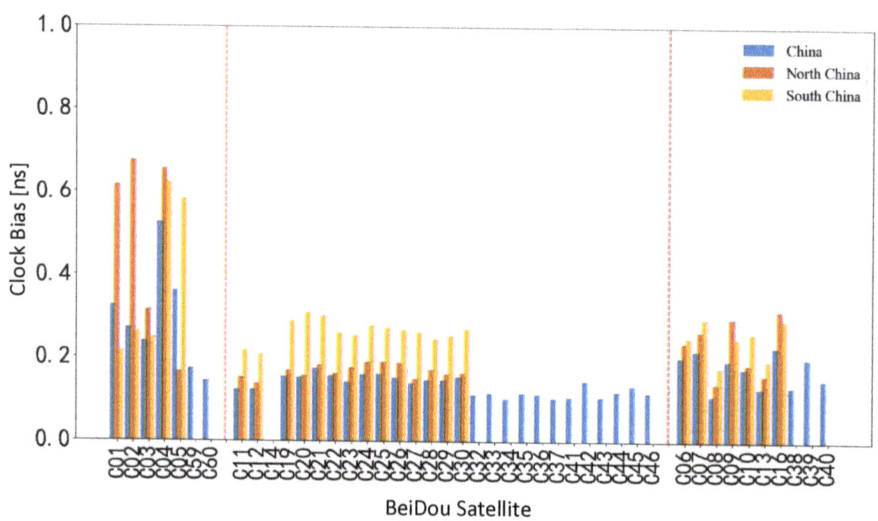


**Figure 9.** The accuracy of SCB in different regions.

### 3.3. *Regional Influence Deviation of Clock Product*

The phase, frequency, and frequency drift of the satellite clocks are obtained by quadratic term fitting to the global clock products calculated in this paper. The theoretical value of the satellite clock is calculated by Equation (12), and the difference is made with the calculated regional satellite clock product, obtaining the mix value of noise and regional influence. Figures 10–12, respectively, show the influence of the China region, North China region, and South China region on SCB calculation (cyan for GEO, red for IGSO, green for MEO). It can be seen from the figure that the level of the mixed value of noise and regional influence is $10^{-7}$ s. In Section 3.2, we have made statistics on the accuracy of regional clock products and know that the accuracy of regional clock products is within $10^{-9}$ s. Based on the above result, it can be seen that noise accounts for a relatively small proportion of the mixed value. We believe that the influence of noise on the region can be ignored for analysis. The regional influence is reflected in the phase offset of the satellite clock calculation. It absorbs some atmospheric parameters, and the influence on the satellite clock bias of all satellites has obviously similar regional characteristics, so it is named as the Region Effect Bias (REB) below. Different from the global SCB product, the regional SCB product with REB can reflect the sudden change of atmospheric delay parameters over the studied area and has a good enhancement effect on the regional PPP.

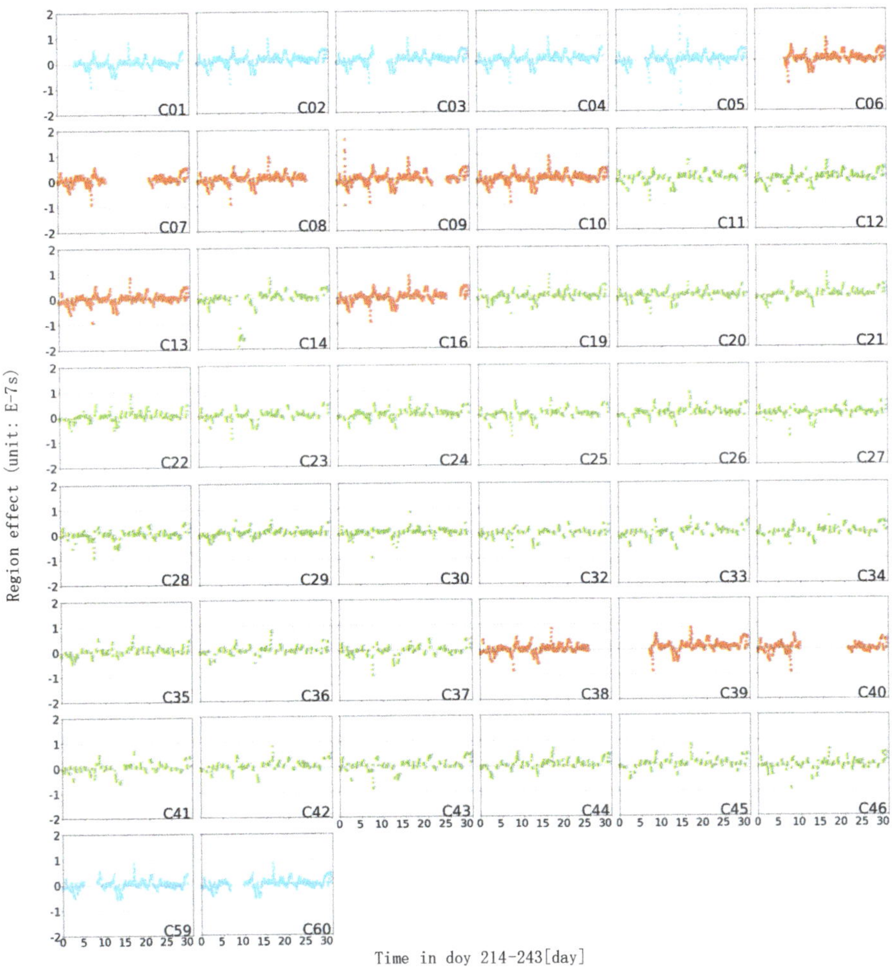


**Figure 10.** China region REB ($10^{-7}$ s).

It can be seen from Figures 10–12 that the time series curves of regional influence bias in the same region are similar. Figure 10 clearly shows that the REB of all solvable SCB products has the same time series characteristics. Although there are many missing epochs in Figures 11 and 12, this situation can also be observed. This indicates that the REB in SCB calculation is related to the distribution region of the station, the REB of the same group of regional networks has the same fluctuation trend on all satellites. Observed in the time series, all satellites in the same area have similar trends in the REB variation, which is irrelevant to the orbits of the satellites. This shows that the REB value of the satellite clock has a strong relationship with the region. The influence factors, such as atmospheric parameters over the region, lead to the rapid change of the REB value in the entire region, and the overall variation range is at the level of $10^{-7}$ s. In the same area, there are slight differences between satellites. It can be seen that the MEO satellites, which are shown as green, are obviously sparser than the IGSO satellites in red, indicating that there are more missing epochs in the clock error calculation of the MEO satellites. Additionally, BDS-2 satellites are denser than BDS-3 satellites, indicating that BDS-3 satellites are missing more epochs. The above two conclusions confirm the discussion of the completeness of satellite clocks in different orbits in Section 3.1. Moreover, in the same area, the fluctuation range of MEO satellites is smaller than that of IGSO and GEO orbiting satellites, and its number of the outliers are less, which is also due to the better orbit determination effect of MEO satellites.

During the study period, the regional influence bias and the range of change in the whole area of China are small. The REB value in the North China region is larger than that in the whole of China, and some values fluctuate significantly. The REB values of China and the North China region are mostly in the range of $\pm 0.5 \times 10^{-7}$ s, whereas the REB values of the South China region fluctuate sharply in the range of $\pm 1 \times 10^{-7}$ s. This shows that the area with a larger range or with a higher latitude has less influence, and the result of this REB value is the SCB parameter absorbing the regional systematic error. However, the ionospheric activity in the lower latitude regions is more active, and it has a greater impact on the regional SCB. The above results provide a reference for station selection in the application of regional clock products; that is to say, when the distribution range of the stations is small, in order to reduce the regional influence, it is appropriate to consider selecting the higher latitude stations to calculate the SCB.

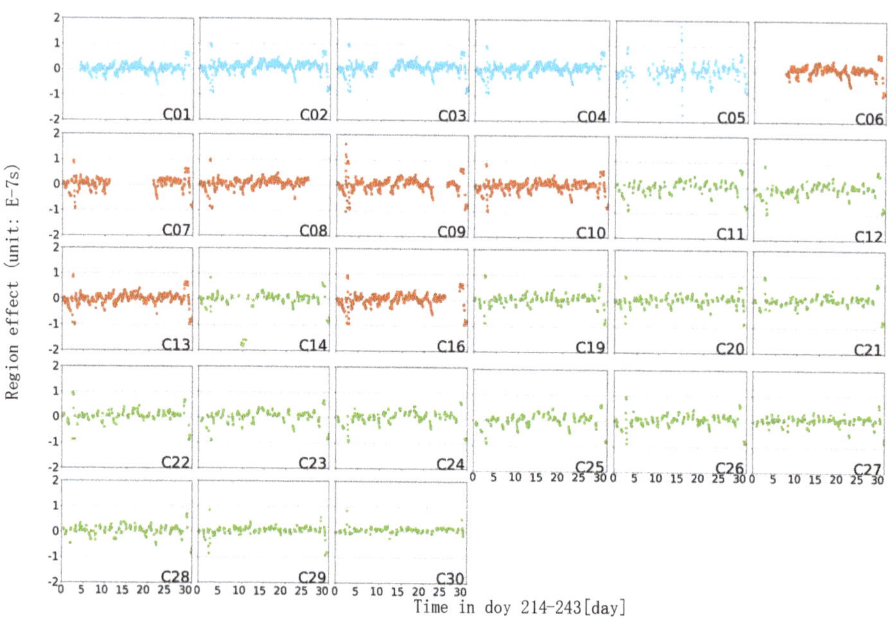


**Figure 11.** North China region REB ($10^{-7}$ s).

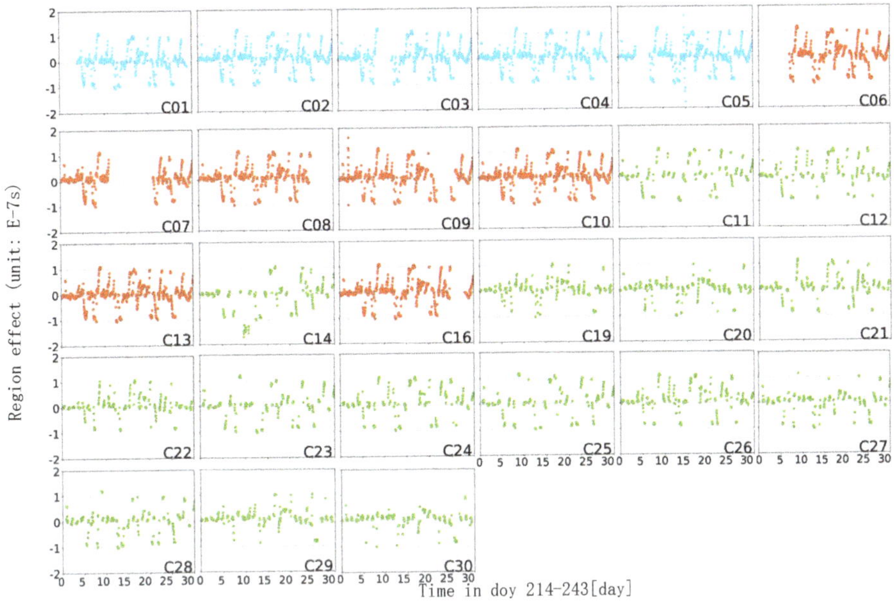


**Figure 12.** South China region REB ($10^{-7}$ s).

## 4. PPP Analysis of Regional Clock Product

We selected IGS stations in the Asia-Pacific region and national stations which had not participated in the calculation in each region to verify the PPP performance of the regional satellite clock product, and the selected stations can receive BDS-2 and BDS-3 signals and use the B1B3 signal for PPP processing. The distribution of IGS stations and national stations is as Figure 13.

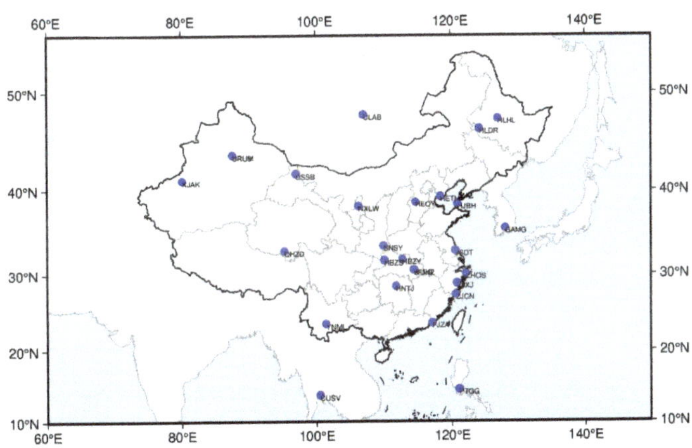


**Figure 13.** PPP station distribution.

The satellite clocks calculated by the global, Chinese, Northern, and Southern China regional stations are used to conduct PPP experiments for the selected stations on days 214–220 of 2021. The first epoch of 20 consecutive epochs when the error is less than 10 cm is used as the convergence time to count the post-convergence accuracy. The seven-day averages of Root Mean Square Error (RMS) after convergence at different IGS stations are counted in Figure 14; the RMS in E, N, and U directions are combined, i.e.,

$$\delta = \sqrt{\delta_E{}^2 + \delta_N{}^2 + \delta_U{}^2} \tag{13}$$

In Equation (13), $\delta$ represents the three-dimensional (3D) RMS in the space; $\delta_E$, $\delta_N$, and $\delta_U$ represent the RMS error in the direction of E, N, and U. It should be noted that in the RMS calculated by the Equation (13), due to the large value of RMS in the U direction, small values in the E and N directions will be submerged, making the overall RMS statistical data seems large.

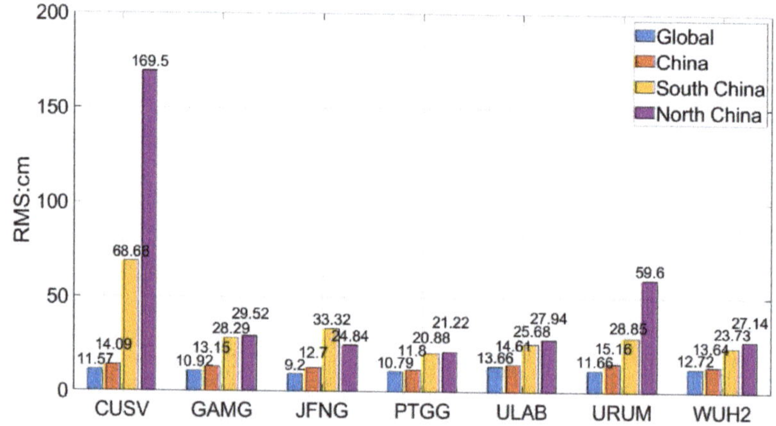

**Figure 14.** Effects of Different Regional Clocks on Different IGS Stations.

From Figure 14, it can be seen that the PPP results of the stations near China solved by the global SCB product have the same accuracy, and the 3D accuracy is at the level of 10 cm. The RMS error of the PPP calculated by the small size regional clock product is higher than that of the global product, which indicates that the reduction of the calculation range of the regional clock has a negative impact on the PPP. The difference between the accuracy of the Chinese regional clock solution and the global solution is insignificant, and they are both at the same level. This indicates that the accuracy of PPP experiments is very close to that of the global clock products if only the SCB products calculated by the national stations in the Chinese region are used. When comparing within the Chinese region, the PPP accuracy of CUSV using the regional clocks in South China in the figure is much higher than that in North China. This phenomenon combined with Figure 8 shows that when performing PPP, the accuracy level of the regional stations calculated far from the clock product increases rapidly with the number of observable satellites. However, the stations closer to the center of the study area, such as PTGG, ULAB, and WUH2, have a larger base of observation satellites, and the continued increase in the number of observation satellites will reduce the contribution to the improvement of PPP accuracy. However, the low-latitude regional SCB product has poor accuracy, as shown in Figure 9, and a large deviation from its regional influence, which prevents further improvement of the accuracy. It will even reduce the PPP accuracy of some stations such as JFNG.

The combination of Figures 14 and 15 shows that the accuracy has reached the level of global clock accuracy when the corresponding satellite clocks are used for PPP in the regional range calculated by the same regional SCB. The stations QHZD, XJAK, YNML, and NXLW in the western region have larger errors, whereas the errors in the southern region are smaller. This phenomenon is similar to the CUSV in Figure 14. This is because the number of satellite observations in the northern region is smaller and the number of satellite observations of the stations far from the satellite clock solution area decrease, and the PPP result accuracy becomes worse. The 3D accuracy of other stations in the North and South China clock calculation regions is at the same level as the global clock, indicating that the SCB products in the smaller mid-latitude range are not suitable for extending the positioning area excessively due to the low number of satellite clock calculations.

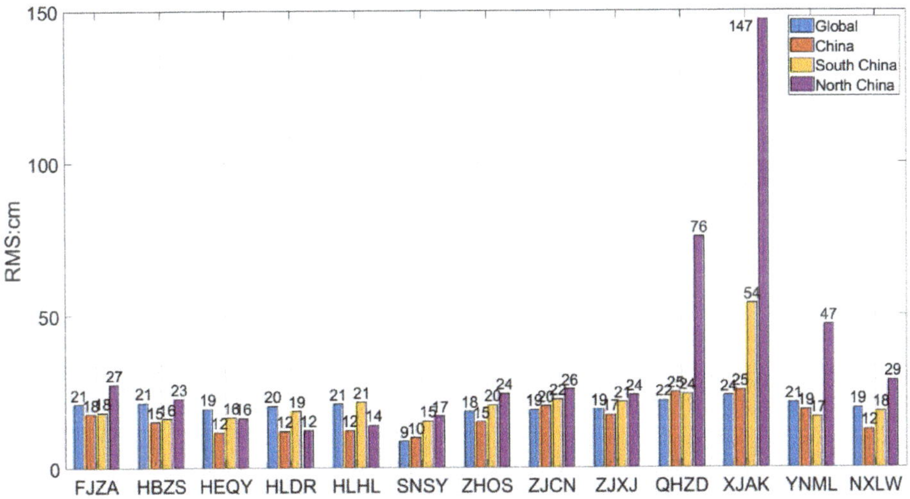


**Figure 15.** Effects of Different Regional Clocks on Different National Stations.

Figures 16 and 17 show the time series of the jump in the PPP results. From the regional clocks of China and North China in Figure 16, in kinematic mode, it can be seen that the PPP results by the regional SCB of China and the global SCB are at the same level, i.e., centimeter level. As shown in kinematic mode, the PPP accuracy of the regional clocks in South China and North China is at the centimeter level in most of the epochs, and there are jumps in the kinematic PPP result sequences of CUSV and URM stations, which shows that some periods after convergence diverge again, and the regional clocks in mid-latitude regions have more jumps and large fluctuations in PPP of stations in low-latitude regions. The regional clock in South China has the same effect on the PPP results of FJZA and CUSV stations, both of which are in the southeast corner of the Chinese region, indicating that the effect here comes from the REB, whereas the jump in the same region using the regional clock in North China has the same obvious consistency as that of the regional clock in South China, and the REB of the regional clock in North China has no obvious effect on the PPP. In the North China region, the low number of observations leads to a decrease in the number of satellite observations in the region of some stations far from the calculation region affecting the accuracy, which leads to the jump in the positioning results of the North China regional clock and affects the positioning accuracy of the regional clock products. In the kinematic mode, the PPP results of the previous epoch are not transferred to the next epoch for iteration to improve the convergence results, whereas in the static PPP mode, the process noise in the state transfer matrix is 0. From the results of static mode positioning, it can be seen that the positioning accuracy of the regional clock reaches the level of the global clock, but the convergence time is significantly longer. From the comparison between Figures 16 and 17, it can be seen that the static mode can effectively eliminate the jump caused by the steep drop in the number of observation satellites in one day, but from the positioning results of the South China clock for FJZA on CUSV, the effect of REB cannot be eliminated completely. Moreover, the convergence time generally reaches more than 2h, which is obviously inferior to the convergence time of the global clock.

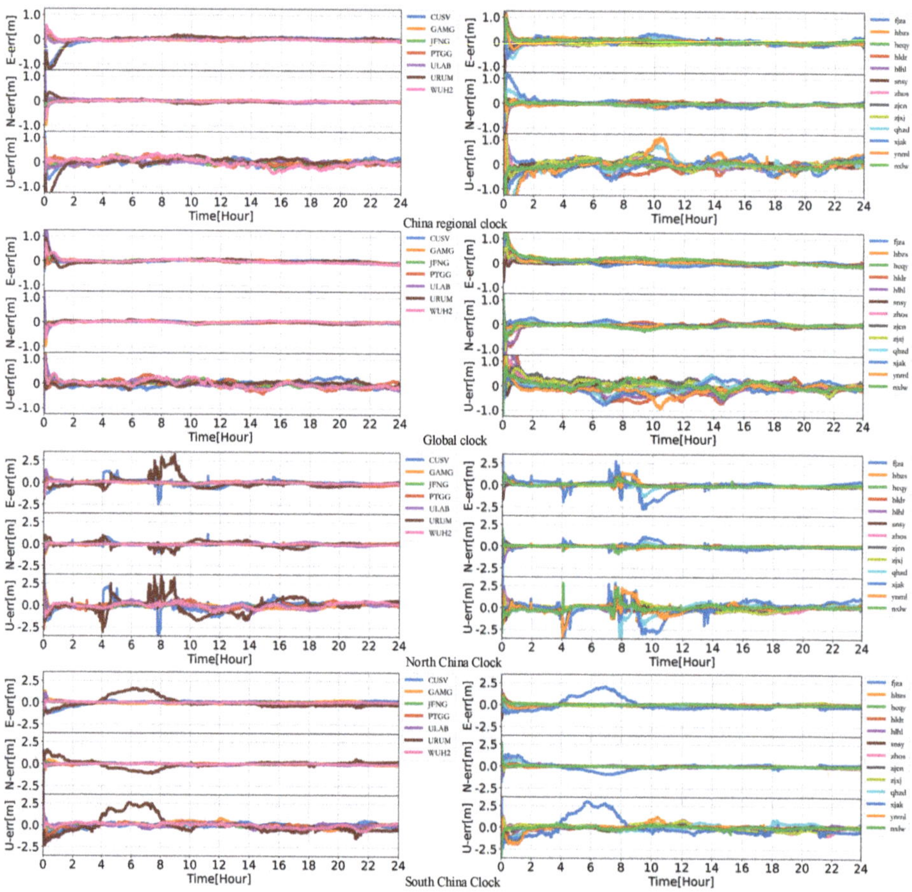


**Figure 16.** Kinematic PPP Time Series Plots of Different Regional Stations on Day 216 of 2021.

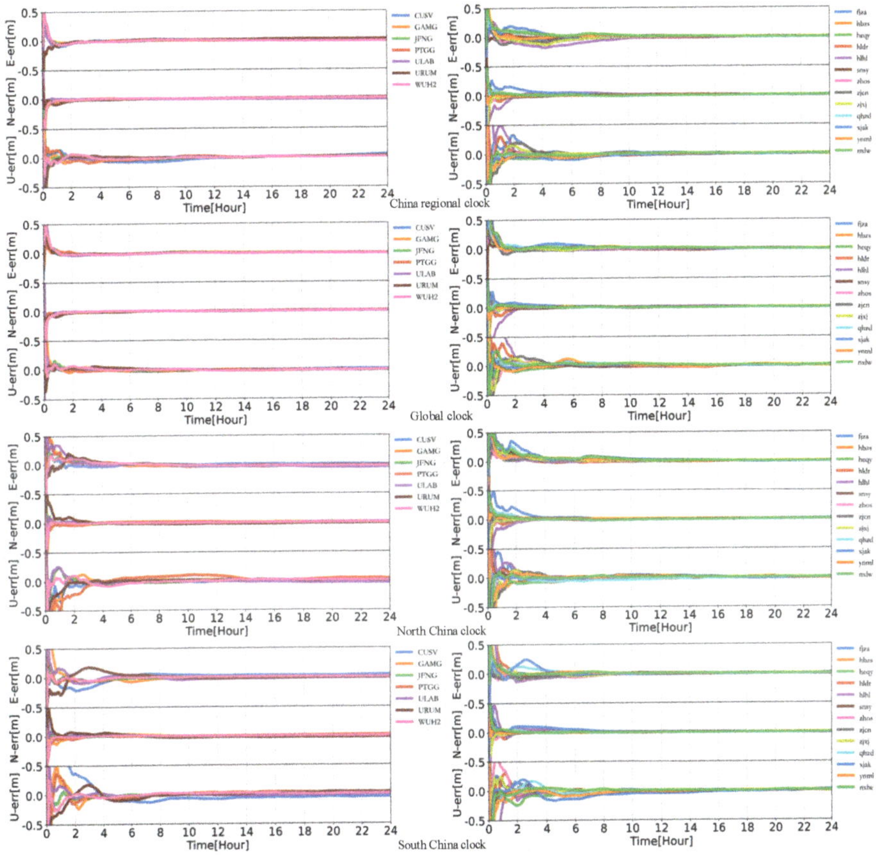


**Figure 17.** Static PPP Time Series Plots of Different Regional Stations on Day 216 of 2021.

## 5. Discussion

We have mentioned the problems of accuracy and regional influence caused by station area and latitude in Sections 3 and 4, and we introduce the concept of regional effect bias (REB). It is found in Section 4 that the stations distributed in the same area have the same fluctuations in the kinematic mode. We believe that the influence of these regional station selections on the BeiDou SCB products comes from the spatial correlation errors such as atmospheric parameters absorbed over the small regional stations. If the research scope is larger, as shown in Figure 2, the difference in spatial correlation error between global and region is small, so the deviation of regional influence is reduced. In Section 3.1, we know that the regional clock has many breakpoints. The calculation after each breakpoint needs to be initialized with the pseudo-range. Therefore, in theory, the regional clock must have a lot of errors due to initialization, and the positioning results should not be as good as the global SCB. However, the positioning accuracy of the regional clock is higher than that of the global SCB; this phenomenon is due to the effect of the SCB parameters absorbing regional system deviation. When the regional SCB calculating, the corresponding range fluctuates when the atmospheric environment changes abruptly and will be absorbed by the regional SCB. These mutations are the main factor leading to the loss of precision in PPP. The atmospheric parameters absorbed by the regional SCB products can offset the effects of these sudden changes during PPP, thereby improving positioning accuracy. Otherwise, some papers also provide another hypothesis that when using globally distributed stations, satellites are always directly above the visible station [33]. Therefore, the clock estimated from the global network cannot compensate for the orbital error of the tangential component, which may also lead to the degradation of the positioning performance of the global network.

Another important question is whether REB will affect the precision statistics of SCB; because Equation (9) of the regional clock should include the regional effect bias REB, it may affect the statistics and analysis of the SCB accuracy after the double-difference. However, according to the conclusions obtained from Figures 10–12 in Section 3.3, the fluctuations of the REBs of each satellite clock calculated in the same area are the same, and the trend term of REB will be greatly weakened after the difference between satellites. The STD level of the final obtained SCB statistics in Figure 9 is consistent with the normal situation, so the SCB products calculated by the inter-satellite difference method adopted in this paper are effective.

## 6. Conclusions

In this paper, the MD method is used to solve the SCB of three regions, China, North China, and South China. The characteristics of the regional clocks are analyzed by using the global clock as the reference. The regional effect bias REB is introduced to measure the deviation of the calculated regional clock from the true value, the accuracy of the regional satellite clock is evaluated, and the PPP is used to verify the regional clock positioning results. The conclusions are as follows:

1. Due to the different designs of the systems, the data completeness of the BDS-2 satellite over the China region is better than that of the BDS-3 satellite. The observation number of the regional BeiDou clocks is ranked from highest to lowest in the Chinese region, the South China region, and the North China region. The decrease of station area has a great influence on the observation of the MEO satellite, whereas the increase of latitude has a great influence on the observation of the GEO and IGSO satellite, which leads to the loss of these regional SCB products, respectively. The contribution of low-latitude stations to regional SCB calculation is higher than that of the mid-latitude stations, which indicates that to increase the number and duration of observation satellites, it is necessary to increase the number of observation stations in the low-latitude region. The available satellite number of SCB in each region satisfies the requirements of the PPP experiment.
2. From the STD of regional clock products, it can be seen that the accuracy of the BeiDou satellite regional clock is from high to low in the China region, North China region, and South China region from high to low. The STD of GEO satellites in all regions are less than 0.7 ns, and that of IGSO and MEO satellites are less than 0.3 ns. The accuracy of SCB products in South China is worse than that in North China, which is due to the low latitude of the stations and the influence of more serious ionospheric errors. Although the ionospheric-free algorithm is used in this experiment, the ionospheric scintillation caused by the active ionosphere will still affect the ambiguity float solutions and the quality of ionospheric elimination, resulting in the instability of SCB observation data in the South China region, and the accuracy is worse than that in the North China region. Therefore, the influence of the ionospheric layer on the accuracy of SCB can be weakened by appropriately choosing the station with higher latitude.
3. In this paper, the regional effect bias REB is introduced to analyze the influence of different regions on the clock error product. For the BeiDou system, the regional influence bias of the same group of regional networks will lead to a similar offset sequence for all satellite clock errors calculated by this group of regional networks, and the REB value is at the level of $10^{-7}$. Among them, the fluctuation range of MEO's offset series is smaller, and there are fewer outliers. During the study period, the larger regional REB deviations and changes are smaller. The regional influence of South China with lower latitude is stronger when the area is similar and leads to worse PPP accuracy. This shows that the regional influence between the larger area and the higher latitude area station-satellite is small. When the distribution range of

the station is small, for weakened regional influence, the higher latitude station can be appropriately considered to calculate the SCB.

4. When using the different regional SCB products calculated in this paper for PPP, the amount of data for SCB-computing using stations in the mid-latitude region is maintained at a low level, resulting in a more serious decrease in the accuracy level of the stations at slightly distant locations near the clock product solving area. When using mid-latitude for regional positioning, the working area for positioning should be controlled as much as possible.
5. The PPP results of the Chinese regional clock and the global clock are at the same level, but the convergence speed of the regional clock is generally inferior to that of the global clock. In the regional mode, the low-latitude clocks are affected by the region, and the mid-latitude clocks are affected by the regional influence, whereas the mid-latitude clocks have a jump in accuracy due to the low amount of observed data. The static mode can improve the fluctuation due to the small amount of data, but it does not completely eliminate the effect of REB.

**Author Contributions:** Conceived the idea, H.W. and W.L.; designed the software, collected the data, and analyzed the experimental data, W.L. and P.L.; collected the related resources and supervised the experiment, W.L. and H.W.; proposed the comment for the paper and experiment, H.M., Y.R., B.L., Y.C. and H.W.; investigation, W.L. All authors have read and agreed to the published version of the manuscript.

**Funding:** This research is supported by the Key Project of China National Programs for Research and Development (No. 2022YFB3903902; No. 2022YFB3903900), the Wenhai Program of the S&T Fund of Shandong Province for Pilot National Laboratory for Marine Science and Technology (Qingdao) (No. 2021WHZZB1000, No. 2021WHZZB1005), the National Natural Science Foundation of China (No. 42274044; No. 41874042), the State Key Laboratory of Geo-Information Engineering and Key Laboratory of Surveying and Mapping Science and Geospatial Information Technology of MNR, CASM (No. 2022-01-09; No. 2021-01-01), and the Scientific Research Project of Chinese Academy of Surveying and Mapping (No. AR2101; No. AR2203; No. AR2214).

**Institutional Review Board Statement:** Not applicable.

**Informed Consent Statement:** Not applicable.

**Data Availability Statement:** Not applicable.

**Acknowledgments:** The authors acknowledge IGS (International GNSS Service) for GNSS observations and CODE (Center for Orbit Determination in Europe), WHU (Wuhan University), and GFZ (Deutsches GeoForschungsZentrum) for GNSS products. Additionally, some of the figures are produced by GMT (Generic Mapping Tools).

**Conflicts of Interest:** The authors declare no conflict of interest.

## References

1. Qu, L.; Du, M.; Wang, J.; Gao, Y.; Zhao, Q.; Qiang, Z.; Guo, X. Precise Point Positioning Ambiguity Resolution by Integrating BDS-3e into BDS-2 and GPS. *GPS Solut.* **2019**, *23*, 63. [CrossRef]
2. Rocken, C.; Johnson, J.; Van Hove, T.; Iwabuchi, T. Atmospheric Water Vapor and Geoid Measurements in the Open Ocean with GPS. *Geophys. Res. Lett.* **2005**, *32*, 12. [CrossRef]
3. Liu, Y.; Sun, X.; Wang, G.; Turco, M.J.; Agudelo, G.; Bao, Y.; Zhao, R.; Shen, S. Current Activity of the Long Point Fault in Houston, Texas Constrained by Continuous GPS (2013–2018). *Remote Sens.* **2019**, *11*, 1213. [CrossRef]
4. Li, X.; Chen, X.; Ge, M.; Schuh, H. Improving Multi-GNSS Ultra-Rapid Orbit Determination for Real-Time Precise Point Positioning. *J. Geod.* **2019**, *93*, 45–64. [CrossRef]
5. Abdi, N.; Ardalan, A.A.; Karimi, R.; Rezvani, M.-H. Performance Assessment of Multi-GNSS Real-Time PPP over Iran. *Adv. Space Res.* **2017**, *59*, 2870–2879. [CrossRef]
6. Cao, X.; Li, J.; Zhang, S.; Pan, L.; Kuang, K. Performance Assessment of Uncombined Precise Point Positioning Using Multi-GNSS Real-Time Streams: Computational Efficiency and RTS Interruption. *Adv. Space Res.* **2018**, *62*, 3133–3147. [CrossRef]
7. Wang, L.; Li, Z.; Ge, M.; Neitzel, F.; Wang, X.; Yuan, H. Investigation of the Performance of Real-Time BDS-Only Precise Point Positioning Using the IGS Real-Time Service. *GPS Solut.* **2019**, *23*, 66. [CrossRef]

8. Ge, M.; Chen, J.; Dousa, J. A Computationally Efficient Approach for Estimating High-Rate Satellite Clock Corrections in Realtime. *GPS Solut.* **2011**, *15*, 9–17. [CrossRef]
9. Liu, Y.; Liu, F.; Gao, Y.; Zhao, L. Implementation and Analysis of Tightly Coupled Global Navigation Satellite System Precise Point Positioning/Inertial Navigation System (GNSS PPP/INS) with Insufficient Satellites for Land Vehicle Navigation. *Sensors* **2018**, *18*, 4305. [CrossRef]
10. Shi, J.; Yuan, X.; Cai, Y.; Wang, G. GPS Real-Time Precise Point Positioning for Aerial Triangulation. *GPS Solut.* **2017**, *21*, 405–414. [CrossRef]
11. Ge, Y.; Zhou, F.; Liu, T.; Weijin, Q.; Wang, S.; Yang, X. Enhancing Real-Time Precise Point Positioning Time and Frequency Transfer with Receiver Clock Modeling. *GPS Solut.* **2018**, *23*, 20. [CrossRef]
12. Tang, X.; Li, X.; Roberts, G.; Hancock, C.; de Ligt, H.; Guo, F. 1 Hz GPS Satellites Clock Correction Estimations to Support High-Rate Dynamic PPP GPS Applied on the Severn Suspension Bridge for Deflection Detection. *GPS Solut.* **2018**, *23*, 28. [CrossRef]
13. Bock, H.; Dach, R.; Jäggi, A.; Beutler, G. High-Rate GPS Clock Corrections from CODE: Support of 1 Hz Applications. *J. Geod.* **2009**, *83*, 1083–1094. [CrossRef]
14. Wu, K.; Shen, Z.; Shen, W.; Sun, X.; Cai, C.; Wu, Y. A Preliminary Experiment of Determining the Geopotential Difference Using Two Hydrogen Atomic Clocks and TWSTFT Technique. *Geod. Geodyn.* **2020**, *11*, 229–241. [CrossRef]
15. Yuan, Z.; Cai, C.; Pan, L.; Kuang, C. An Improved Multi-Satellite Method for Evaluating Real-Time BDS Satellite Clock Offset Products. *Remote Sens.* **2020**, *12*, 3638. [CrossRef]
16. Zhang, X.; Li, X.; Guo, F. Satellite Clock Estimation at 1 Hz for Realtime Kinematic PPP Applications. *GPS Solut.* **2011**, *15*, 315–324. [CrossRef]
17. Fu, W.; Huang, G.; Zhang, Q.; Gu, S.; Ge, M.; Schuh, H. Multi-GNSS Real-Time Clock Estimation Using Sequential Least Square Adjustment with Online Quality Control. *J. Geod.* **2019**, *93*, 963–976. [CrossRef]
18. Li, X.; Xu, Y.; Wang, L. Undifferenced Precise Satellite Clock Error Estimation and Precision Analysis. *Geomat. Inf. Sci. Wuhan Univ.* **2010**, *35*, 661–664. [CrossRef]
19. Dixon, K. StarFire: A Global SBAS for Sub-Decimeter Precise Point Positioning. In Proceedings of the 19th International Technical Meeting of the Satellite Division of The Institute of Navigation (ION GNSS 2006), Fort Worth, TX, USA, 29 September 2006; pp. 2286–2296.
20. Hauschild, A.; Montenbruck, O. Kalman-Filter-Based GPS Clock Estimation for near Real-Time Positioning. *GPS Solut.* **2009**, *13*, 173–182. [CrossRef]
21. Melgard, T.; Vigen, E.; de Jong, K.; Lapucha, D.; Visser, H.; Oerpen, O. G2—The First Real-Time GPS and GLONASS Precise Orbit and Clock Service. In Proceedings of the 22nd International Technical Meeting of the Satellite Division of The Institute of Navigation (ION GNSS 2009), Savannah, GA, USA, 22–25 September 2009; pp. 1885–1891.
22. Chen, L.; Song, W.; Yi, W.; Shi, C.; Lou, Y.; Guo, H. Research on a Method of Real-Time Combination of Precise GPS Clock Corrections. *GPS Solut.* **2017**, *21*, 187–195. [CrossRef]
23. Pan, S.; Chen, W.; Jin, X.; Shi, X.; He, F. Real-Time PPP Based on the Coupling Estimation of Clock Bias and Orbit Error with Broadcast Ephemeris. *Sensors* **2015**, *15*, 17808–17826. [CrossRef] [PubMed]
24. Mervart, L.; Lukes, Z.; Rocken, C.; Iwabuchi, T. Precise Point Positioning with Ambiguity Resolution in Real-Time. In Proceedings of the 21st International Technical Meeting of the Satellite Division of The Institute of Navigation (ION GNSS 2008), Savannah, GA, USA, 16–19 September 2008; pp. 397–405.
25. Zhang, Q.; Moore, P.; Hanley, J.; Martin, S. Auto-BAHN: Software for near Real-Time GPS Orbit and Clock Computations. *Adv. Space Res.* **2007**, *39*, 1531–1538. [CrossRef]
26. Yao, Y.; He, Y.; Yi, W.; Song, W.; Cao, C.; Chen, M. Method for Evaluating Real-Time GNSS Satellite Clock Offset Products. *GPS Solut.* **2017**, *21*, 1417–1425. [CrossRef]
27. Wu, Y.; Liu, X.; Liu, W.; Ren, J.; Lou, Y.; Dai, X.; Fang, X. Long-Term Behavior and Statistical Characterization of BeiDou Signal-in-Space Errors. *GPS Solut.* **2017**, *21*, 1907–1922. [CrossRef]
28. Chen, J.; Zhao, X.; Hu, H.; Ya, S.; Zhu, S. Comparison and Assessment on Long-Term Performance of BDS-2/BDS-3 Satellite Atomic Clocks. *Meas. Sci. Technol.* **2021**, *32*, 115021. [CrossRef]
29. Guo, J.; Xu, X.; Zhao, Q.; Liu, J. Precise Orbit Determination for Quad-Constellation Satellites at Wuhan University: Strategy, Result Validation, and Comparison. *J. Geod.* **2016**, *90*, 143–159. [CrossRef]
30. Wu, J.T.; Wu, S.C.; Hajj, G.A.; Bertiger, W.I.; Lichten, S.M. Effects of antenna orientation on GPS carrier phase. *Astrodynamics* **1992**, *1991*, 1647–1660.
31. Schmid, R.; Dach, R.; Collilieux, X.; Jäggi, A.; Schmitz, M.; Dilssner, F. Absolute IGS antenna phase center model igs08. atx: Status and potential improvements. *J. Geod.* **2016**, *90*, 343–364. [CrossRef]
32. Dilssner, F.; Springer, T.; Schönemann Enderle, W. Estimation of satellite antenna phase center corrections for BeiDou. *IGS Workshop* **2014**, 23–27.
33. Lou, Y.; Zhang, W.; Wang, C.; Yao, X.; Shi, C.; Liu, J. The Impact of Orbital Errors on the Estimation of Satellite Clock Errors and PPP. *Adv. Space Res.* **2014**, *54*, 1571–1580. [CrossRef]

MDPI
St. Alban-Anlage 66
4052 Basel
Switzerland
www.mdpi.com

*Remote Sensing* Editorial Office
E-mail: remotesensing@mdpi.com
www.mdpi.com/journal/remotesensing

Disclaimer/Publisher's Note: The statements, opinions and data contained in all publications are solely those of the individual author(s) and contributor(s) and not of MDPI and/or the editor(s). MDPI and/or the editor(s) disclaim responsibility for any injury to people or property resulting from any ideas, methods, instructions or products referred to in the content.

www.ingramcontent.com/pod-product-compliance
Lightning Source LLC
Chambersburg PA
CBHW061549210926
51850CB00009B/2044
*9783036589862*